Bioenergy for Sustainable Development in Africa

Rainer Janssen • Dominik Rutz
Editors

Bioenergy for Sustainable Development in Africa

Springer

Editors
Dr. Rainer Janssen
WIP – Renewable Energies
Sylvensteinstraße 2
81369 Munich
Germany
Rainer.Janssen@wip-munich.de
www.wip-munich.de

Dipl.-Ing. Dominik Rutz M.Sc.
WIP – Renewable Energies
Sylvensteinstraße 2
81369 Munich
Germany
Dominik.Rutz@wip-munich.de
www.wip-munich.de

ISBN 978-94-007-2180-7 e-ISBN 978-94-007-2181-4
DOI 10.1007/978-94-007-2181-4
Springer Dordrecht Heidelberg London New York

Library of Congress Control Number: 2011941607

Printed on acid-free paper

Springer is part of Springer Science+Business Media (www.springer.com)

Preface

The book "Bioenergy for Sustainable Development in Africa" is a response to the current global discussion on sustainability of bioenergy which often neglects the needs and perspectives of developing countries. In five parts on "Biomass Production and Use", "Biomass Technologies and Markets", "Biomass Policies", "Sustainability of Biomass Production and Use", and "Financing and Socio-Economic Issues" the book addresses in 31 chapters bioenergy development opportunities for Africa and related risks. Contributions to this book are based on the experience of selected authors from Africa, Europe, and other continents, including researchers, investors, policy makers and other stakeholders such as representatives from NGOs.

This publication builds upon the results of the "COMPETE Bioenergy Competence Platform for Africa" which was supported by the European Commission in the Sixth Framework Programme for Research from January 2007 to December 2009.

Acknowledgements

The editors would like to thank all authors that contributed to this book on Bioenergy for Sustainable Development in Africa as well as all partners of the COMPETE Bioenergy Competence Platform for Africa, which was supported by the European Commission in the Sixth Framework Programme for Research from 2007 to 2009.

We would also like to thank our colleagues at WIP Renewable Energies for their continuous support of our activities within the current global discussion about the sustainability of bioenergy in Africa.

Contents

Editors' Biography

Rainer Janssen graduated in Physics (Dr. rer.nat.) at the Technical University of Munich, Walter Schottky Institute, Germany and performed studies at the University of Toronto, Canada. He is Head of the Biomass Department at WIP Renewable Energies and Senior Expert in the Biomass field. He is involved in the production, distribution and market penetration of bioenergy (solid biomass, biogas) and biofuels for transport (e.g. bioethanol, biodiesel, vegetable oil) with special emphasis on the development of supportive framework conditions and policy regulations in the EU, Latin America, Africa and other emerging economies.

Since 2000, Dr. Janssen is involved in the coordination of a variety of international and European projects. Recent projects include the FP7 project Global-Bio-Pact focussing on socio-economic impacts of biomass and bioproducts on a global level, the FP7 project BioTop on biofuels assessment on technical opportunities and research needs for Latin America, and the international competence platform COMPETE (FP6) on the sustainable use of energy crops and agro-forestry systems in Africa.

In recent years, Dr. Janssen was invited expert in the field of bioenergy for the European Commission (DG RTD, DG ENER), IEA (International Energy Agency) Bioenergy, and the GIZ (Deutsche Gesellschaft für Technische Zusammenarbeit). Since 2009, Dr. Janssen is member of working group 4 on "Sustainability" of the European Biofuels Technology Platform and member of working group 4 on "Market & Policies Development" within the Biomass Panel of the European Technology Platform on Renewable Heating and Cooling.

Dominik Rutz graduated in Environmental Science (Dipl.-Ing.) and Consumer Science (M.Sc.) at the Technical University of Munich, Germany, and the Institut National d'Horticulture, France. He is Senior Expert at WIP Renewable Energies and his main field of experience includes market support and international cooperation on bioenergy in developing countries and emerging economies worldwide.

Mr. Rutz coordinated the European Union supported projects BioTop (Biofuels Assessment on Technical Opportunities and Research Needs for Latin America), Global-Bio-Pact (Global Assessment of Biomass and Bioproduct Impacts on Socio-economics and Sustainability), BiG>East (Biogas for Eastern Europe) and

UrbanBiogas (Urban Waste for Biogas Production). He is author of three handbooks on biogas, liquid biofuels, and biomass heating systems, and furthermore invited expert and lecturer e.g. for a course on second generation biofuels at UNAM (Universidad Nacional Autónoma de México) and REMBIO (Red Mexicana de Bioenergía).

While travelling to more than 40 countries worldwide and stimulated by his diploma thesis on vegetation ecology in the Okavango Delta in Botswana in 2004, his main fascination is dedicated to the beauty of nature and cultural spirit of the African continent. Nature conservation on the one hand and social equity on the other hand are the main drivers for his engagement in decentralised renewable energy supply. Thereby his interest goes far beyond the bioenergy field, towards a holistic and sustainable energy system.

Contributors

Greg Austin AGAMA Energy, The Green Building, 9b Bell Crescent Close, Westlake Business Park, 7945 Westlake, South Africa, greg.austin@agama.co.za

Dina Bacovsky BIOENERGY 2020+, Gewerbepark Haag 3, 3250 Wieselburg-Land, Austria, dina.bacovsky@bioenergy2020.eu

Grant Ballard-Tremeer Eco Ltd, P.O. Box 900, BR1 9FF London, Bromley, UK, grant@ecoharmony.com

Kirongo Balozi Department of Forestry & Wood Science, Chepkoilel University College, Moi University, P.O. Box 1125, Eldoret 30100, Kenya, balozibk@hotmail.com

Bothwell Batidzirai Copernicus Institute, Utrecht University, Heidelberglaan 2, 3584 CS Utrecht, The Netherlands, B.Batidzirai@uu.nl

Raffaella Bellanca Eco Ltd, P.O. Box 900, London, Bromley BR1 9FF, UK, info@ecoharmony.com

Kemjika Benaiah Ajoku Raw Materials Research and Development Council, Abuja, Nigeria, ajokukemji@yahoo.co.uk

Harro von Blottnitz Chemical Engineering Department, University of Cape Town, P. Bag X3, Rondebosch 7701, South Africa, harro.vonblottnitz@uct.ac.za

Clarietta Chagwiza University of Fort Hare, Private Bag X1314, Alice 5700, South Africa, clarrytee@yahoo.co.uk

Touria Dafrallah Environnement et Développement du Tiers Monde (ENDA TM), 54, rue Carnot, Dakar BP 3370, Senegal, touria.dafrallah@hotmail.com

Mamadou Dianka Union Monétaire et Economique Ouest Africaine (UEMOA), BP 543 Ouagadougou, Burkina Faso, mdianka@uemoa.int

Rocio A. Diaz-Chavez Centre for Environmental Policy, Imperial College London, Exhibition Road 313A, Mech Eng Bld., South Kensington, London SW7 2AZ, UK, r.diaz-chavez@imperial.ac.uk

Janske van Eijck Department of Science, Technology and Society, Copernicus Institute, Utrecht University, Heidelberglaan 2, 3584 CS, Utrecht, The Netherlands, J.A.J.vanEijck@uu.nl

Lazare Etiégni Department of Forestry & Wood Science, Chepkoilel University College, Moi University, P.O. Box 1125, Eldoret 30100, Kenya, lazetiegni@amatala.org

André P.C. Faaij Copernicus Institute, Utrecht University, Heidelberglaan 2 3584 CS, Utrecht, The Netherlands, a.p.c.faaij@chem.uu.nl

Francesca Farioli Interuniversity Research Centre on Sustainable Development, (CIRPS)-SAPIENZA Università di Roma, Piazza San Pietro in Vincoli 10, 00184, Rome, Italy, francesca.farioli@uniroma1.it

Gavin Fraser Rhodes University, P.O. Box 94, Grahamstown 6140, South Africa, g.fraser@ru.ac.za

Michael Hofmann Camco, 172 Tottenham Court Road, London W1T 7NS, UK, michael.hofmann@camcoglobal.com

Steven Hunt The Schumacher Centre for Technology & Development, Practical Action Consulting, Bourton on Dunsmore, Rugby CV23 9QZ, United Kingdom, steven.hunt@practicalaction.org.uk

Moses Imo Department of Forestry & Wood Science, Chepkoilel University College, Moi University, P.O. Box 1125, Eldoret 30100, Kenya, imomoses@yahoo.com

Rainer Janssen WIP Renewable Energies, Sylvensteinstr. 2, 81369 Munich, Germany, rainer.janssen@wip-munich.de

Francis X. Johnson Stockholm Environment Institute, Kräftriket 2B SE-106 91, Stockholm, Sweden, francis.johnson@sei.se

Charles B.L. Jumbe Centre for Agricultural Research & Development, Bunda College of Agriculture, University of Malawi, P.O. Box 219, Lilongwe, Malawi, charlesjumbe@yahoo.com

Donald L. Kgathi Harry Oppenheimer Okavango Research Institute, University of Botswana, Shorobe Road, Matlapana, Maun, Botswana, kgathi@mopipi.ub.bw

Kaysara Khatun Basque Centre for Climate Change (BC3), Alameda Urquijo 4, 4° – 1a, Bilbao 48008, Spain, kaysara.khatun@bc3research.org

Thalia Konaris Eco Ltd, P.O. Box 900, London, Bromley BR1 9FF, UK, info@ecoharmony.com

Thapelo Letete Energy Research Centre, University of Cape Town, P. Bag X3, Rondebosch, 7701 South Africa, t.letete@uct.ac.za

Michael Madjera Evangelical Church in Middle Germany, P.O. Box 1424, 39004 Magdeburg, Germany, michael.madjera@onlinehome.de

Sumedha Malaviya Center for Sustainable Technologies, Indian Institute of Science, Malleshwaram, Bangalore 560 012, India, sumedhamalaviya@gmail.com

Isaac Mazonde Office of Research and Development, University of Botswana, 4775 Notwane Rd, Gaborone, Botswana, drd@mopipi.ub.bw

Sarah Mohamed Rhodes University, P.O. Box 94, Grahamstown 6140, South Africa, sarahm31@hotmail.com

Glynn Morris AGAMA Energy, The Green Building, 9b Bell Crescent Close, Westlake Business Park, Westlake 7945, South Africa, glynn.morris@agama.co.za

Kalaluka Munyinda Centre for Energy Environment and Engineering Zambia (CEEEZ), 176 Parirenyatwa Road, Suite B, Private Bag E721, Lusaka, Zambia, ceeez@coppernet.zm; ceeez@zamnet.zm; munyinda_kalaluka@yahoo.com

Michael Murray-Hudson Okavango Research Institute, University of Botswana, Shorobe Road, Matlapana, Maun, Botswana, mmurray-hudson@orc.ub.bw

Stanford Mwakasonda Engen Petroleum Ltd, Engen Court, Thibault Square, P.O. Box 35, Cape Town 8000, South Africa, Stanford.mwakasonda@engenoil.com

Barbara N. Ngwenya Okavango Research Institute, University of Botswana, Shorobe Road, Matlapana, Maun, Botswana, bntombi@orc.ub.bw

Odipo Osano School of Environmental Studies, Moi University, P.O. Box 3900, Eldoret 30100, Kenya, odipo@africaonline.co.ke

Josef Rathbauer Francisco Josephinum – Federal Secondary School and Research Institute for Agriculture, Agricultural Engineering and Food Technology, BLT – Biomass, Logistics, Technology, Rottenhauserstrasse 1, Wieselburg, AT 3250, Austria, josef.rathbauer@fjblt.bmlfuw.gv.at

Nijavalli H. Ravindranath Center for Sustainable Technologies, Indian Institute of Science, Malleshwaram, Bangalore 560 012, India, ravi@ces.iisc.ernet.in

Dominik Rutz WIP Renewable Energies, Sylvensteinstr. 2, 81369 Munich, Germany, dominik.rutz@wip-munich.de

Estomih N. Sawe TaTEDO, Mpakani A, Plot No. KJM/MPA/98, Near Institute of Social Works, Kijitonyama, P.O. Box 32794, Dar es Salaam, Tanzania, energy@tatedo.org; edirector@tatedo.org

Anna Segerstedt Institut für Umweltökonomik und Welthandel, Leibniz Universität Hannover, Königsworther Platz 1, 30167 Hannover, Germany, segerstedt@iuw.uni-hannover.de

Mogodisheng B.M. Sekhwela Office of Research and Development, University of Botswana, 4775 Notwane Rd. Gaborone, Botswana, Sekhwela@mopipi.ub.bw

Kingiri Senelwa Department of Forestry & Wood Science, Chepkoilel University College, Moi University, P.O. Box 1125, Eldoret 30100, Kenya, ksenelwas@yahoo.co.uk

Thomson Sinkala Thomro Biofuels, Lusaka, Zambia, tsinkala@thomro-zambia.com

Edward M.W. Smeets Copernicus Institute, Utrecht University, Heidelberglaan 2, Heidelberglaan 2, 3584 CS, Utrecht, The Netherlands, esmeets@hotmail.com

Department of Science, Technology and Society, Copernicus Institute, Utrecht University, Heidelberglaan 23584 CS, Utrecht, The Netherlands, e.m.w.smeets@uu.nl

Takeshi Takama Japanese International Development Agency, Jl. Angkasa I No. 2, Kemayoran, Jakarta Pusat 10720, Indonesia, takeshi.takama@sei.se

Kirsten Ulsrud Department of sociology and human geography, University of Oslo, Molkte Moes vei 31, P.O. Box 1096, Blindern 0317, Oslo, Norway, kirsten.ulsrud@sosgeo.uio.no

Hartley Walimwipi Centre for Energy, Environment and Engineering Zambia, Parirenyatwa road, 176, Private Bag E 721, Lusaka, Zambia, hartleykabunda@yahoo.co.uk; ceeez@coppernet.zm

Arnaldo Walter University of Campinas (Unicamp) and Brazilian Bioethanol Science and Technology Laboratory (CTBE), Mendeleyev, 200, Campinas 13083-860, Brazil, awalter@fem.unicamp.br

Jeremy Woods Porter Institute, Imperial College London, Exhibition Road 313A Mech Eng Bld South Kensington, London SW7 2AZ, UK, j.woods@imperial.ac.uk

Manfred Wörgetter Francisco Josephinum - Federal Secondary School and Research Institute for Agriculture, Agricultural Engineering and Food Technology, BLT – Biomass, Logistics, Technology, Rottenhauserstrasse 1, Wieselburg, AT 3250, Austria, manfred.woergetter@fjblt.bmlfuw.gv.at

Francis Davison Yamba Centre for Energy Environment and Engineering Zambia (CEEEZ), 176 Parirenyatwa Road, Suite B, Private Bag E721, Lusaka, Zambia, ceeez@coppernet.zm; ceeez@zamnet.zm; yambafd@yahoo.com

Abbreviations

ABPP	Africa Biogas Partnership Program
ABREF	African Biofuels Renewable Energy Fund
AIDS	Acquired immune deficiency syndrome
ANADEB	National Agency for the Development of Biofuels
ARI	Acute respiratory infections
ASC	Alternative specific constant
AUC	African Union Commission (www.africa-union.org)
Bio-SNG	Synthetic natural gas
BtL	Biomass-to-Liquid
CAFTA	Central America free trade agreement
CAG	Consortium advisory group
CAPEX	Capital expenditure
CASP	Comprehensive Agricultural Support Programme (www.nda.agric.za/docs/CASP/casp.htm)
CBI	Caribbean Basin Initiative
CDM	Clean development mechanism
CEF	Central Energy Fund (www.cef.org.za)
CER	Carbon emission reductions
CFB	Circulation fluidised bed
CH_4	Methane
CHP	Combined heat and power
CO	Carbon monoxide
CO_2	Carbon dioxide
COMPETE	Competence Platform for Bioenergy in Arid and Semi-arid Ecosystems in Africa (www.compete-bioafrica.net)
DCA	Discrete choice analysis
DFID	UK Department for International Development (www.dfid.gov.uk)
DME	Dimethylether
DNA	Designated national authorities
DRI	Drought risk index
EC	European Commission (www.ec.europa.eu)

ECOWAS	Economic Community of West African States (www.ecowas.int)
EDC	Energy Development Corporation (www.cef.org.za)
EF	Entrained flow
ENERGIA	International Network on Gender and Sustainable Energy (www.energia.org)
ENSO	El Niño Southern Oscillation
ESDA	Energy for Sustainable Development Africa (www.esda.co.ke)
ESIA	Environmental and social impact assessment
ESMAP	Energy Sector Management Assistance Programme
ESSP	Earth System Science Partnership (www.essp.org)
EU	European Union (www.europa.eu)
FAEE	Fatty acid ethyl ester
FAME	Fatty acid methyl ester
FANR	Food, Agriculture and Natural Resources Agency (www.fanrpan.org)
FAO	Food and Agriculture Organization of the United Nations (www.fao.org)
FCFA	Franc des Colonies Françaises d'Afrique (currency in West Africa)
FSC	Forest Stewardship Council (www.fsc.org)
FT	Fischer-Tropsch
GBEP	Global Bioenergy Partnership (www.globalbioenergy.org)
GCM	General circulation models
GDP	Gross domestic product
GHG	Greenhouse gases
GIZ	Deutsche Gesellschaft für Internationale Zusammenarbeit (German Agency for International Cooperation) (www.giz.de)
GTZ	Gesellschaft für Technische Zusammenarbeit (German Technical Cooperation) (now GIZ) (www.giz.de)
GWC	Global warming commitment
GWP	Global warming potential
HHEA	Household energy economic analysis
HHV	Higher heating value
HIV	Human immunodeficiency virus
ICRISAT	International Crops Research Institute for the Semi-arid Tropics (www.icrisat.org)
IDC	South Africa's Industrial Development Corporation (www.idc.co.za)
IEA	International Energy Agency (www.iea.org)
IFAD	International Fund for Agricultural Development (www.ifad.org)
IFEU	Institute for Energy and Environmental Research Heidelberg (www.ifeu.de)
IIED	International Institute for Environment and Development (www.iied.org)
ILUC	Indirect land use change
IPCC	Intergovernmental Panel on Climate Change (www.ipcc.ch)
IPP	Independent power producer
IRR	Internal rate of return

ISCC	International Sustainability and Carbon Certification System (www.iscc-system.org)
IUCN	International Union for Conservation of Nature (www.iucn.org)
KBDA	Kenya Biodiesel Association
LCA	Life cycle assessment
LDC	Least developed country
LPG	Liquefied petroleum gas
MAFISA	Micro Agricultural Financial Industrial Scheme of South Africa
MAR	Mean annual rainfall
MDGs	Millennium Development Goals (www.un.org/millenniumgoals)
MFP	Multifunctional platform
MTBE	Methyl tertiary butyl ether
MWTP	Marginal willingness to pay
N_2O	Nitrous oxide
NCAR	National Center for Atmospheric Research (www.ncar.ucar.edu)
NCEP	National Centers for Environmental Research (www.epa.gov/ncer)
NEB	Net energy balance
NERSA	National Energy Regulator of South Africa (www.nersa.org.za)
NGO	Non governmental organisation
NMHC	Non-methane hydrocarbons
NO_x	Nitrogen oxides
OECD	Organisation for Economic Co-operation and Development (www.oecd.org)
OMVG	Organisation pour la Mise en Valeur du Fleuve Gambie
OMVS	Organisation pour la Mise en Valeur du fleuve Sénégal
OPEC	Organization of the Petroleum Exporting Countries (www.opec.org)
PANPP	Pays Africains Non-Producteurs de Pétrole
PDSI	Palmer Drought Severity Index
PIC	Products of incomplete combustion
PISCES	Policy Innovation Systems for Clean Energy Security (www.pisces.or.ke)
PM	Particulate matter
PPA	Power purchase agreement
PPO	Pure plant oil
ProBEC	Programme for Biomass Energy Conservation (www.probec.org)
R&D	Research and development
REC	Renewable energy certificates
RED	Renewable Energy Directive (Directive 2009/28/EC of the European Parliament and of the Council of 23 April 2009 on the promotion of the use of energy from renewable sources and amending and subsequently repealing Directives 2001/77/EC and 2003/30/EC)
REFIT	South Africa Renewable Energy Feed-in Tariff Programme
REN21	Renewable Energy Policy Network for the 21st Century (www.ren21.net)
RET	Renewable energy technology

RSB	Roundtable on Sustainable Biofuels (www.rsb.epfl.ch)
RSPO	Roundtable on Sustainable Palm Oil (www.rspo.org)
RTD	Research and technical development
RTRS	Round Table on Responsible Soy (www.responsiblesoy.org)
SADC	Southern Africa Development Community (www.sadc.int)
SANEDI	South African National Energy Development Institute (www.saneri.org.za)
SEI	Stockholm Environment Institute (www.sei-international.org)
SHS	Solar home systems
SIDA	Swedish International Development Cooperation Agency (www.sida.se)
SNG	Synthetic natural gas
SNV	Netherlands Development Organisation (www.snvworld.org)
SODEGO	Solidary and community development
SO_x	Sulphur oxides
SRES	Special report on emissions scenarios
SRF	Short rotation forest
SVO	Straight vegetable oil
SWOT	SWOT analysis (strengths, weaknesses, opportunities, threats)
TaTEDO	Tanzanian Traditional Energy Development Organisation (www.tatedo.org)
TJ	Tera joule
TPES	Total primary energy supply
UEMOA	Union Économique et Monétaire Ouest-Africaine (www.uemoa.int)
UN	DESA United Nations Department of Economic and Social Affairs (http://www.un.org/en/development/desa/index.html)
UNEP	United Nations Environment Programme (www.unep.org)
UNICEF	United Nations Children's Fund (www.unicef.org)
UNIDO	United Nations Industrial Development Organisation (www.unido.org)
UNFCCC	United Nations Framework Convention on Climate Change (www.unfccc.int)
VEETC	Volumetric ethanol excise tax credit
VER	Verified emission reduction
WAM	West African monsoon
WARDA	West Africa Rice Development Association (http://www.warda.cgiar.org/)
WBGU	Wissenschaftliche Beirat der Bundesregierung Globale Umweltveränderungen (German Advisory Council on Global Change) (www.wbgu.de)
WEO	World Energy Outlook (www.worldenergyoutlook.org)
WHO	World Health Organisation (www.who.int)
WIP	Wirtschaft und Infrastruktur GmbH & Planungs Co KG (WIP Renewable Energies) (www.wip-munich.de)
WMO	World Meteorological Organisation (www.wmo.int)
WWW	World weather watch (stations)

List of Figures

List of Tables

Part I
Biomass Production and Use

Chapter 1
Keynote Introduction: Traditional and Improved Use of Biomass for Energy in Africa

Edward Smeets, Francis X. Johnson, and Grant Ballard-Tremeer

Abstract Traditional biomass energy systems are widely used in Africa, mainly because of the low cost and lack of available alternatives in rural areas. Projections indicate that the (relative) contribution of traditional bioenergy will decrease, but that the total use of traditional biomass energy systems will increase during the coming decades. The efficiencies of wood-fuel (firewood and charcoal) energy systems are usually low and the use of these systems has serious negative consequences, such as indoor air pollution and related health effects, deforestation and the labour intensive and sometimes dangerous process of firewood collection. Improvements in stoves, charcoal production efficiency and switching fuels can increase the efficiency by several tens of percent points and thereby reduce the demand for labour for the collection of firewood and the costs. Other advantages of improved traditional bioenergy systems are reduced greenhouse gas emissions, reduced indoor air pollution and reduced deforestation. Various initiatives have been successful in implementing the use of improved household stoves, although the results suggest that the success of improved traditional biomass systems depends on the local conditions and socio-economic impacts of these systems.

E. Smeets (✉)
Copernicus Institute, Utrecht University, Budapestlaan 6, 3584 CD, Utrecht, The Netherlands
e-mail: esmeets@hotmail.com

F.X. Johnson
Stockholm Environment Institute, Kräftriket 2B, SE-106 91 Stockholm, Sweden
e-mail: francis.johnson@sei.se

G. Ballard-Tremeer
Eco Ltd, P.O. Box 900, BR1 9FF Bromley, London, UK
e-mail: grant@ecoharmony.com

R. Janssen and D. Rutz (eds.), *Bioenergy for Sustainable Development in Africa*, DOI 10.1007/978-94-007-2181-4_1, © Springer Science+Business Media B.V. 2012

Keywords Wood-fuel • Fuel-wood • Firewood • Charcoal • Traditional energy systems • Fuel-wood stoves • Charcoal kilns • Efficiency • Indoor air pollution • Black carbon • Incomplete combustion • Energy ladder • Dedicated woody energy • LPG • Kerosene • Biogas • Ethanol • Ethanol-gelfuel

1.1 Introduction

Biomass currently accounts for about 10% of the total global energy supply, making it by far the most important renewable energy source (IPCC 2007; IEA 2008; IEA 2009). Most of this biomass (70–80%) consists of wood, agricultural residues and urban waste and is used in traditional stoves for cooking and heating (IPCC 2007).

Biomass sources for traditional bioenergy systems include fuel-wood and charcoal (wood-fuels), animal dung and crop residues. In many countries, fuel-wood accounts for 80–100% of biomass use, although the percentage is lower in east South-East Asia, where there is significant use of agricultural residues and/or dung (Table 1.1). Typical devices used in households are the three-stone fire-place, or simple cooking enclosures. Traditional charcoal production is typically carried out with a simple 'earth mound kiln', and charcoal burned in simple braziers. Apart from the household sector, biomass is also used by small and medium-sized enterprises in a variety of traditional commercial applications as well as small and medium enterprises of developing countries. Since charcoal and fuel-wood account for more than 90% of traditional biomass in Sub-Saharan Africa, the focus in this chapter is especially on fuel-wood and charcoal.

Traditional bioenergy systems have several important disadvantages. First, the quality of the energy services provided is extremely low in comparison to the energy and economic expenditure. The efficiencies of such systems are often low (7–20%) and they provide only low grade heat that cannot be easily regulated. Large volumes of biomass are needed to meet the daily energy demand. Women and children are mostly responsible for the labour intensive collection of fuel-wood, which can take up to 8 h per day (World Bank 2008). Further, traditional bioenergy systems typically have no chimneys and, combined with poor ventilation, lead to high levels of indoor

Table 1.1 Estimated use of biomass for cooking in various world regions (Mm^3)

	Fuel-wood	Crop residues	Dung	Charcoal
North America	41	0	0	0
Latin America	80	0	0	16
Africa	371	52	0	14
Europe	147	0	0	0
South Asia	344	76	75	3
East Asia	193	323	0	0
South-East Asia	164	43	0	6
Oceania	10	0	0	0
Global total	1,351	495	75	39

Source: Fernandes et al. (2007)

air pollution. In poorly ventilated dwellings, indoor smoke can exceed acceptable levels for small particles in outdoor air by 100-fold. Exposure is particularly high amongst women and children, who spend the most time near the domestic hearth. Every year, indoor air pollution in Sub-Saharan Africa is responsible for the death of an estimated 350,000 children due to lower respiratory infections and 34,000 adult women due to chronic obstructive pulmonary disease (Bailis et al. 2005).

Improving the quality of energy supply in Africa is central to sustainable development and poverty reduction efforts. At this moment, some 583 million people, or 76% of the population, in Sub-Saharan Africa depend solely upon traditional bioenergy (IEA 2002). The contribution of traditional bioenergy to the energy supply is projected to decrease during the coming decades. But the total number of people in Africa that will rely on traditional biomass for cooking and heating is expected to increase from 583 million in 2002 to 823 million (IEA 2002). Access to modern energy sources, such as natural gas and electricity, allows people to generate income and improve their living conditions. So even modest improvements in the efficiency of traditional energy systems and in access to modern energy services can have a positive impact on the livelihoods of people.

In this section the possibilities, constraints and impacts of improving traditional bioenergy systems are discussed as a way to increase the efficiency of the use of scarce resources, to combat energy poverty and to contribute to rural development.

1.2 Conventional Traditional Bioenergy Systems

1.2.1 Firewood and Charcoal

The percentage of the population in rural areas relying on firewood as the primary source of energy is above 85% in most countries in Sub-Saharan Africa (UNDP 2009). The contribution of firewood to the energy supply is relatively constant in time, due to the low costs and the lack of available alternatives. In urban regions charcoal is the main source of energy. In many respects, charcoal is more suitable for household use compared to firewood. Charcoal emits fewer pollutants, it has higher energy content, and it is lighter and therefore easier to transport. Because of its advantages over firewood, there have been a number of efforts to promote its use in household cooking. The use of charcoal in Africa is projected to increase by more than 50% between 2010 and 2030 (IEA 2002), as a result of rapid urbanization and population growth.

The traditional three-stone stove is widely used in Sub-Saharan Africa (Fig. 1.1). An important disadvantage of this type of stove is the relative low efficiency of about 14% (Table 1.3). Charcoal is typically used in traditional metal stoves (Fig. 1.1), which have slightly higher end-use efficiencies of 15–18% (Malimbwi and Zahabu 2007). Nevertheless, due to the loss of energy during charcoal production about five times the amount of wood is needed when cooking on charcoal as compared to firewood when assuming a charcoal kiln efficiency of 15% (Table 1.3).

Fig. 1.1 A traditional three stone stove (*left*) and a charcoal stove (*right*) (Source: Ballard-Treemer)

1.2.2 Impacts of Conventional Traditional Bioenergy Systems

The use of traditional bioenergy systems has several serious negative socio-economic and environmental impacts. Traditional bioenergy systems lead to high greenhouse gas emissions. Because of the incomplete combustion of wood-fuels, 10–20% of the carbon released is in the form of methane (CH_4), nitrous oxide (N_2O), carbon monoxide (CO) and non-methane hydrocarbons (NMHC). These compounds are referred to as products of incomplete combustion (PIC) and have much higher global warming potential than carbon dioxide (Smith et al. 2000). The combined impact of the five emissions is called the Global Warming Commitment (GWC) (see Table 1.2). This table also shows that the GWC of traditional bioenergy applications depend heavily on the source of the biomass. With renewable harvesting of biomass, carbon dioxide (CO_2) missions are completely recycled and thus there is no net increase in GWC from CO_2. For PICs, renewable harvesting only affects the GWC by eliminating the portion of the global warming potential of each gas owing to its eventual conversion to CO_2 in the atmosphere. In non renewable harvesting, however, all the carbon in biomass is a net addition to the atmosphere, as for fossil fuels. Further, Table 1.2 shows that traditional systems have very high emissions, compared to improved traditional systems, but also that improved traditional systems using biomass from renewable sources have emissions that are much lower than modern energy carriers.

Table 1.2 Global warming commitment (grams of carbon as CO_2 per MJ of delivered energy) of various cooking fuels

Fuel	Renewable	Non-renewable
Biogas	148	148
LPG	139	139
Kerosene wick stove	154	154
Kerosene press stove	173	173
Improved metal stove	576	171
Improved vented ceramic	472	150
Three stone	740	206

Source: Smith et al. (2000)

The inefficient and incomplete combustion of wood-fuels in traditional bioenergy systems releases a number of hazardous pollutants, including carbon monoxide (CO), sulphur and nitrogen oxides (NO_x, SO_x) and particulate matter (PM). Poor ventilation exacerbates the indoor air pollution and leads to acute respiratory infections (ARI), as well as causing chronic bronchitis, emphysema, and chronic obstructive pulmonary disease. Several studies have also linked childhood exposure to the smoke with asthma, though others have concluded that there is no interrelation.

The collection of firewood is a labour intensive and dangerous process and the burden often falls in rural areas in Sub-Saharan Africa upon women and, to a lesser extent, young girls. Women gather firewood by foot, often walking long distances with an average load of 20 kg (IEA 2006). A survey of 30 households near Lake Malawi found a mean distance to a viable firewood source of 2.1 km, resulting in a mean trip length of 241 min and mean time spent collecting wood per day of 63 min (Biran et al. 2004). Also the production of charcoal is a labour intensive process, although production is often commercialized. Because of the low efficiencies, large quantities of wood are needed. In Nairobi, for example, a household that relies exclusively upon charcoal consumes 240–600 kg of charcoal per year, which equals 1.5–3.5 tons of biomass (Kammen 2006). As a result, the production and trade of charcoal is an important economic activity, which provides income, tax revenue and employment especially in rural areas. Charcoal production creates employment of 200–350 person-days per TJ energy, compared to 100–170 for firewood (where commercially supplied), 80–110 for coal, 10–20 for LPG and 10 for kerosene, based on figures from South-East Asia (RWEDP 1997).

Further, the collection and use of biomass for traditional bioenergy systems can in some cases contribute to the overexploitation and degradation of ecosystems. Examples are forest degradation or deforestation, deterioration of watersheds, and loss of soil fertility and biodiversity. Although traditional biomass use is not a major driver of deforestation, population pressures and the impacts of urbanization have resulted in increased pressure on forest health and ecosystems. The wood-fuel crisis that was widely discussed in the 1970s and 1980s gave way to a recognition that the impacts of traditional biomass use have been and can be addressed through a variety of institutional changes and policy reforms (Leach and Mearns 1988). Another concern that has arisen is the impact on GHG emissions due to the decrease in above ground biomass and soil degradation that results from unsustainable use of traditional wood-fuels. Addressing such concerns require more analysis on the GHG balances associated with traditional biomass use.

Another climate-related concern associated with traditional biomass use that has become better understood in recent years is the climate impact of black carbon (soot). Black carbon is a product of incomplete combustion and can have significant impacts on both regional and global climatic processes, and thus contributing to climate change (Ramanathan and Carmichael 2008). It can be difficult to separately calculate the impacts from combustion of biomass as compared to combustion of fossil fuels or to differentiate amongst the diverse types of biomass burning (Gustafsson et al. 2009). Nevertheless, a consensus has emerged that black carbon is a major contributor to climate change and that biomass burning in households is a significant source of black carbon (Bond et al. 2004).

1.3 Improved Traditional Bioenergy Systems

Existing traditional bioenergy systems can be improved by using more efficient stoves and charcoal kilns and by switching to other fuels. Table 1.3 gives an overview of the performance of conventional and improved traditional cooking stoves.

Key drivers for improved stove design are combustion efficiency, heat transfer efficiency, safety, costs, durability and local cooking practice. A wide variety of improved fuel-wood cooking stoves have been designed during the past decades; from relative inefficient (23%) stoves that are constructed from clay and grass at very low costs, but having a short lifespan, to more efficient (29%) stoves made from bricks or metal, having a long lifespan, but considerable investment cost. Furthermore, some types of improved stoves are portable, which is important because the traditional three-stone cooking stove can be easily replaced. Improved stoves are currently in use in many countries, but penetration has generally been small in terms of the percentage of households. This applies especially for the more advanced stoves that have been designed during the past few years. Examples of such stoves are the First Energy Oorja Stove, Envirofit and StoveTEC stoves. Some countries aimed at training the users to build their own stoves. Although this has in

Table 1.3 Costs, efficiencies and lifetimes of various cooking stoves and the related cost of energy, in terms of utilized heat

Energy carrier	Stove type	Efficiency (%)	Cost (US$)	Life-time	Cost of heat (US$/GJ$_H$)
Wood	Three stone	14	Free	–	14
Wood	Mud	23	1.4	2 month	9
Wood	Brick	29	33.2	5 years	7
Charcoal	Traditional	17	1.7	3 years	35 legal 21 illegal
Charcoal	Improved	45	8.0	3 years	13 legal 8 illegal
Kerosene	Kerosene	38	12.5	3 years	95

Sources: Wiskerke et al. (2008), Ballard-Tremeer and Jawurek (1996)

some cases resulted in localized economic and social benefits, this approach will not reach the existing millions of biomass stove users in Sub-Saharan Africa in the near-term.

The use of improved charcoal kilns is also an approach to greatly increase the overall system efficiency for the delivered energy services. Traditional earth and earth mound kilns have an efficiency of 10–20%, depending on the skills of the charcoal producer and on the tree species. A skilled charcoal producer who uses well-dried wood can reach efficiencies of up to 30% (Wiskerke et al. 2008). Improved charcoal kilns can increase the efficiency of charcoal production to 45%, as well as enhancing the quality characteristics of charcoal. Whilst most of the low-cost improved charcoal kilns have demonstrated high efficiencies under test conditions, none of the developed designs have attained substantive dissemination, largely because of the surprisingly high efficiency of traditional kilns under field conditions. A large proportion of charcoal production in Africa is carried out as a semi-illegal part-time activity since the wood used is often illegally procured. Consequently, few charcoal makers are willing to construct in-situ kilns since they would be vulnerable to punitive official measures such as imposition of tax and seizure. Consequently, dissemination of improved charcoal techniques to the informal sector has proven to be a difficult undertaking. Improved charcoal production technologies have been more successful in areas where production is undertaken on a commercialized basis as in the case of Malawi. Another focus area is the transportation of charcoal. Due to the fragility of charcoal, excessive handling and transporting over long distances can increase the amount of fines (i.e. charcoal dust that cannot be used or sold without further processing) to about 40% and thus greatly reducing the value of the charcoal. Distribution in bags can help to limit the amount of fines produced in addition to providing a convenient measurable quantity for both retail and bulk sales.

There are also a number of alternatives to improve the efficient production of biomass feedstock. The production of fuel-wood from dedicated woody energy crops can reduce the demand for wood from natural resources and related overexploitation. Use of degraded or low productivity areas to grow woody energy crops can take advantage of lower land costs and offer the potential to increase the soil quality as a bonus to the resulting energy production. Integrated agro-forestry and multi-purpose systems are especially promising. Wiskerke et al. (2010) analyzed and compared the costs and benefits of three biomass energy supply systems for rural households in north Tanzania, namely (1) a small-scale forestation project for carbon sequestration under the Clean Development Mechanism (CDM) of the Kyoto Protocol, (2) a short rotation woodlot for the production of fuel-wood or charcoal, optionally with intercropping, and (3) a jatropha plantation that generates oil used as cooking fuel, as a diesel substitute for off-grid household electrification, or for soap production. The results show that rotational woodlots are the most economically attractive option, especially when producing poles from stem wood and charcoal from branches, whilst practising intercropping of maize.

It can be concluded that there is a huge potential to increase the efficiency of existing traditional bioenergy systems by tens of percentage points in Africa.

The use of such improved technologies can also reduce indoor air pollution and contribute to reducing the costs and labour associated with wood-fuel systems. Especially women and children are expected to benefit from such a transition, because women are typically responsible for the collection of fuel-wood and women and children are exposed to indoor air pollution. The lessons learnt and trends point towards considerable markets for improved biomass-burning products and fuels.

1.4 Fuel Switching

Several projections indicate that traditional bioenergy systems will slowly be replaced by modern fuels as households become wealthier, following the 'energy ladder' (Fig. 1.2). The theory suggests that when the income of households increases, they ascend the ladder from low quality fuels to more convenient, cleaner and modern fuels. In the first stage, households depend solely upon solid biomass, deriving energy from the combustion of residues, waste and firewood. In the intermediate stage, households shift to fuels that burn more efficiently, but still result in significant emissions, such as charcoal and kerosene. In the most advanced stage, households move to the most convenient and cleanest commercial fuels, usually liquefied petroleum gas (LPG) and electricity. Aside from their low greenhouse gas impact, some of these fossil fuels have several other advantages over wood-fuels, such as a higher energy density and higher combustion efficiency. Further, both LPG and kerosene produce less greenhouse gas emissions per unit of useful energy than wood-fuel. But, if the wood-fuels are produced from renewable sources, the carbon emissions of LPG and kerosene would be of comparable magnitude.

Recently, there has been an increasing attention on liquid and gaseous biofuels for cooking, such as biogas, ethanol and ethanol-gelfuels. Several countries in Africa are currently producing ethanol from sugarcane at significant scales, including Malawi, Zimbabwe, Ethiopia, and Kenya. Other potentially suitable crops are cassava,

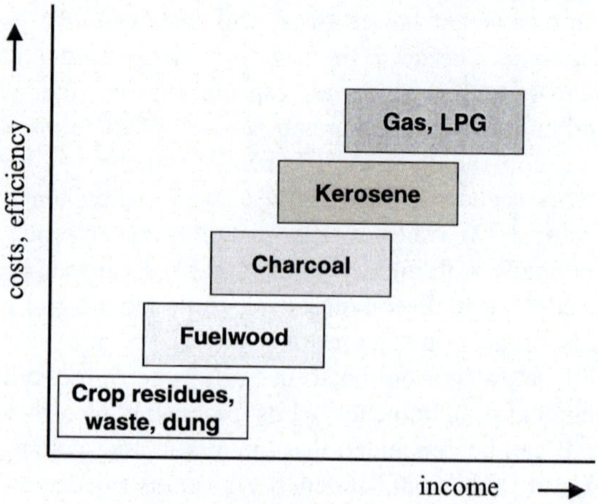

Fig. 1.2 Schematic representation of the fuel ladder

sweet sorghum, maize, and wheat. Ethanol can be burned directly in specialized stoves, though further conversion to gelfuel is a simple process that offers notable advantages. Gelfuel has a much higher viscosity than ethanol, making it easier to handle and a safe alternative, however, gelfuel has also several disadvantages. Biogas and ethanol offer the greatest potential for the reduction in GHG emissions, as both can be burned close to completion and produced sustainably. Assuming sustainable production, carbon output due to the use of biogas in household cooking would be on the order of 100 times less than for unsustainable wood-fuels (Smith et al. 2000). This is because the k-factor (factor showing how incomplete the combustion is) is so low that almost all of thc carbon is released as carbon dioxide, emissions that are offset by carbon uptake due to the sustainable production of fuel.

A disadvantage of the fuel ladder theory is that it implies perfect substitution of one fuel for another and that income is the decisive factor. However, recent studies suggest that there are many factors besides income that are important in determining the fuel choice. Social, economic, and technological barriers prevent the linear progression towards clean cooking fuels put forth by the energy ladder. More specifically factors such as fuel availability, affordability, cultural norms, and cooking practice impact fuel/stove adoption and continued use (IEA 2002; Pachauri and Spreng 2003; Masera et al. 2000). During the past years the 'fuel-stacking' model has gained ground. This model suggests that households integrate modern fuels slowly into existing energy use patterns, resulting in the use of multiple cooking fuels simultaneously (Masera et al. 2000). The fuel-stacking model is an important aspect to consider when designing and implementing policies to increase the use of improved traditional energy systems.

1.5 Conclusions

The overwhelming majority of biomass used for energy in Sub-Saharan Africa consists of fuel-wood and charcoal that is burned in low-efficiency stoves for cooking, resulting in significant socio-economic, health, and environmental impacts. A numbcr of alternatives exist to provide higher quality energy services at lower lifecycle costs and with significant reductions of negative impacts. The challenge is to find ways to make these more efficient options more affordable and widely available, but also to ensure that they are suitable for the usage patterns in different regions and localities.

References

Bailis R, Ezzati M, Kammen D (2005) Mortality and greenhouse gas impacts of biomass and petroleum energy futures in Africa. Science 308:98–103

Ballard-Tremeer G, Jawurek HH (1996) Comparison of five rural, wood-burning cooking devices: efficiencies and emissions. Biomass Bioenerg 11(5):419–430

Biran A, Abbot J, Mace R (2004) Families and firewood: a comparative analysis of the costs and benefits of children in firewood collection and use in two rural communities in Sub-Saharan Africa. Hum Ecol 32:1–25

Bond T, Venkataraman C, Masera O (2004) Global atmospheric impacts of residential fuels. Energ Sust Dev 8(3):20–32

Fernandes SD, Trautman NM, Streets DG, Roden GA, Bond TC (2007) Global biofuel use, 1850–2000. Global Biochem Cycles 21:2019

Gustafsson Ö, Kruså M, Zencak Z, Sheesley RJ, Granat L, Engström E, Praveen PS, Rao PSP, Leck C, Rodhe H (2009) Brown clouds over South Asia: biomass or fossil fuel combustion? Science 323:495–498

IEA (2002) World energy outlook 2002. International Energy Agency, Paris

IEA (2006) World energy outlook 2006. Energy for cooking in developing countries. International Energy Agency, Paris, p 488

IEA (2008) World energy outlook 2008. International Energy Agency (IEA), Paris, p 578

IEA (2009) Key world energy statistics 2009. International Energy Agency, Energy Statistics Division, Paris, p 82

IPCC (2007) Climate change 2007: mitigation. Contribution of working group III to the fourth assessment report of the Intergovernmental Panel on Climate Change. Cambridge University Press, Cambridge

Kammen D (2006) Bioenergy in developing countries: experiences and prospects. Focus Brief 10. IFPRI, Washington DC, p 2

Leach G, Mearns R (1988) Beyond the wood-fuel crisis: people, land, and trees in Africa. Earthscan, London, p 309

Malimbwi RE, Zahabu E (2007) Situation analysis of charcoal sector in Dar es Salaam; charcoal supply and consumption. Department of Forest Mensuration and Management, Faculty of Forestry and Nature Conservation, Sokoine University of Agriculture, Morogoro, Tanzania

Masera OR, Saatkamp BD, Kammen DM (2000) From linear fuel switching to multiple cooking strategies: a critique and alternative to the energy ladder model. World Dev 28(12):2083–2103. doi:10.1016/S0305-750X(00)00076-0

Pachauri S, Spreng D (2003) Energy use and energy access in relation to poverty. Economics Working Paper 25. Center for Energy Policy and Zurich, Switzerland, p 20

Ramanathan V, Carmichael G (2008) Global and regional climate changes due to black carbon. Nat Geosci 1:221–227

RWEDP (1997) Regional study on wood energy today and tomorrow in Asia. FAO Regional Wood Energy Development Programme, Bangkok, p 179

Smith KR, Uma R, Kishore VVN, Zhang J, Joshi V, Khalil MAK (2000) Greenhouse implications of household stoves: an analysis for India. Annu Rev Energ Environ 25:741–763

UNDP (2009) The energy access situation in developing countries. A review focusing on the least developed countries and Sub-Saharan Africa. United Nations Development Programme, New York, USA, 142 p

Wiskerke W, Dornburg V, Faaij APC, Malimbwi RE, Rubanza CDA (2008) Towards a sustainable biomass energy supply for rural households in semi-arid Shinyanga, Tanzania. Department of Science, Technology and Society, Copernicus Institute, Utrecht University, Utrecht, The Netherlands, 110 p

Wiskerke WT, Dornburg V, Rubanza CDK, Malimbwi RE, Faaij APC (2010) Cost/benefit analysis of biomass energy supply options for rural smallholders in the semi-arid eastern part of Shinyanga Region in Tanzania. Renew Sust Energ Rev 14(1):148–165

World Bank (2008) The welfare impact of rural electrification: a reassessment of the costs and benefits. World Bank Interdependant Evaluation Group, Washington DC, p 178

Chapter 2
Biomass Potential in Arid and Semi-Arid Regions in Botswana

Mogodisheng B.M. Sekhwela and Donald L. Kgathi

Abstract The development of new biomass energy technologies to reduce greenhouse gas emissions and mitigate climate change is opening new opportunities in Botswana. Cleaner technologies have increasingly led to the transformation of biomass into various bioenergy carriers for the use in modern energy systems that have lower GHG emissions. Whilst the use of biomass for energy purposes in rangelands has been as problematic as overgrazing, an increasing trend towards bush encroachment is observed in the dry-lands of southern Africa as a result of climate change. This chapter explores the potential of bioenergy development on arid lands in Botswana based on species associated with bush encroachment, invading species, and other potential energy sources such as oil plants. Bush encroachment and invasive species could be assessed for sustainable production of wood pellets either under management systems of the naturally occurring resources or by established plantations. Naturally occurring oil plants in Botswana such as *Ximenia sp.* are identified as potential sources for bioenergy production. Further research on the possibilities of using this plant species for bioenergy production is recommended.

Keywords Potential • Botswana • Bush encroachment • Climate change • Dry-lands • Oil plants • *Ximenia*

M.B.M. Sekhwela (✉)
Office of Research and Development, University of Botswana, 4775, Notwane Rd
Gaborone, Botswana
e-mail: Sekhwela@mopipi.ub.bw

D.L. Kgathi
Harry Oppenheimer Okavango Research Centre, University of Botswana,
Shorobe Road, Matlapana, Maun, Botswana
e-mail: dlkgathi@orc.ub.bw

R. Janssen and D. Rutz (eds.), *Bioenergy for Sustainable Development in Africa*,
DOI 10.1007/978-94-007-2181-4_2, © Springer Science+Business Media B.V. 2012

2.1 Introduction

The development of new biomass energy technologies to reduce greenhouse gas emissions and to mitigate climate change is opening new opportunities in Botswana. The consequences of the changing climate and its associated hazardous and extreme weather conditions have increased commitment to the development of less polluting energy sources. As predicted by Hall et al. (1993), communities are now seeking to exploit the high potential of biomass energy. Cleaner technologies have increasingly led to the transformation of biomass into various bioenergy carriers for the use in modern energy systems that have lower greenhouse gas (GHG) emissions. As an encompassing term, bioenergy includes traditional biomass energy, energy of the poor in developing countries, and higher value and more efficient biomass-based energy carriers such as electricity, biogas and bioethanol (Kartha et al. 2005). Bioenergy is associated with environmental and ecological impacts that positively or adversely affect environmental sustainability (Sekhwela 1997). Therefore, the sustainability of bioenergy has become increasingly important globally in recent years and forms the basis for assessing the potential for the development of future energy production systems (Fehrenbach et al. 2008). Sustainability criteria are already developed for biofuels, particularly in the European Union (Schlegel and Kaphengst 2007), and several certification systems have already been introduced for instance for forestry products (e.g. Forestry Stewardship Council – FSC).

With respect to sustainability the main issues of concern addresses greenhouse gas balances, biodiversity, environment, food security, welfare of the community, and wellbeing of workers and of the local population (Fehrenbach et al. 2008; Royal Society 2008; Smeets 2008; Rutz et al. 2010). In order to ensure that biofuels are produced in a sustainable way, a number of countries are now introducing sustainability certification systems, whereby the buyers of biofuels ensure that their production complies with certain standards or sustainability criteria (Smeets 2008). Potential markets for bioenergy products and services are developing within these frameworks defined by certification processes based on sustainability criteria. Collectively, the frameworks provide critical guidance as well as a basis for evaluating new initiatives in the development of bioenergy systems.

The sustainability criteria are based on the experience of unsustainable bioenergy production and supply, including that of some imported products and services. An example is the palm oil imported into the European Union from Asia, where its production may have caused deforestation and loss of biodiversity (Wicke et al. 2008). Hence, there is an increasing pressure for the development and adoption of sustainability criteria in countries that produce bioenergy as the EU considers the import of bioenergy products and services to be part of the options for meeting its GHGs reduction targets (de Pous 2008).

Sub-Saharan Africa is considered a high potential region for the production of bioenergy products for export to EU markets under these globally evolving sustainability criteria (Bekunda et al. 2008). This could create opportunities for a number of countries in the region, including Botswana which has started looking at the potential for bioenergy production as an alternative for substituting fossil fuels (Kedikilwe 2008).

This chapter explores the potential of bioenergy production on arid lands in Botswana, based on woody biomass from bush encroachment and invading species as well as from other potential energy sources such as oil plants. The specific objectives of this chapter are (1) to describe the potential for sustainable production and supply of bioenergy products and (2) to highlight opportunities and challenges that exist and could enable better decisions on bioenergy pathway choices.

2.1.1 Analytical Framework for Woody Biomass Sustainability

The predominant form of bioenergy in developing countries is traditional biomass energy. Its harvesting has been associated with deforestation and land degradation, particularly around major villages and urban centres in Africa (Krutilla 1995). In addition, increased removal of the preferred species and the overgrazing of rangelands have subsequently led to bush encroachment, which is associated with impenetrable thickets that have no immediate economic value (Skarpe 1990 and 1991). Whilst this bush encroachment process has been largely attributed to overgrazing by livestock, there is also increasing evidence that the woody vegetation component in the dry-lands of southern Africa is likely to increase as a result of climate change (Bond and Midgley 2000). There are also problems associated with invading woody species that are now expanding uncontrollably in Botswana and South Africa such as *Prosopis sp.* (Zimmerman and Pasiecznik 2005). However, in South Africa the economic value of these species is being derived by using the wood for timber and the fruits as fodder. Additional benefits may include the use for bioenergy systems (Zimmerman and Pasiecznik 2005). Hence, there is a potential economic value associated with the increasing woody biomass in the dry-lands of Botswana which can sustain rural livelihoods, currently adversely affected by the changing vegetation resources. Examples of these invading woody plant species include many of the *Acacia* species and *Dichrostachys cinerea*, particularly in overgrazed areas, and *Prosopis sp.* that is taking over the Kalahari sand areas, outcompeting indigenous species along water courses and lowering the water tables in Botswana. However, some of the invading species like *Acacia mellifera* and *Dichrostachys cinerea* were found to have high economic value as fuel-wood and construction wood, particularly when they have reached the shrub and tree stages (Sekhwela et al. 2007).

Whilst studies on land degradation mainly focused on the risks of increasing woody vegetation, few studies have looked at the opportunities, especially in dry-lands such as those of Botswana and other southern African countries. The economic potential has already been identified by some communities, but it can be increased by the use in bioenergy systems. This will further mitigate climate change and could contribute to the attainment of the millennium development goals (MDGs). Thus, bush encroachment may lead to increased incomes that improve the economic status of the poor. Further potential of such resources could emerge beyond the energy dimensions to include other wood products and reduce pressure on the harvest of natural forest resources and destruction of wildlife habitat and biodiversity.

2.1.2 *Study Area and Sources of Data*

The study area, Botswana, is a semi-arid country with an annual rainfall ranging from a minimum of 250 mm in the south to a maximum of 650 mm in the north. Botswana is a middle income country with a gross national income per capita of US$ 5,840 (for 2007) (World Bank 2008). Although fuel-wood is the main source of energy in rural areas, modern energy sources are also used. Electricity and petroleum based products are the main sources of energy in urban areas. The 2005 energy balance of Botswana shows that the contribution of modern renewable energy to final energy supply is insignificant. Petroleum was the most important source of energy (34%), followed by coal (29%), traditional biomass energy (27.5%) and electricity (8.8%) (Mabowe 2009). Botswana has large resources of coal, estimated at 212 billion tons. There are currently no known petroleum reserves and Botswana is therefore a net importer of petroleum products (Mabowe 2009).

Data for this chapter were synthesised from secondary sources including studies of bush encroachment and woody biomass carried out in Botswana and in the region. Biomass production potential was estimated from ongoing long-term studies in Botswana and in the region, whenever it was possible. Export potential was assessed on the basis of existing international trade on woody products. Information on oil plants was derived mainly from secondary sources, and where data on plant occurrence and population distribution was available, potential supply areas have been highlighted in maps.

2.2 Results and Discussions

The plant species identified for potential bioenergy production have been categorised into woody biomass resources (mainly composed of the highlighted invader species) and oil plants.

2.2.1 *Woody Biomass Resources*

Woody plant species that could be harvested for bioenergy processing are found all over Botswana. Some of the species occur in the whole country from the sandy Kalahari soils to the loamy and relatively fertile soils along the eastern part of the country (Fig. 2.1). Such species also span the different rainfall regimes indicating

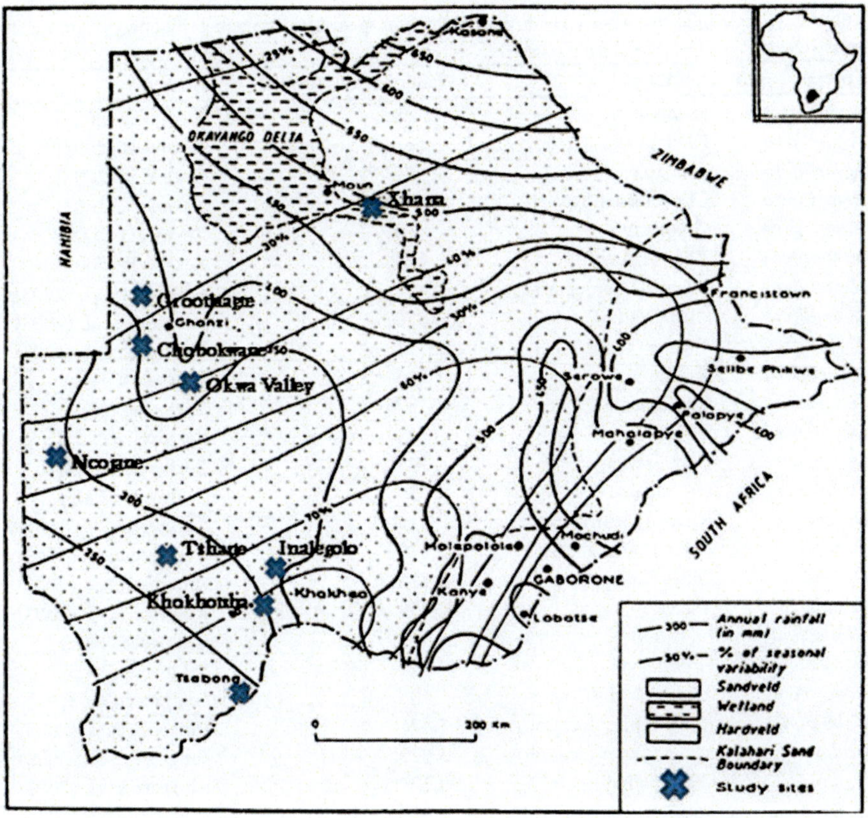

Fig. 2.1 Map of sites with different woody plant species identified as potential bioenergy species in Botswana

high adaptation potential. The quantitative data derived from various studies show variable plant densities with higher figures in some areas reflecting formation of impenetrable thickets of some species such as *A. mellifera* (Table 2.1) (Skarpe 1991; Sekhwela 2003; Ringrose et al. 2003). This particular species occurs all over the country due to overgrazing. It is invasive and forms close canopies. However, it grows to medium tree layer. It is thus a suitable source of fuel-wood and construction poles (Sekhwela 2003). A similar pattern of invasiveness and dominance in overgrazed areas is also exhibited by *D. cinerea,* a species with a high fodder value (fruits) (Moleele 1998). It is also used as fuel-wood and construction poles for fencing of arable fields as observed in the Kweneng District in Botswana. Farmers mentioned that the poles produced from this species are termite resistant (Fig. 2.2).

The potential of these invasive species could be relatively high as they regenerate easily and are often multiple stemmed, allowing for selective harvesting of preferred stem sizes (Table 2.2). The measured production rates were found to be comparatively high, given the generally low rates of slow growing trees in the arid and

Table 2.1 Examples of woody bush encroachment species as potential bioenergy source in Botswana

Species	Location	Plant densities (no/ha)	Source
A. mellifera	Tsabong	98	Sekhwela (2003)
A. mellifera	Khokhotsha	33	Sekhwela (2003)
A. mellifera	Inalegolo	91	Sekhwela (2003)
A. mellifera	Chobokwane	181	Sekhwela (2003)
A. mellifera	Grootlaagte	10	Sekhwela (2003)
A. mellifera	Xhana (Maun)	129	Sekhwela (2003)
A. mellifera	Maun (Other location)	74	Ringrose et al. (2003)[a]
A. mellifera	Okwa Valley	3,556	Ringrose et al. (2003)
A. mellifera	Tshane	2,383	Ringrose et al. (2003)
A. mellifera	Ncojane	3,765	Skarpe (1991)
D. cinerea	Khokhotsha	7	Sekhwela (2003)
D. cinerea	Inalegolo	13	Sekhwela (2003)
D. cinerea	Chobokwane	135	Sekhwela (2003)
D. cinerea	Grootlaagte	39	Sekhwela (2003)
D. cinerea	Xhana (Maun)	515	Sekhwela (2003)
D. cinerea	Maun (Other location)	2,839	Ringrose et al. (2003)
D. cinerea	Okwa Valley	3,185	Ringrose et al. (2003)
D. cinerea	Tshane	259	Ringrose et al. (2003)
G. bicolour	Inalegolo	142	Sekhwela (2003)
G. bicolour	Grootlaagte	94	Sekhwela (2003)
G. bicolour	Xhana (Maun)	105	Sekhwela (2003)
G. bicolour	Maun (Other location)	1,876	Ringrose et al. (2003)
G. flava	Maun (Other location)	5,629	Ringrose et al. (2003)
G. flava	Okwa Valley	20,629	Ringrose et al. (2003)
G. flava	Tshane	9,667	Ringrose et al. (2003)

[a]Plant densities were obtained by expressing the total plant counts for each species in belt transects of 3 m by 90 m

semi-arid regions (Tietema 1993; Sekhwela 2000). The vigour exhibited during invasion and coppice regeneration, particularly of *Grewia bicolor* which is common in the sandy Kalahari (Maun, Okwa Valley and Tshane), and as measured in the Tsabong area (southern Kalahari) (Table 2.1 and Fig. 2.1), showed significant annual production (Sekhwela 2000, unpublished data). Other species like *Colophospermum mopane*, which is also a vigorous secondary coloniser and coppicer, also form exclusive stands and have high production rates that cause small farmers to cut back and burn the regrowth yearly in fields before ploughing, also called 'slash-and-burn' practice (Table 2.3). Similarly, the massive root system and multiple stem nature of *G. bicolor* cause strong re-growth after cutting, showing high potential for continuous production with selected stem harvest in a given growth cycle (Sekhwela 2000, unpublished data). Whilst it is cherished for its sweet berries that are harvested and sold at local markets, it was also found highly popular for the production of courtyards in the Kalahari. Hardly any record was made of dead *Grewia* plants in the woodland assessments (Sekhwela 2003).

Fig. 2.2 Poles of *D. Cinerea* used in arable fields fencing in the Kweneng District, Botswana

Table 2.2 Regeneration through coppice of potential bioenergy woody plant invaders in Botswana

Species	Annual coppice	Source
Acacia mellifera	Vigorous	Sekhwela (2000)
Dichrostachys cinerea	Vigorous	Sekhwela (2000) (unpublished data)
Grewia bicolor	Vigorous	Sekhwela (2000) (unpublished data)
Colophospermum mopane	Vigorous	Mushove and Makoni (1993)
Acacia tortilis	Vigorous	Tietema (1993)

Table 2.3 Indicative biomass production rates of potential bioenergy woody plant species in their natural growth environment (fresh weight)

Species	Standing stock (t/ha per year)	Annual production (t/ha per year)	Source
Acacia mellifera	5–8	0.7	Sekhwela (2000)
Dichrostachys cinerea	0.5[a]	–	Sekhwela (2000) (unpublished data)
Grewia bicolor	0.2[a]	–	Sekhwela (2000) (unpublished data)
Colophospermum mopane	117	10.7	Tietema (1993)
Acacia tortilis	24	1.8	Tietema (1993)

[a]Encroaching plants still bushed at the time of measurements in Maun (*D. cinerea*) and Tsabong (*G. Bicolor*)

Whilst these species are considered here for potential bioenergy resources, their current importance underlines their multiple use and potential for diversifying the economic opportunities of rural communities. The fodder value of the fruits of the vigorous invader *D. cinerea*, is considered high in Zimbabwe, whilst cattle were observed to ingest its fruits in a monitoring study in Botswana (Abbot 1993; Moleele 1998). Similarly, Sekhwela (2003) recorded a heavy reliance on *A. mellifera* flowers and sprouting twigs by goats in the Kalahari during the dry season that saves livestock during the period of high food stress. The fruits of *A. tortilis* that have in the past been ingested by livestock in the rangelands are nowadays collected by rural inhabitants for sale to livestock farmers. This creates a conflict with livestock owners of cattle which ingest the fruits of this tree species in the area (Dube and Sekhwela 2008). The foregoing results highlight the interactive nature of the use of natural resources in rural economies for energy and other purposes as both *A. mellifera* and *D. cinerea* are valuable for the production of fuel-wood and construction poles. However, the oversupply of these tree species, due to their invasiveness and vigorous growth, makes them a nuisance, hence they have the potential to be used for the development of modern energy systems.

The harvested bushes/wood could also be chipped and packaged for export as feedstock for biomass energy conversion systems (Heinimo and Junginger 2009). The harvesting and packaging would create employment for the local communities together with resource management systems to be put in place for sustainable supply. This would provide income for the poor whose livelihoods are currently compromised by these invading species. The harvesting of these resources would increase carbon sequestration with new replacement growth of the regeneration processes, or restoration of grasslands for normal economic activities. Opportunities for producing other forms of bioenergy such as liquid biofuels could be explored, given the large variety of 'nutty' plant species that are found in the dry-lands such as *Sclerocarya birrea*, *Ximenia sp.* and other plant species that store energy in the form of oils (Taylor 2000; Bekunda et al. 2008). These species are within the reach of the rural communities and can contribute to economic development by providing adaption strategies to potential impacts of climate change in Botswana (Dube and Sekhwela 2008).

2.2.2 Oil Plant Resources

There are generally few studies that have explored the oil potential of plants found in Botswana (Table 2.4). Such studies have generally been left to the cosmetic industry until recently with increasing desire to find alternative liquid fuel resources. The existing information is largely anecdotal for the few species that are now being exploited for cosmetics and food, such as *Sclerocarya birrea* and *Ximenia sp.*, for which production assessments have been carried out (Phytotrade 2007). For instance, the oil content of *Ximenia sp.* was found to be 70% of the kernel weight, with the quality close to that of olives (Phytotrade 2008). The oil content is even higher than

Table 2.4 Oil plant species found in Botswana (Source: Taylor 2000)

Species	Oil source	Uses of other plant parts
Cleome gynandra	Seeds	Leaves and flowers edible
Croton megalobotrys	Seeds	Bark and seeds medicinal
Croton menyharttii	Leaves	Fruit edible
Xeminia species	Seeds	Fruit edible, bark and leaves medicinal
Sclerocarya birrea	Seeds	Fruit edible
Pluchea leubnitziae	Whole plant	Essential oils
Ricinus commins	Seeds	–
Pappea capensis	Seeds	Fruits edible, seeds and leaves medicinal

that of *Jatropha curcas*, an energy crop with 40% oil content (Azam et al. 2005). However, the short acid chains of *Ximenia* oil lead to gumming in the engine combustion chambers (Azam et al. 2005). Due to its high oil content, *Ximenia sp.* shows good potential as energy plant, but there is need for more research on the possibilities of commercial production as well as on oil extraction and processing for energy products and services.

Whilst *Sclerocarya birrea* has a larger fruit than *Ximenia sp.*, it has a lower oil component relative to the size of the fruit. However, the fruit of *S. birrea* has been largely valued for its flesh which is used as food and also for drinks and liquors. Whilst the same can be said about *Ximenia sp.*, only the flesh that is high in vitamin C has been traditionally used as food, and the kernel is usually thrown away. There is hardly any information on the oil potential of other species in Botswana (Table 2.4) and in many other southern African countries (Bekunda et al. 2008).

As indicated in Table 2.4, many of the species have other economic uses that raise their potential for diversifying rural economies. In particular, *Ximenia sp.*, which is a good browse species that coppices well and that has a fast growth rate. Therefore it could be valuable for small farmers (Abbot 1993). *Sclerocraya birrea* which is valued for its wood is a vigorous and fast coppicing plant. All the above-mentioned plants occur naturally. They are beneficial to humans, livestock and wild animals. Their use for various applications has been identified. The potential impacts of these species for bioenergy development at a larger scale than the current use are not yet known. It is however notable that the current uses of these species by rural communities have been in existence for millennia being sustainable in terms of resource availability, biodiversity and related ecosystem services.

2.2.3 Trade in Bioenergy Products and Services

According to Heinimo and Junginger (2009), the development of international trade in biomass for energy purposes is still in its early stages, but grows rapidly as influenced by international climate agreements. Sub-Saharan Africa is considered to become one of the important producers in the long-term, whilst the main demand will be in the OECD countries and South-East Asia. Heinimo and Junginger (2009) indicate that wood pellets will be a successfully traded commodity due to their low moisture content, high calorific value, stability over time and easy handling which allows for long distance transportation. Pellets would therefore be the most favoured option for using invading woody plant species growing in remote areas with poor infrastructure. Conversion of this resource into tradable pellets could contribute to the creation of small-scale rural industrial enterprises and a diversification of income generating activities for the rural poor.

2.2.4 Sustainability Potential

Available information on both woody and oil plant resources indicates a high potential for meeting sustainability criteria, particularly from production systems that could 'mimic' natural ecosystems. The invasive species are mostly secondary colonisers that exhibit the behaviour due to the imbalances created by improper land use systems that result in overgrazing and consequent land degradation. Where these species are managed properly, the production can be sustainable and provide ecosystem services as part of the natural habitat of the areas. The vigorous growth exhibited by some of the species shows their ability to grow in semi-arid and arid conditions that are predicted to further degrade due to climate change (Hulme 1996). The adaptation would increase the potential to meet the water use and preservation criteria, whilst the natural occurrence of the species will positively influence its biodiversity contribution. For instance, *Ximenia sp.* was found to be valuable as a habitat for the pupal stage of the *Imbrasia belina* moth of which the host is *Colophospermum mopane* in north-eastern Botswana (Sekhwela et al. 2007).

There is generally no reliable long-term data and information to determine the sustainability of potential production systems of oil feedstock on a commercial basis. The processing of oil from these indigenous plant species is also not adequately known. Hence, Bekunda et al. (2008) noted with concern that despite the fact that the potential of oil plants in South Africa, has long been identified, hardly any research has been carried out. It also became apparent that even the exotic plants such as *Jatropha curcas* have not been sufficiently studied for their suitability for growing in the arid and semi-arid conditions such as those found in southern Africa (Bekunda et al. 2008). Nevertheless, some countries such as Namibia have embarked on production of bioenergy based on *Jatropha curcas* without further research (Namibian Agronomy Board 2007). Whilst Botswana has also identified the same species as potentially suitable for production, it has embarked on studies to determine the suitability and sustainability of production systems (Botswana Government 2009a). Botswana has also recently completed a study on the development of a biomass energy strategy which is supposed to create a sound development of bioenergy resources in the country (Botswana Government 2009b). It is important that local species are included in such studies. For instance, anecdotal information shows that the softly coated seed of *C. mopane* contains oil, highlighting again the potential multiple benefits from indigenous species.

In reviewing the current approach to the development of alternative fuels, the Earth System Science Partnership (ESSP) noted that it is not yet clear how the identified bioenergy production systems are positively contributing to sustainable climate change mitigation (Klepper et al. 2008). It was also realised that invariably bioenergy systems are dependent on ecosystems and vulnerable to climate change, like food production systems. ESSP also noted that whilst the production of bioenergy crops has recently been linked to food shortages, the issue of land availability for bioenergy production has not been settled and there is need for further research. Klepper et al. (2008) also noted that it is not yet clearly known how much biofuels can contribute to mitigate climate change, and how sustainability standards for bioenergy systems should be defined.

2.2.5 *Implications for Bioenergy Policy*

As highlighted in the introduction, the existing sustainability criteria are guiding principles for new energy initiatives that have to be in line with the global driver to reduce GHG emissions and mitigate climate change. The criteria also include aspects on the protection and enhancement of environment and ecosystem services, whilst maintaining socio-economic development that has both local and off-site benefits. Therefore, the current draft on bioenergy policies in Botswana can benefit from the sustainability criteria. The approach adopted by the Government of Botswana to undertake studies to determine the suitability and potential of various bioenergy production systems and crops is a step in the right direction. Research also includes indigenous plants that have potential for bioenergy production and other applications for food, fodder and medicine in Botswana and other southern African countries. As already suggested, indigenous species have the advantage of serving multiple uses and forming part of the existing natural ecosystem services that include habitat for fauna.

2.3 Conclusions

The introduction of sustainability criteria could provide a robust framework to guide the development of suitable and sustainable bioenergy systems in Botswana. As bioenergy production systems rely on plant-based feedstock, locally available energy resources can provide opportunities for diversifying rural economies.

The identified woody plant species that have low economic value due to their overproduction relative to local use can be used for the production of modern energy systems locally or off-site. The production of wood chips or pellets could be a suitable option for biomass trade. Similarly, oil plants are largely unexplored for biofuel purposes, and currently little information is available and often limited to trees that serve cosmetic industries. The utilisation of invading woody species for bioenergy production is likely to contribute to the mitigation of climate change and to enhance the attainment of the MDGs. Therefore, further research is recommended, particularly on production systems to use plants for the production of bioenergy. Specifically, research is urgently needed on the following topics:

- Potential for sustainable reduction of invasive species and their use for bioenergy feedstock and markets
- Impact of invasiveness on biodiversity and harvesting for bioenergy purposes
- Potential for plantations of invasive species and oil plants
- Determination of suitable processing technologies for both woody and oil containing resources

References

Abbot PG (1993) Research and development of simple silvicultural systems for community management of Miombo in Malawi. In: Piearce GD, Gumbo DJ (eds) The ecology and management of indigenous forests in southern Africa. The Forestry Commission, Harare

Azam MM, Warris A, Nahar NM (2005) Prospects and potential of fatty acid methyl esters of some non-traditional seed oils for use as biodiesel in India. Biomass Bioenerg 29:293–302

Bekunda M, Palm CA, De Fraiture C et al (2008) Biofuels in developing countries. In: Howarth RW, Bringezu S (eds) Biofuels: environmental consequences and interactions with changing land use. Proceedings of the Scientific Committee on Problems of the Environment (SCOPE), 22–25 September 2008, Gummersbach, Germany. Cornell University, Ithaca

Bond WJ, Midgley GF (2000) A proposed CO_2- controlled mechanism of woody plants invasion in grasslands and savannas. Glob Change Biol 6:865–869

Botswana Government (2009a) A tender for provision of consultancy services for the development of national biofuels guidelines and research on the effects of *Jatropha sp* farming on local farmers. Ministry of Minerals, Energy and Water Affairs, Gaborone

Botswana Government (2009b) Botswana biomass energy strategy. Final report, Ministry of Minerals, Energy and Water Affairs, Gaborone

de Pous P (2008) EEB policy briefing on EU biofuels and bioenergy policy. European Environmental Bureau (aisbl), Brussels. www.eeb.org

Dube PO, Sekhwela MBM (2008) Indigenous knowledge, institutions and practices for coping with variable climate in the Limpopo Basin of Botswana. In: Leary N, Adejuwon J, Barros V, Burton I, Kulkarni J, Lasco R (eds) Climate change and adaptation. Earthscan, London, pp 71–89.

Fehrenbach H, Glegrich J, Reinhardt G et al (2008) Criteria for a (include all authors) sustainable use of bioenergy on a global scale. Research report 206-41-112, Federal Environmental Agency, Federal Ministry of the Environment, Netherlands. http://www.umweltbundesamt.de

Hall DO, Rosillo-Calle F, Williams RH et al (1993) Biomass for energy: supply prospects. In: Johansson HK, Reddy AKN, Williams RH (eds) Renewable energy: sources for fuels and electricity. Island Press, Washington D.C

Heinimo J, Junginger M (2009) Production and trading of biomass for energy – an overview of the global status. Biomass Bioenerg 33:1310–1320

Hulme M (1996) Climate change and southern Africa: an explotation of some potential impacts and implications for SADC Region. Climate Research Unit, University of East Anglia, Norwich, UK

Kartha S, Leach G, Rajan SC (2005) Advancing bioenergy for sustainable development: guidelines for policymakers and investors. Energy Sector Management Assistance Programme (ESMAP) report 300/05, Advancing Bioenergy for Sustainable development Word Bank, Washington

Kedikilwe K (2008) The Minister speech, annexure 1. Proceedings of the Botswana biofuels workshopm, 15th–16th July 2008. Department of Energy, Ministry of Minerals, Energy and Water Affairs, Gaborone

Klepper G, Canadell P, Leemans R et al (2008) ESSP Research on bioenergy and earth sustainability: tapping GEC programme-wide expertise for the benefit of science and society. IHDP Update Science Dialogue Extra, Special Edition 2008: 31–36. www.ihdp.org

Krutilla K, Hyde WF, Barnes D (1995) Periurban deforestation in developing countries. Forest Ecol Manag 74:181–195

Mabowe BR (2009) A proposal for biodiesel development from Jatropha seeds. Ministry of Minereals, energy and water resources, Gaborone

Moleele NM (1998) Encroacher woody plant browse as feed for cattle: cattle diet composition for three seasons at Olifants Drift South East Botswana. J Arid Environ 40:255–268

Mushove PT, Makoni JT (1993) Coppicing ability of *Colophospermum mopane*. In: Piearce GD, Gumbo DJ (eds) The ecology and management of indigenous forests in southern Africa. The Forestry Commission, Harare, Zimbabwe, pp 226–230

Namibian Agronomic Board (2007) Annual report number 20. Ministry of Agriculture, Water and Forestry, Windhoek

Phyto Trade Africa (2007) Morula Oil. http://www.phytotradeafrica.com/products/MarulaOil.html

Phyto Trade Africa (2009) Xemenia Oil. http://www.phytotradeafrica.com/products/xeminiaoil

Ringrose S, Matheson W, Wolski P, Huntsman-Mapila P (2003) Vegetation cover trends along the Botswana transect. J Arid Environ 54:297–317

Royal Society (2008) Sustainable biofuels: prospects and challenges. Royal Society, London

Rutz D, Van Dam J, Janssen R, Hiegl W, Diaz Chavez RA, Woods J, Faaij APC, Smeets EMW, Vos J, Vis M, Bottriell K, Rettenmaier N, Reinhardt G, Nussbaum R, Gozzi A, Haye A, da Silva Walter AC, Hilbert JA, Coto C, Fallot A, Sawe EN, Togola I, Ouattare O, Jaax R (2010) Global socio-economic impact assessment of biofuel and bioproduct chains. Proceedings of the 18th European biomass conference and exhibition; ISBN 978-88-89407-56-5

Schlegel S, Kaphengst T (2007) European Union policy on bioenergy and the role of sustainability criteria and certification systems. J Agric Food Ind Organ 5(7):1–17

Sekhwela MBM (1997) Environmental impact of wood biomass use in Botswana: the case of fuelwood. In: Kgathi DL, Hall DO, Hategeka A, Sekhwela MBM (eds) Biomass energy policy in Africa. Zed Books Ltd, London, pp 64–144

Sekhwela MBM (2000) Woody biomass production ecology in Kalahari communal areas of Botswana. PhD Thesis, Botany Department, University of Queensland, Brisbane

Sekhwela MBM (2003) Woody vegetation resource changes round selected settlement along aridity gradient in the Kalahari, Botswana. J Arid Environ 54:469–482

Sekhwela MBM, Ntseane PG, Dube PO et al (2007) Towards an integrated sustainable harvesting, management and conservation of Phane and Mophane Woodland. Phase I final report, Office of Research and Development, University of Botswana, Gaborone

Skarpe C (1990) Shrub layer dynamics under different herbivore densities in arid savanna, Botswana. J Appl Ecol 27:873–885

Skarpe C (1991) Spatial patterns and dynamics of woody vegetation in arid savanna. J Veg Sci 2:565–572

Smeets E (2008) Possibilities and limitations for sustainable bioenergy production systems. PhD Thesis, Copernicus Institute for Sustainable Development and Innovation, Utrecht University, The Netherland

Taylor FW (2000) Changes in the utilisation and management of renewable natural resources (veld products) in the Kalahari. In: Ringrose S, Chanda S (eds) Towards sustainable management in the Kalahari region: some essential background and critical issues. Directorate of Research and Development, University of Botswana, Gaborone

Tietema T (1993) Possibilities for the management of indigenous woodlands in southern Africa: a case study from Botswana. In: Piearce GD, Gumbo DJ (eds) The ecology and management of indigenous forests in southern Africa. The Forestry Commission, Harare

Wicke B, Dornburg V, Junginger M et al (2008) Different palm oil production systems for energy purposes and their greenhouse gas implications. Biomass Bioenerg 32(12):1322–1337

World Bank (2008) World development report. World Bank, Washington D.C

Zimmerman H, Pasiecznik NM (2005) Realistic approaches to the management of *Prosopis* species in South Africa. Plant Protection Research Institute, Agricultural Research Council, Pretoria

Chapter 3
Jatropha: A Promising Crop for Africa's Biofuel Production?

Janske van Eijck, Edward Smeets, and André Faaij

Abstract Jatropha has often been proposed as a miracle crop for the production of oil, because of the high yields and low requirements in terms of land quality, climate and crop management. A large number of companies have started with jatropha production in Africa which is projected to increase rapidly. Yet, the sector is not fully developed and therefore the economic viability is unclear. Crucial issues for the economic performance are the crop management system, level of inputs and thereby yield and labour requirements, the price of jatropha seeds, and the business model used (e.g. farmer-centred, plantation model). Other factors influencing the sustainability of jatropha production and use are land use conversions and their resulting impacts on GHG emissions, as well as socio-economic impacts which depend largely on the combination of local socio-economic circumstances and on the business model. Plantations have generally larger negative effects on biodiversity and land issues than farmer-centred models, but larger positive effects on employment levels. Farmer-centred models are generally more pro-poor due to technological spillovers and the larger number of farmers involved. Especially when jatropha products are used to increase energy access, local communities can benefit. More research is required to determine optimised agricultural practices, long-term effects on food security, local prosperity and gender issues and technological development of equipment that can use jatropha products. It should be avoided to replace food crops with jatropha to avoid negative impacts on food security.

J. van Eijck (✉) • E. Smeets • A. Faaij
Department of Science, Technology and Society, Copernicus Institute,
Utrecht University, Budapestlaan 6, 3584 CD, Utrecht, The Netherlands
e-mail: J.A.J.vanEijck@uu.nl; www.chem.uu.nl/nws; esmeets@hotmail.com
www.chem.uu.nl/nws; A.P.C.Faaij@uu.nl

R. Janssen and D. Rutz (eds.), *Bioenergy for Sustainable Development in Africa*,
DOI 10.1007/978-94-007-2181-4_3, © Springer Science+Business Media B.V. 2012

Keywords Jatropha • Cultivation • Yields • Labour requirement • Plantation • Outgrowers • Crop management system • Business model • Socio economic impacts • Farmer-centred model • Contract farming • Outgrower • Government-centred model • Corporate-centred model

3.1 Introduction

Sub-Saharan Africa is frequently mentioned as a region with large potential for bioenergy production (Marrison and Larson 1996; Hoogwijk et al. 2005; Smeets et al. 2007). However, the production and use of bioenergy does not necessarily contribute to sustainable development. The environmental and socio-economic impacts depend, amongst others, on natural conditions (climate, soil), socio-economic settings (employment, poverty, governance), and crop production systems (low vs. intermediate or high inputs) (Dornburg et al. 2010; Schut et al. 2010b).

Jatropha has been promoted as a potential renewable energy source. Initially it was claimed to simultaneously reclaim waste land, enhance rural development as well as producing biofuels (Francis et al. 2005). Large investments in jatropha (*Jatropha curcas* L.) are ongoing, a total of 900,000 ha were planted globally up to 2008, and projections for 2015 show a total of 12.8 million ha of which large parts are in Asia (±80%) and Africa (±15%) (GEXI 2008). But, agronomy, socio-economic and technical aspects of the jatropha value chain and its implications on the sustainable livelihoods of people are still unknown. The uncertainties will be described in this chapter focusing on cultivation management, yields, labour requirements, price of jatropha seeds, different business models, and on socio-economic impacts.

3.2 Cultivation Management and Yields

Most of the *Jatropha curcas* L. grown in Africa is from the Cape Verde variety. Other two varieties that are identified are the Nicaraguan and Mexican varieties, with larger, but with fewer fruits and less toxic seeds compared to the Cape Verde variety (IFAD/FAO 2010). The plant requires relatively little management and starts to produce seeds after 1–3 years, depending on soil and climate conditions (van Eijck and Romijn 2008). Jatropha can be productive for 30–50 years (Achten et al. 2008; van Eijck and Romijn 2008; IFAD/FAO 2010).

Field preparation required before planting jatropha consists of ploughing, marking and digging holes. Jatropha does not require high quality soils, but cannot stand water logging and therefore heavy soils like heavy clay soils are less suitable (Achten et al. 2008). *Planting* of jatropha uses different propagation techniques, namely direct sowing, seedlings raised in nurseries or via vegetative propagation by plant cuttings. Cuttings have a much quicker growth path, but the plants will not develop tap roots which may be problematic in extremely dry areas. Plant spacing can vary, the optimum density is determined by the purpose of planting, e.g. as fence, erosion stopper, support tree for vanilla, or to produce biofuels on monoculture plantations. If other crops are planted on the same plot (i.e. in intercropping systems), a distance of at least 3–5 m is required between jatropha rows. This allows other crops to obtain enough sunlight. *Weeding* is normally only required in the first 4 years after planting jatropha. After this period jatropha plants have grown enough to compete favourably for sunlight with weeds (Loos 2008; Mitchell 2008). Especially after rainy periods weeds can grow vigorously and compete for nutrients in the soil. Therefore, weeding is one of the most important tasks during the initial stages of cultivation. *Pruning* should be carried out every year, since this practice induces branching, which in turn stimulates higher yields. Pruning should be performed before the start of rainy periods (Achten et al. 2008; Behera et al. 2010). Periodic thinning is advised for plantations, to enable sunlight to reach the jatropha plants (Jongschaap et al. 2007). *Fertilisation* can increase yields significantly. This is an important aspect because nutrient levels may decrease when jatropha seeds are harvested and if these nutrients are not replenished. Based on the nutrient composition of jatropha fruits taken from Jongschaap et al. (2007) it can be estimated that from harvesting one ton of seeds 14.3–34.3 kg N, 0.7–7.0 kg P and 14.3–31.6 kg K are removed from the soil (Achten et al. 2008). However, existing fertilisation rates are often much lower (Mwangi 1996; IAC 2004). At the moment several tests are being executed to identify the value of seedcake, which is obtained after pressing of jatropha seeds, as fertiliser. Furthermore, *pest and disease control* reduces fungal and insect attacks, although the effectiveness on the long-term for large plantations is not (yet) known. Jatropha can be sensitive during the first years, in Kenya for example many farmers reported damage by insects (GTZ 2009b).

When jatropha plants are mature, fruits will be generated after every rainy period. Several fruit development stages (flowering and fruiting) occur on the same plant, so repeated harvesting is required over a period of several months (Mitchell 2008). Each fruit contains three seeds, although occasionally two or four seeds can be found. The fruits need to be dehulled to obtain the seeds. This demands larger labour efforts (see Sect. 3.3) or machinery, if available. Drying of the seeds is required if the moisture content of the seeds is too high. Packing and transport to the processing unit are the final steps before the oil is produced.

The level of inputs varies with different cultivation systems. Inputs refer to the use of agro-chemicals (pesticides, herbicides), fertilisers and other management practices like pruning and weeding. Three levels of inputs can be distinguished. In case of a *low input system* no agro-chemicals or fertilisers (synthetic or manure) are used and no pruning and irrigation is applied. Only weeding is performed. This crop management is common practice in smallholder agricultural systems (Loos 2008; Mitchell 2008; Fermont et al. 2009). In an *intermediate input system* limited use is made of agro-chemicals, fertilisers (synthetic or manure), and weeding and pruning is carried out, but no irrigation is applied. Lastly, a *high input* system refers to a high level of fertilising, weeding, pruning, and irrigation. This last system is not common practice for smallholders, but is practised by more capital intensive investments on plantation scale. The different levels of input have a significant impact on the yield, although the number of field data is limited, especially with respect to the performance on the long-term.

Other inputs required for jatropha cultivation are: tools for field preparation, weed control and pruning (hoes and machetes), the initial seeds or seedlings for planting, fertilisers, pesticides and irrigation systems (if applicable), and potentially land rent. Also packaging material expenses can be part of input expenses. Depending on the year, total expenses excluding labour in a low or intermediate input system lie between 21 and 68 US\$ ha^{-1} $year^{-1}$.

There is large variability in jatropha yields due to differences in climate, soil characteristics and crop management (Achten et al. 2008, FACT Foundation 2009; van Eijck et al. 2011). Systematic yield monitoring for jatropha started only recently, so there are still large uncertainties about long-term yields. Older jatropha literature mentions a large range of 0.5–12 t ha^{-1} $year^{-1}$ (Francis et al. 2005). Based on recent data the bandwidth is narrowed, the upper range of 12 t ha^{-1} $year^{-1}$ has been reduced. For example, Jongschaap et al. (2007) projected a maximum of 7.8 t ha^{-1} $year^{-1}$for mature stands with a range of 1.5–7.8 t ha^{-1} $year^{-1}$. And Trabucco et al. (2010) modelled yields based on a maximum of 5 t ha^{-1} $year^{-1}$.

Jongschaap et al. (2007) also mention observed yields for 1 or 2 year old plantations, ranging from 0.6–4.0 t ha^{-1} $year^{-1}$, whilst Heller (1996) and Tewari (2007) state 2.0–3.0 t ha^{-1} $year^{-1}$ as a good range for semi-arid wastelands. The report by Endelevu Energy on jatropha in Kenya projected a much lower maximum yield in very dry areas of 0.86 kg per tree, which relates to 1.4 t ha^{-1} $year^{-1}$ if 1,600 trees are planted ha^{-1}(GTZ 2009b). A recent report that combined literature of jatropha related studies includes a range of verified observed data of 0–2.5 t ha^{-1} $year^{-1}$ (van Eijck et al. 2010). The yield is determined largely by the input system. Low input systems will generate a yield in the lower ranges (like the GTZ study), whilst intermediate and high input systems will generate yields closer to the maximum.

According to Nielsen (2009a), the growth of jatropha follows an S-shaped curve with slow growth in the beginning, an increased growth in later years levelling off towards the end. And according to observations of Nielsen (2009a) and van Eijck et al. (2011) maximum yield under poor conditions can be reached as late as in year 8. However, this is greatly influenced by climate conditions.

The report by van Eijck et al. (2010) furthermore concluded that scientific data related to cultivation practices and yields is still lacking. Often yields are published without additional information like plant spacing, rainfall data, soil condition, fertiliser application and composition, and other specific agronomic data. Partly due to this reason, cultivation management techniques are still not optimised for local conditions in the regions where jatropha is grown. Although there is no doubt that considerable yields can be obtained from older stands of jatropha (15 years and older) planted in very wide spacings, it is unclear how long this will take when jatropha is planted commercially. This is a crucial limiting factor in the determination of the potential for jatropha.

3.3 Labour Requirements

The total manual labour requirements for jatropha cultivation and harvesting were investigated by several studies as shown in Table 3.1.

The total number of labour days varies with the input system and the production year. It is between 17 and 107 days ha^{-1} $year^{-1}$ or 34–214 US$ ha^{-1} $year^{-1}$ (with a labour rate of 2 US$ day^{-1}) (van Eijck et al. 2011).

The major time consuming task in the cultivation of jatropha is harvesting, until now done manually. Depending on the density of jatropha cultivation, one person can harvest on average around 40 kg seeds per day. Data from 12 jatropha sites showed that when jatropha seeds are picked from fences or 'wild' jatropha, the total amount of seeds that can be picked per day is on average 20–30 kg, whilst plantations that are well maintained could make it possible to pick 40–60 kg per day (FACT Foundation 2009). Empirical evidence in Mozambique suggests that one person can collect seeds at 8–24 kg per day or 1–3 kg per hour, assuming an 8 h working day, including dehulling, and 40 kg per day or 5 kg per hour without dehulling (Nielsen 2009b). This shows the relatively large labour requirement for dehulling. Electrical and manually operated dehullers are developed, though usually dehulling takes place at the field.

Mechanised harvesting systems are under development, but are problematic, because both ripe and immature seeds as well as flowers grow simultaneously at the shrubs (Jongschaap et al. 2007). This could reduce yields. Mechanised harvesting systems are nevertheless being developed, but will most probably not be feasible for smallholders. However, several harvesters have been developed and are being tested. Mechanical harvesters have not been optimised yet, but one example that is on the market (BEI harvester) can harvest 1.5 ha per hour. The labour requirement for jatropha is significantly reduced by approximately 40% if mechanised harvesters are used.

Other tasks that require labour occur mostly in the initial stages of jatropha cultivation. Field preparation and planting for example only have to be executed once in the lifetime of jatropha. Weed control is especially important during the first years of establishing the plantation (Loos 2008; Mitchell 2008).

Table 3.1 Labour requirements for jatropha cultivation and harvesting

Days ha^{-1} $year^{-1}$	Details	Source
70	From year 6 onwards	Jongschaap et al. (2007)
200	First year	Francis et al. (2005)
50	Year 2+	
91–107	First year	van Eijck et al. (2011)
17–82	Year 2+ (low or intermediate input system)	

Table 3.2 Prices paid for jatropha seeds reported by various studies

Country	Type of project	Price per kg (US$)	Details	Study
Tanzania	Outgrowers (2008)	0.07	Guaranteed price to farmer	Struijs (2008)
Tanzania	Outgrowers (2009)	0.08–0.17	0.08 US$ is guaranteed	van Eijck (2009)
Tanzania	Small farmers	0.08–0.25	For soap production	Messemaker (2008), Altenburg et al. (2009)
Mozambique	Small-scale	0.08–0.18	0.18 US$ is paid to participating farmers, 0.08 US$ to other farmers	Nielsen (2009b)

3.4 Price of Jatropha Seeds

Jatropha oil is an alternative for fossil diesel and therefore from a market perspective, jatropha oil has to be competitive with diesel fuel. Jatropha oil can also be used for soap production, but that market is relatively small. Jatropha oil (Straight Vegetable Oil or SVO) can be chemically modified by the transesterification process to obtain jatropha biodiesel. However, this will increase the costs, especially at small-scale. The focus of this chapter is on SVO.

The consumer price of jatropha SVO is determined by the costs of feedstock, transport, processing and potential taxes. For feedstock, around 4 kg of seeds are needed per litre oil. The conversion will add around 0.20 US$ per litre to the total expenses, based on motor press costs in Zimbabwe (Openshaw 2000). Average fossil diesel prices in Africa (in 2008) are around 1 US$ per litre (GTZ 2009a). Therefore, without transport costs or taxes, the maximum feedstock price could amount to 0.80 US$ l^{-1} or 0.20 US$ kg^{-1}. But, as transport expenses have to be taken into account as well, current prices paid for seeds are between 0.06–0.20 US$ kg^{-1} (see Table 3.2). The prices paid for jatropha seeds could increase in the future if the oil price rises, production systems become more efficient and cost effective, e.g. by increased yields, optimising the value of the by-products, and by establishing more efficient transport systems.

3.5 Business Models

Various business models can be applied for the production of jatropha. A possible distinction is made by Altenburg et al. (2009) who discusses a *government*-centred model (cultivation on communal and government land), a *farmer*-centred model (cultivation on private land) and a *corporate*-centred model (large-scale cultivation). Furthermore, a farmer-centred model can have different business models as described by Vermeulen and Cotula (2010) and Bijman et al. (2009). Bijman et al.

differentiate the following farmer-centred systems: (1) centralised model (processor sources from a large number of (small) farmers), (2) nucleus estate model (processor sources from farmers and own production facilities) and (3) multipartite model (joint venture between the state, a private company and farmers). The last model allows a state agency to provide technical support and inputs whilst the processor guarantees market access. In this way the risks and expenses are divided between the state and private companies. This model is implemented and tested in India, for example by the use of Rural Business Hubs or the Rural Employment Guarantee Scheme (GRAIN 2008).

Which business model is applied depends on local circumstances and project goals. Sometimes practical issues determine the choice of business models. In Mozambique, a project designed for the energy needs of local people decided to press the seeds at a central location because quality control was only feasible with larger quantities. It made transport expenses higher and seedcake could not be returned to farmers easily (Nielsen 2009b). In Tanzania similar factors contributed to the decision to establish a central processing unit (van Eijck 2009).

In this chapter two business models are highlighted, namely the "farmer-centred model" and the "corporate-centred model". As the name of the corporate-centred model is not regarded sufficiently clear, the term "plantation model" is applied in this chapter.

3.5.1 Farmer-Centred Model

Farmer-centred models are also called contract farming models or outgrowers models. Production is done by cultivation of biofuel crops on family owned land and family operated farms aimed at self-sufficiency with respect to the production of food. Bijman et al. (2009) analysed contractual arrangements for jatropha smallholders in Mozambique. They state that the transaction costs between farmers and jatropha processors are high, thereby increasing risks. This is due to the long time lag between planting and harvesting, the lack of knowledge and experience as well as access to inputs and supporting services. Processors also risk abuse of inputs (if provided) and sideselling of the crop. Finally, formal contractual arrangements were found to be potentially difficult to maintain due to weak property rights enforcement (Bijman et al. 2009). Organising farmers in a group facilitates access of the members to credit schemes and makes the implementation of trainings easier. Farmer clubs are for example set up in Mozambique (Nielsen 2009b). The advantage of outgrower models is that start-up costs are less than for large-scale corporate models, as no land has to be acquired before start-up. Land acquisition is often very politically sensitive (see Sect. 3.6).

Experiences from setting up farmer-centred models (outgrower network) show clear benefits for local people. However, such models are very time consuming, as it takes effort to convince outgrowers, requires adequate funding to cover for the long pay back period. In addition, market distortions reduce reliability of feedstock supply (van Eijck 2009). Finally, in outgrowers systems the distances are an important factor as certain regions may be more easily accessible than others (e.g. due to bad infrastructure), and hence, transport costs can be high.

3.5.2 *Plantation Model*

Large-scale operations are characterised by the use of large areas of land and advanced crop management techniques in combination with wage labourers. Large demand for biofuel drives investors to use large-scale corporate models. But a number of companies that applied this model have currently financial problems or went bankrupt. Part of these problems may be explained by the financial crisis in 2009, but another part could be explained by inflated yield figures and a lack of long-term funding. For example, a jatropha company in Tanzania, that requested 8,000 ha of land, spent 3 years to complete the land acquisition process (Gordon-Maclean et al. 2008; FAO 2010). In Tanzania at least one plantation company went bankrupt (Bioshape) and another in Mozambique (Energem). So far, to the authors' knowledge, no successful large-scale jatropha plantations exists producing commercial amounts of jatropha seed. Thus, it will take a few more years before this model can proof its suitability for jatropha. Today, large plantations are being developed, for example 160,000 ha in Kenya (Christian 2010). Senegal aims at the development of approximately 300,000 ha jatropha plantations in 2012 (Ministry of Agriculture Senegal 2007).

3.5.3 *Comparison Between the Farmer-Centred and the Plantation Models*

The financial viability of jatropha varies with the business model used. A large influencing factor is the value of the by-products. Without the use of by-products, the jatropha system is currently hardly economical. This is also true as current yields are relatively low. More research is required to develop agronomic practices and plants with higher yields. For plantation models, the high initial investment and the relatively long time before cash is generated, involves cash flow risk. A farmer-centred system does not have these high initial costs, but providing training and extension services is rather costly and time consuming. The price of seeds determines the price of jatropha oil which in turn should be competitive with fossil diesel to ensure local markets. Jatropha oil is currently not viable as commercial substitute for fossil diesel. Carbon credits can help investors to obtain additional revenues, but not many initiatives have yet been successfully. In farmer-centred models transport expenses are important.

The size of the plantations has high impacts on the region. In the south of Tanzania for example, a large-scale plantation of 80,000 ha expected to employ 10,000 people.

If minimum wages of 65 US$ per month would be considered, a total of 156 million US$ would be paid to local employees over a period of 20 years. This is a higher number compared to farmer-centred business models. For example, if seeds are bought from local communities at a factory gate price of 0.17 US$, this price accounts for 0.08 US$ paid to farmers, around 0.03 US$ commission to collectors, and around 0.06 US$ to local transport. All these payments are done at local level, creating a total money flow of around 72 million US$ in 20 years, if 80,000 tons of seeds from 80,000 ha are considered. This means 80,000 farmers (if the average size is 1 ha) will have an additional income of around 80 US$ per year. So, although the total amount of money brought into local communities is less in farmer-centred business models, it reaches more people. Furthermore, farmer-centred models can provide additional sources of income for smallholders including the use of by-products as energy, food or wood-fuel (Achten et al. 2007; Achten et al. 2010).

3.6 Socio-Economic Impacts Related to Business Models

Many different studies have been published that list potential socio-economic impacts. Since many projects are still in early phases, the long-term effects are not yet clear. This chapter addresses food security, local prosperity, land rights, and gender issues based on the review of literature analysing existing projects.

3.6.1 Food Security

Food security impacts are difficult, if not impossible, to determine at project level. It is influenced by a variety of factors such as food availability, food access, food utilisation and food stability (UN 2008). Food availability relates to the crop production whilst food access relates to food prices and income level. The other two factors are less directly linked to the production of biofuels. Food utilisation relates to the ability of the population to absorb nutrients and food stability relates to events that can cause reduced access to food such as conflicts, disasters, etc. A study by FAO investigated the linkages between biofuel crop production and food security in Tanzania and found no significant negative impact (FAO 2010). The largest contributor to food insecurity in Tanzania is the currently low agricultural yields. It was concluded that even a slight increase of current yields will offset any effect on food security. Furthermore, FAO also emphasised the welfare gains from additional income and employment. Food security can, however, be negatively impacted when the cultivation of food crops is replaced by jatropha. This so far only happened in two cases, namely in Honduras and Brazil (Ariza-Montobbio and Lele 2010; Finco and Doppler 2010). In Brazil, a company that worked with contract farmers, promoted the cultivation as monoculture. When this company faced financial problems, the market could no longer be guaranteed and the farmers lost income and land availability, leading to a loss of food security. Such problems can be avoided if markets for jatropha seeds are well established with a variety of competing market actors. However, such cases show that converting areas used for food production can have negative impacts on food security and should therefore be avoided.

3.6.2 *Local Prosperity*

Local prosperity is determined by the benefits of the local population, either income or non-financial benefits like access to energy, capacity building and education. Especially the availability of jatropha oil for local communities and its use in special equipment has positive effects. If jatropha seeds are processed and the products are used locally, energy access levels can be increased. Jatropha production and processing offers opportunities for a wide variety of products that can be used locally:

- Soap
- Seedcake for biogas
- Oil for special stoves
- Oil for special lights
- Electricity by using oil in adapted generators

For domestic purposes, jatropha oil could mainly be used as cooking fuel in adapted cooking stoves, and as lighting source in adapted lamps. Another use is in electricity generators, potentially connected to a local electricity grid. However, to date only few stoves and lamps have been developed which are working properly with jatropha SVO. As SVO has a higher viscosity than fossil diesel or kerosene, it can not be used in standard equipment.

Employment levels vary according to the business model used. Plantations normally generate more direct labour; whilst farmer-centred models reach more people. Arndt et al. (2009) compared outgrower and plantation models, and concluded that the outgrower (smallholder) approach is more pro-poor. This is due to the differences in labour and capital intensity. Outgrower approaches use more unskilled labour resulting in technology spillovers. This result is supported by FAO claiming that all biofuel production scenarios improve household welfare and local prosperity, but small-scale outgrower schemes, especially for cassava and jatropha, are most effective at raising poorer households' incomes (FAO 2010).

Skills and attitudes are other important determinants of local prosperity. Technological spillovers ensure capacity building and lead to increased yields. However, financial problems (e.g. difficulties in cash flow), as currently experienced by some jatropha companies, can have a negative impact. In Mozambique, it has been reported that the bankruptcy of a jatropha company has led to decreased trust of people (Schut et al. 2010a). This is a negative change in attitude by local communities that has to be avoided. Thus, taking precautions and developing clear exit strategies should be part of a project design. A gradual scaling up of the organisation could lead to a more sustainable operation. It will enable local communities to absorb the changes, and in case of displacement of people will have less sudden impacts. It will also allow time for internal learning process.

3.6.3 Land Rights

Land conflicts are very common in Africa. In most cases land can only be leased, but lease times can be as long as 99 years (Tanzania). In plantation models, it is required to acquire land titles whereas for farmer-centred models this does not apply. Reoccurring issues when obtaining administrative land rights of large plots of land are: previous land use, compensation and transparency. Even though land may be registered as unused, local communities often obtain fuel and fodder from these areas or use it for grazing cattle. Furthermore, if previous land use was forest, converting this biomass into jatropha will generate a carbon debt. The carbon sequestered by jatropha will not necessarily offset this effect, depending on the density of the forest (Romijn 2010). Secondly, compensation paid to local communities for land acquisition is difficult since the determination of the land value is a complicated issue. Often the villagers are satisfied with any amount, because the land was not generating income before (Cotula et al. 2009). Land administration processes take very long time, in one case in Tanzania it was reported to take 3 years (Gordon-Maclean et al. 2008; Sulle and Nelson 2009). Transparency is often lacking and in some cases investors by-pass official (often very unclear) procedures. Promises made by investors often are not laid down in written agreements, which, in addition, are rarely in local language. This aspect does not only apply for jatropha, but in general for land acquisition processes. Neutral brokers serving as a liason between investor and community can help creating good deals for the community and a clear understanding on both sides. In Mozambique it has been observed that infrastructure was developed by jatropha investors leading to an improvement of the situation of the local population. In farmer-centred systems, land issues have less direct impacts. It has been reported, however, that in some cases the situation of vulnerable groups deteriorated, mostly caused by land pressure (Salfrais 2010). It is important to investigate the situation of land pressure before promotion of jatropha systems. But, on smaller scales, planting jatropha as fences can help to reduce land boundary conflicts (Salfrais 2010).

3.6.4 Gender

Gender issues for jatropha have not been investigated in great detail so far. Harvesting jatropha is normally done by women. More technical tasks, such as driving tractors and processing, are usually executed by men. However, the impact of a labour force of e.g. 10,000 (mainly women) workers, is unclear. Women could as well be trained for higher skilled jobs. In one case in Mozambique, it was observed that favourable working hours at the plantation enabled women to keep tending their household plots where they often cultivate food crops for their own consumption (Peters 2009). The availability of jatropha oil for domestic use (cooking and lighting) could benefit especially women.

3.7 Conclusion

The jatropha sector is not yet fully developed. Therefore it is difficult to assess the actual potential or long-term effects. Several projects demonstrate that jatropha can contribute to rural development, particularly when jatropha was present in the region and income is generated by selling seeds. Other projects have demonstrated that it is possible to establish jatropha in dry areas. However, some jatropha projects have failed financially due to different reasons including the lack of sufficient experience to compile a solid and accurate business plan. As no company is producing commercial quantities of jatropha oil it is currently difficult to determine the actual economic viability of jatropha production. Large-scale plantations will have larger negative impacts on biodiversity and land rights, but larger positive effects on employment and the local economy. Large-scale projects require gradual implementation to avoid sudden and large effects on the local environment, economy and population. Farmer-centred models have less negative effects on biodiversity and land rights. Employment levels are lower, but reach more people and are generally considered to be more pro-poor. Jatropha can clearly have benefits for local communities, especially when energy access is increased. This can be done by using jatropha oil or seedcake in special equipment for lighting or cooking. Jatropha will have a positive environmental effect when planted as additional crop, but not when natural vegetation is cleared. It can help to reduce soil and wind erosion. More research is required on agronomic practices to ensure increased yields. Since it is a perennial crop and since agronomic research is time consuming, it is expected to take at least 5 years before better genotypes are available. Finally, long-term effects on food security, local prosperity and gender have to be monitored. In order to minimise impact on food security, it should be avoided to replace food crops with jatropha.

References

Achten WM, Mathijs E, Verchot L, Singh SP, Aerts R, Muys B (2007) Jatropha bio-diesel fueling sustainability? Biofuel Bioprod Bior 1(4):283–291

Achten WMJ, Verchot L, Franken YJ, Mathijs E, Singh VP, Aerts R, Muys B (2008) Jatropha bio-diesel production and use. Biomass Bioenerg 32(12):1063–1084

Achten WMJ, Maes WH, Aerts R, Verchot L, Trabucco A, Mathijs E, Singh VP, Muys B (2010) Jatropha: from global hype to local opportunity. J Arid Environ 74(1):164–165

Altenburg T, Dietz H, Hahl M, Nikolidakis N, Rosendahl C, Seelige K (2009) Biodiesel in India, Value chain organisation and policy options for rural development, vol 43, Studies. German Development Institute (Deutsches Institut für Entwicklungspolitik), Bonn

Ariza-Montobbio P, Lele S (2010) Jatropha plantations for biodiesel in Tamil Nadu, India: viability, livelihood trade-offs, and latent conflict. Ecol Econ 70(2):189–195

Arndt C, Benfica R, Tarp F, Thurlow J, Uaiene R (2009) Biofuels, poverty, and growth: a computable general equilibrium analysis of Mozambique. Environ Dev Econ 15:81–105

Behera SK, Srivastava P, Tripathi R, Singh JP, Singh N (2010) Evaluation of plant performance of Jatropha curcas L. under different agro-practices for optimizing biomass – a case study. Biomass Bioenerg 34(1):30–41

Bijman JM, Slingerland M, van Baren S (2009) Contractual arrangements for smallholders in biofuel chains: a case study of Jatropha in Mozambique. VII international PENSA conference, Sao Paulo, Brazil

Christian C (2010) Natural oil discovered in "diesel tree". Fort McMurray Today. Fort McMurray

Cotula L, Vermeulen S, Leonard R, Keeley J (2009) Land grab or development opportunity? Agricultural investment and international land deals in Africa. IIED/FAO/IFAD, London/Rome

Dornburg V, van Vuuren V, van de Ven G, Langeveld H, Meeusen M, Banse M, van Oorschot M, Ros J, van de Born GJ, Aiking H, Londo M, Mozaffarian M, Verweij P, Lysen E, Faaij A (2010) Bioenergy revisited: key factors in global potentials of bioenergy. Energy Environ Sci 3:258–267

FAO (2010) Bioenergy and Food Security. The BEFS analysis for Tanzania. Environment and natural resources management working paper 35. Maltsoglou I and Khwaja Y, Rome, FAO, p 240

Fermont AM, van Asten PJA, Tittonell P, van Wijk MT, Giller KE (2009) Closing the cassava yield gap: an analysis from smallholder farms in East Africa. Field Crop Res 112(1):24–36

Finco MVA, Doppler W (2010) Bioenergy and sustainable development: the dilemma of food security and climate change in the Brazilian savannah. Energ Sust Dev 14(3):194–199

Foundation FACT (2009) The Jatropha handbook. FACT foundation, Eindhoven

Francis G, Edinger R, Becker K (2005) A concept for simultaneous wasteland reclamation, fuel production, and socio-economic development in degraded areas in India: need, potential and perspectives of Jatropha plantations. Nat Resour Forum 29:12–24

GEXI (2008) Global market study on Jatropha. London, Berlin, GEXI LLP

Gordon-Maclean A, Laizer J, Harrison PJ, Shemdoe R (2008) Biofuel industry study, Tanzania: an assessment of the current situation. Tanzania & Sweden, World Wide Fund for Nature (WWF)

GRAIN (2008) Agrofuels in India, private unlimited. GRAIN's Quarterly Magazine: Seedling April 2008:15–23

GTZ (2009a) International fuel prices 2009, 6th edn. GTZ, Eschborn

GTZ (2009b) Jatropha reality-check, a field assessment of the agronomic and economic viability of Jatropha and other oilseed crops in Kenya, study conducted by Endelevu Energy, World Agroforestry Centre, Kenya Forestry Research Institute

Heller J (1996) Physic nut: Jatropha curcas L, Promoting the conservation and use of underutilized and neglected crops. International Plant Genetic Resources Insitute, Rome

Hoogwijk MA, Faaij A, Eickhout B, de Vries B, Turkenburg W (2005) Potential of biomass energy out to 2100, for four IPCC SRES land-use scenarios. Biomass Bioenerg 29(4):225–257

IAC (2004) Realizing the promise and potential of African agriculture. InterAcademy Council, Amsterdam

IFAD, FAO, Brittaine R, Lutaladio N (2010) Jatropha: a smallholder bioenergy crop, the potential for pro-poor development, vol 8, Integrated crop management. FAO, Rome

Jongschaap REE, Corré WJ, Bindraban PS, Brandenburg WA (2007) Claims and facts on Jatropha curcas L. Wageningen University and Research Centre, Wageningen

Loos TK (2008) Socio-economic impact of a Jatropha-project on smallholder farmers in Mpanda. Thesis, Hohenheim, Tanzania

Marrison CI, Larson ED (1996) A preliminary analysis of the biomass energy production potential in Africa in 2025 considering projected land needs for food production. Biomass Bioenerg 10(5–6):337–351

Messemaker L (2008) The green myth? Assessment of the Jatropha value chain and its potential for pro-poor biofuel development in Northern Tanzania. Thesis, Faculty of Geosciences, Utrecht University, Utrecht

Ministry of Agriculture Senegal (2007) New orientation for the Agriculture Sector Policy, REVA Plan, Special Biofuels Program. March 2007, ENDA, Energy, Environment, Development Biofuels in Senegal Jatropha Programme. 2007–2012

Mitchell A (2008) The implications of smallholder cultivation of the biofuel crop, Jatropha curcas, for local food security and socio-economic development in northern Tanzania. Thesis, Anthropology and Ecology of Development, University of London

Mwangi WM (1996) Low use of fertilizers and low productivity in sub-Saharan Africa. Nutr Cycl Agroecosys 47(2):135–147

Nielsen F (2009a) Trip report: FACT/ADPP project Mozambique

Nielsen F (2009b) Jatropha curcas oil production for local development in Mozambique. African Crop Sci Conf Proc 6:71–75

Openshaw K (2000) A review of Jatropha curcas: an oil plant of unfulfilled promise. Biomass Bioenerg 19(1):1–15

Peters F (2009) Socio-economic impact study of biofuel plantation on farm households in Mozambique. Thesis, Department of Social Sciences, Wageningen University

Romijn HA (2010) Land clearing and greenhouse gas emissions from Jatropha biofuels on African Miombo Woodlands. Energy Policy In Press, Corrected Proof

Salfrais N (2010) Small is beautiful? The impacts of small-scale bio fuel production on people's access to land in the Koulikoro Region, Mali.Thesis, Geosciences, Utrecht University

Schut M, Slingerland M, Locke A (2010a) Biofuel developments in Mozambique. Update and analysis of policy, potential and reality. Energy Policy 38(9):5151–5165

Schut MLW, Bos S, Machuama L, Slingerland MA (2010b) Working towards sustainability. Learning experiences for sustainable biofuel strategies in Mozambique. Wageningen University and Research Centre, Netherlands, in collaboration with CEPAGRI, Mozambique, Wageningen, p 103

Smeets EMW, Faaij APC, Lewandowski IM, Turkenburg WC (2007) A bottom-up assessment and review of global bioenergy potentials to 2050. Prog Energy Combust Sci 33(1):56–106

Struijs, J (2008) Shinda Shinda, Option for sustainable bioenergy: a Jatropha case study. Bilthoven, RIVM (National Institute for Public Health and the Environment)

Sulle E, Nelson F (2009) Biofuels, land access and rural livelihoods in Tanzania. L. Cotula, London

Tewari D (2007) Jatropha and Biodiesel. Ocean Books Ltd, New Delhi

Trabucco A, Achten WMJ, Bowe C, Aerts R, van Orshoven J, Norgrove L, Muys B (2010) Global mapping of Jatropha curcas yield based on response of fitness to present and future climate. GCB Bioenerg 2(3):139–151

UN (2008) Energy security and sustainable development in Asia and the Pacific. Environment and Development Division (EDD), United Nations ESCAP, Bangkok

van Eijck J (2009) Case study: the smallholder model of biofuel production in Tanzania, Commissioned by GTZ and ProBEC

van Eijck J, Romijn H (2008) Prospects for Jatropha biofuels in Tanzania: an analysis with Strategic Niche Management. Energy Policy 36(1):311–325

van Eijck J, Smeets E, Faaij A (2011) The economic performance of Jatropha, Cassava and Eucalyptus production systems for energy in an East African smallholder setting. Submitted to Agricultural Systems

van Eijck J, Smeets E, Jongschaap R, Romijn H, Balkema A (2010) Jatropha assessment; agronomy, socio-economic issues and ecology, facts from literature. Copernicus Institute, Utrecht University, Eindhoven University of Technology and Wageningen PRI, Utrecht

Vermeulen S, Cotula L (2010) Making the most of agricultural investment: a survey of business models that provide opportunities for smallholders. IIED, FAO, IFAD, SDC, London, Rome, Bern

Chapter 4
Small-Scale Production of Jatropha in Zambia and its Implications for Rural Development and National Biofuel Policies

Thomson Sinkala and Francis X. Johnson

Abstract Concerns about energy security and the need to promote rural development have been key factors in the promotion of biofuels in many developing countries in Africa. At the same time, the low cost of labour and plentiful land in some regions of Africa has motivated many foreign investors to set up biofuels schemes that are aimed at export markets. Small-scale production of biofuels in a Least Developed Country (LDC) such as Zambia offers a potentially more viable alternative, or in some cases a complement, to large-scale schemes. The lower capital investment required and the fact that households and communities can use by-products allows for value-added at the local level. The case of jatropha exhibits a number of benefits if there is a willingness to experiment with various production schemes and develop different products. In this chapter small-scale jatropha production in Zambia is assessed using a case study at Thomro farms. The relation of small-scale schemes to national priorities and policies is reviewed and the future role of jatropha at local and national levels is discussed.

Keywords Jatropha • Zambia • Agriculture • Household energy • Household economy • Co-products • Rural development • Biofuels • Energy policy • Family farm • Small-scale production

---

T. Sinkala (✉)
Thomro Biofuels, Lusaka, Zambia
e-mail: tsinkala@thomro-zambia.com

F.X. Johnson
Stockholm Environment Institute, Kräftriket 2B, SE-106 91, Stockholm, Sweden
e-mail: francis.johnson@sei.se

R. Janssen and D. Rutz (eds.), *Bioenergy for Sustainable Development in Africa*,
DOI 10.1007/978-94-007-2181-4_4,

4.1 Introduction

Bioenergy and its associated co-products can be a valuable part of the sustainability transition in developed and developing countries alike. New emerging markets can contribute to agricultural and rural development, energy security and environmental restoration. When practised at small scale in the context of rural households, bioenergy crops can improve household and community economies and contribute to social stability and lower migration to crowded urban centres. Small-scale bioenergy schemes can offer an alternative to large-scale bioenergy and liquid biofuels, which have raised a variety of environmental and social concerns, such as displacement of food crops and land degradation.

Zambia is a country with a significant amount of land and resources per capita, but still suffers from high levels of poverty in rural areas. At the same time, it has no domestic fossil fuels and must devote a large share of its foreign exchange to oil imports. As a landlocked country, Zambia faces even higher costs than some of its neighbours in getting its products to and from markets, thereby further increasing energy costs (Johnson and Matsika 2006). With rising oil prices in recent years, biofuels have emerged as a means for Zambia to reduce its dependency on imported oil. Those biofuel crops that can provide a variety of other products and services in addition to energy are especially appealing in some developing countries where the co-products can displace imports, generate cash revenue and exports, and improve health and life quality in rural areas.

Amongst the biofuel crops that have received special attention in Zambia is *Jatropha curcas*, an oil-bearing plant that can grow in a variety of tropical and sub-tropical climates and provides many co-products, which are often valuable for rural communities. This chapter looks at the case of jatropha in Zambia and examines how small family-based farms can benefit from its cultivation. Such local uses and markets are generally quite distinct from, although not necessarily in conflict with, the role envisioned for jatropha as part of Zambia's national energy policies and biofuels strategies. The linkages, synergies and conflicts across different scales and applications are explored in this chapter through special reference to a case study from Thomro farms in Zambia.

4.2 Energy Markets and Policies in Zambia

In Zambia, the major sources of energy include fuel-wood, hydropower, and petroleum. Fuel-wood accounts for about 70% of the nation's energy needs whilst hydropower contributes about 14%. The estimated hydropower potential in Zambia is 6,000 MW, and the current installed capacity is 1,760 MW. Hydroelectric plants provide 99% of electrical energy, nearly all of which is from the large plants at Kafue Gorge, Kariba North Bank and Victoria Falls (MEWD 2008). Petroleum products account for 12% of the energy mix, all of which is imported.

Zambia's electricity is consumed mainly by the mines whilst only about 22% of Zambia's population have access to electricity (Republic of Zambia 2008). The remainder use woodfuel, charcoal and other traditional energy sources for their household energy needs. The charcoal delivered to urban centres is derived from peri-urban areas, resulting in deforestation and associated land degradation. In some provinces, this deforestation is exacerbated by poor agricultural practices, such as slash and burn agriculture and a shifting cultivation method known as *citimene*. Such approaches, maintained at subsistence levels, have been combined with social distribution methods. However, they can also stifle innovation in improving resource use and creating new livelihoods (Kakeya et al. 2006).

In Zambia's Northern Province, the level of deforestation has been estimated at 200,000 ha per year during the last several decades, which may have contributed to a changing climate and decreased rainfall in the region (Muloshi 2007). By 2006, deforestation had affected more than 4.3 million hectares (35%) of forested land over a period of 40 years, a situation which is unsustainable, even for a sparsely populated country like Zambia.

Therefore, without alternative energy and sustainable agricultural practices, both poverty and forest cover loss in Northern Province and in other areas of Zambia will increase. The high levels of poverty will have a negative effect on development, by leading to further inefficiency in the utilisation of natural resources and greater environmental degradation.

Considering the large and burdensome fuel import bill in Zambia, lack of indigenous petroleum production and the country's available vast arable land suitable for growing energy crops, the prospects of local biofuels production in Zambia are considerable. The biofuels industry would not only bring general economic benefits, but could also help to reduce disparities in fuel prices which, until September 2010 when Government issued uniform prices throughout the country, were unfavourable to areas further away from Lusaka and Copperbelt (see Table 4.1).

Only 14% of the total arable land of 42 million hectares is currently used for agricultural production. With its considerable land resources, water resources and agricultural land per person (5.79 ha/capita), Zambia could reduce poverty and

Table 4.1 National price disparities at fuel pumps across different regions of Zambia (1 US$ = K5000, May 2010)

	Petrol		Diesel		Kerosene	
Provincial capital	Kwacha	US$	Kwacha	US$	Kwacha	US$
Kasama	8,390	1.68	7,717	1.54	5,598	1.12
Livingstone	8,030	1.61	7,355	1.47	5,288	1.06
Chipata	8,229	1.65	7,556	1.51	5,460	1.09
Solwezi	7,825	1.57	7,150	1.43	5,111	1.02
Ndola	7,461	1.49	6,786	1.36	4,798	0.96
Lusaka	7,573	1.51	6,898	1.38	4,893	0.98
Kabwe	7,543	1.51	6,868	1.37	4,868	0.97
Mansa	8,252	1.65	7,579	1.52	5,480	1.10
Mongu	8,337	1.67	7,664	1.53	5,553	1.11

Table 4.2 Comparison of population, resources, and GDP for selected countries

Country	Population (Million)	Arable land used (%)	Area/Person (ha/capita)	Total renewable water resources (km^3/capita)	GDP (1,000 USD/capita)	Population below poverty Line (% (year measured))
Zambia	13.0	7	5.79	8.69	1.6	86 (1993)
Brazil	201.1	7	4.23	40.94	10.1	26 (2008)
USA	310.2	18	3.17	9.89	46.5	12 (2004)
S. Africa	49.1	12	2.48	1.02	10.3	50 (2000)
France	64.1	33	1.00	2.95	41.6	6.4 (2004)
Switzerland	7.6	10	0.54	7.01	64.5	7.4 (2009)
Germany	82.3	33	0.43	2.28	39.8	11 (2001)
Netherlands	16.8	22	0.25	5.34	47.6	10.5 (2005)

increase its economic wealth. However, as Table 4.2 shows, Zambia not only has low GDP per capita, but also high poverty levels (86%). If Zambia decided to use, for example, 20 million hectares for biofuels, the country would still have available 4.25 ha/capita, which remains higher than any of the countries shown in Table 4.2.

In recognition of the future role of biofuels, the National Energy Policy of 2008 included the following goals or measures:

- Expansion of the role of biofuels in the national fuel mix
- Ensuring security of supply and stable prices of fuels by promotion of biofuels for transport as an alternative to petroleum
- Ensuring availability of data and information on market demand, resource assessment and applicability of biofuels
- Providing a legal and institutional framework for the biofuels sub-sector
- Supporting investment in the biofuels industry through appropriate incentives, standards and research

To ensure security of supply and stable fuel prices, Zambian policies promote biofuels for transport as an alternative to petroleum by supporting:

- Cultivation of energy crops
- Investment in biofuels through appropriate incentives
- Participation of Zambians in the biofuels industry as shareholders
- Maintenance of food security and environmental sustainability

The policies also address the fact that traditional biomass energy (fuel-wood, charcoal, agricultural and forestry wastes) forms the largest part of Zambia's energy mix. To improve the standard of living the policy articulates that there is need to switch from low quality household energy sources to modern energy resources such as electricity, petroleum products, biofuels and biogas.

4.3 Assessing Feedstock Options in Zambia

A number of feedstocks for biodiesel and bioethanol production were assessed to evaluate their relative contributions towards societal goals. The parameters used to assess target biofuels feedstocks in Zambia include:

- Scope of wealth ownership at national and individual levels
- Appropriate production technology
- Job creation
- Resilience against external disturbances
- Diversity of products
- Size of investment
- Market scope
- Land requirements
- Water requirements
- Food security
- Geographical coverage
- Environment

These parameters determine the sustainability of the biofuels industry in Zambia including impacts on other sectors. The parameters are briefly explained below and an evaluation matrix is given in Table 4.3.

Table 4.3 Biofuels feedstock assessment/ranking in Zambia

Parameters	Jatropha	Oil palm	Castor	Sugarcane	Sweet sorghum	Cassava
Yield (kg/ha)	2,000	5,000	1,500	6,000	4,000	4,500
Wealth ownership (individual / national)[a]	3	3	3	3	3	3
Production technology[a]	3	3	2	1	1	1
Job creation[a]	3	2	3	3	3	3
Resilience[a]	3	3	3	3	3	3
Geographical coverage[a,b]	3	1	3	1	3	2
Environmental damage[a]	3	3	3	2	3	3
Investment[a,c]	3	2	2	1	2	3
Diversity of products[a]	3	2	3	2	2	3
Food security[a]	3	2	2	1	2	3
Market[a]	3	1	3	1	2	2
Land/water requirement[a]	2+3	3+1	2+3	3+1	3+3	3+3
SCORE (Rank)	35 (1)	26 (3)	31 (2)	23 (3)	30 (2)	32 (1)

[a]All parameters (except technology, environmental damage, land, water and investment) have: high = H = 3; average = A = 2; low = L = 1; for technology, environmental damage, land, water and investment, the key is: high = H = 1; average = A = 2; low = L = 3

[b]Low rank of 1 for geographical coverage for sugarcane and palm oil is due to the fact that growth is limited to areas with sufficient rainfall

[c]Investment requirements depend mainly on scale (sugarcane is highest) but also on whether crop is perennial (e.g. Jatropha)

Wealth ownership (individual/nation): It is preferable if the wealth derived from Zambia's biofuels activities is fully owned and controlled by Zambian citizens. A feedstock which facilitates local ownership would create a base for further developments and promote sustainability of the industry.

Appropriate production technology: As Zambia lacks own developments of innovative technologies, it is preferable to target biofuels feedstocks suitable for simple and proven technologies without losing the intended quality of products. This would ensure greater participation and biofuels industry ownership by Zambians.

Job creation: Zambia is experiencing high poverty levels and unemployment. Policies should target biofuels feedstocks that create jobs for Zambians, especially in rural areas. This could help to reduce, and perhaps reverse, the migration of rural dwellers to cities. Unemployed urban dwellers could have new opportunities for steady employment on a plantation or the possibility to develop their own small plot or farm.

Resilience against external disturbances: There should be a high degree of local production and consumption and few external production inputs to minimise the risk of shocks on Zambia's biofuels industry due to external pressures or price changes.

Diversity of products: Policies should target feedstocks that have more products in addition to liquid biofuels, thereby facilitating low and stable production costs as well as increased business opportunities along the value chain.

Size of investment: The smaller the size of initial investment required for biofuels feedstock, the larger the number of Zambians that could become involved. High upfront investment is often the greatest barrier for economic development in Zambia and in LDC in general.

Scope of the Market: Many exotic crops and their by-products are targeted towards the export market. It is preferable to promote biofuel feedstocks that can be also used by the grower himself/herself. In good times, growers could use products in the primary (biofuels) business line whereas in bad times they could use products in secondary business lines (own use, local market), thus minimising external shocks.

Land requirements: The biofuels programme should preferably target feedstocks with high yield per hectare and requiring low quality of soils. This would reduce land competition and input requirements.

Water requirements: It is preferable to target feedstocks that do not require irrigation. This would allow biofuels development further away from water bodies, reduce conflicts over water use and minimise the risk due to droughts.

Food security: Policies should target feedstocks that do not compromise food security and do not result in net displacement of land for food production.

Geographical coverage: Policies should target feedstocks with broad geographical coverage suitable for many regions of Zambia. This would ensure broader participation and ownership and enhance political stability.

Environmental Damage: It is preferable to use feedstocks that have low or even net positive environmental impacts. Land, water and other resources must be used prudently to safeguard the interests of future generations.

As shown in Table 4.3, jatropha ranks high in many respects and had the highest overall score. It is a plant that favours participation at small-scale, and thus has special benefits for the rural poor with access to land. Consequently, there are already many small-scale participants in the jatropha industry in Zambia.

4.4 Small-Scale Case Study: Thomro Biofuels Farm

Thomro Biofuels' jatropha plantation is located about 12 km from Lusaka city centre. The jatropha plants cover about 12 ha out of a total of 105 ha. The first plantings were done in 2005/2006 and the plantation used seedlings from various sources in D.R. Congo, Tanzania and Zambia. Some key lessons learned in terms of the overall physical and economic viability of operations at Thomro Biofuels farm are summarised below. The discussion focuses on non-energy products and characteristics, since these are more closely related to local value-added, and can be developed alongside the expected market for biodiesel at national level. Experiences in Kenya, for example, have illustrated the economic risks of focusing on jatropha production only for biodiesel (Endelevu Energy 2009).

4.4.1 Weeding of Jatropha Fields

At the moment, a significant share of the work in the jatropha plantation involves weeding, as grass is probably the worst enemy of jatropha. For weeding, Thomro has used various approaches including: hoes, herbicides, livestock and slashing. Using goats adds economic and ecological value to the plantation operations whilst providing a solution to the weed problem. As Thomro must maintain a maximum number of goats for weeding purposes, the excess goats can be sold, thus earning additional income. Livestock also adds manure to the field during grazing time. The manure from the goat shelter is used for crop production. A 25 kg bag of manure costs 0.6 US$ in the Thomro farm area.

In Zambia, a goat costs 16–90 US$ depending on the source, size and breed. The cropping of 10 average size goats of local breed is currently estimated to provide income of 300 US$. Thomro is breeding to increase the number of goats; from the six female and one male goat acquired in 2008, Thomro now has 25 heads of goats (as of September 2010).

4.4.2 Improving Pollination in Jatropha Fields

Jatropha is pollinated by insects. At Thomro Biofuels farm, the insects observed are primarily honey bees and septic flies. To increase the insect life for pollination,

Thomro Biofuels has created and installed a trial beehive. Honey produced from jatropha flowers is a high-value product that adds to income. By producing honey with jatropha pollen, bees enhance jatropha fruit production, which leads to higher oil yields per tree and hectare. The harvest from this trial beehive yielded a retail value of 134 US$ of honey. If there was a beehive per each hectare of the 12 ha under jatropha, this would have yielded 1,608 US$.

According to Zambia Agribusiness Forum, Zambia has emerged as Africa's largest exporter of honey and bee products to the European Union and the USA, with supply to those markets projected at 1,000 tons by the end of 2010. Zambia had 50,000 people deriving their livelihood from bee products in 2010. The global market for honey and other products related to honey bees is worth over 200 billion US$. The existence of a global market suggests a large potential to derive synergies between jatropha and bee industries.

4.4.3 *Soap Production from Jatropha*

At present, Thomro Biofuels is extracting oil from jatropha seeds for soap production whilst the residue cake is being utilised as organic fertiliser for crop production. From one litre of jatropha oil, about six pieces of jatropha soap of 100 g each can be produced. At a retail price of 0.6 US$ each, a litre of jatropha fetches 3.60 US$. This is several times more than the price for a litre of jatropha oil or biodiesel. As Table 4.1 shows, the most expensive price for diesel in Zambia is at Kasama, where diesel costs 1.54 US$. In other countries such as Madagascar (where its main use is for soap-making) the price of jatropha oil is 1.50 US$, thus illustrating the higher value-added that can be obtained where market prices can rise closer to their value relative to alternative uses.

The soap market in Zambia is more than 50 million US$ per year. Soap production with jatropha oil has been promoted in Mali, Tanzania and Madagascar where it has gained recognition in the market as an anti-septic natural soap. There is therefore potential for this in Zambia's market.

4.4.4 *Use of the Jatropha Cake*

The jatropha cake has relatively high levels of nitrogen (N), phosphorus (P) and potassium (K). Two studies have reported the following nutrient levels:

- 6.0% nitrogen, 2.75% phosphorus and 1.0% potassium (Chungu 2007)
- 3.2–4.44% nitrogen, 1.4–2.09% phosphorus, 1.2–1.68% potassium (van Eijck 2007).

These levels are reasonably high and therefore jatropha cake can be used as organic fertiliser. The increased availability of organic fertilisers in Zambia has the potential to minimise shifting cultivation, leading to a net reduction in deforestation and environmental degradation.

4.5 Greenhouse Gas Balance for Biodiesel from Jatropha

Zambia has no obligation to reduce its GHG emissions and indeed it is expected that Zambia will increase its emissions. It has a similar ethical right to develop its economy as other Least Developed Countries in their efforts to promote growth and development. Nevertheless, the GHG balance of jatropha biofuels must be taken into consideration, since financing opportunities may be associated with emissions, and also because all countries of the world will have to develop low-carbon paths of development.

The main GHG emissions are CO_2, CH_4 and N_2O. CO_2 is less problematic with biofuels since most of the CO_2 is recycled during plant growth. For example, a full grown jatropha shrub absorbs around 8 kg of CO_2 per year, which translates into CO_2 sequestration of 20 tons ha^{-1} year^{-1}, assuming a plantation with 2,500 shrubs ha^{-1} (Muok and Källbäck 2008). CH_4 and N_2O are more problematic as they originate from soils, fertilisers, and land clearing or burning.

Emissions from cultivated soils associated with fertiliser decomposition and N_2O emissions are a main concern due to their high global warming potential (GWP-100). Emissions are highly dependent on soil conditions (moisture, nitrate, etc.) and cultivation practices. N_2O emissions associated with biofuels are not different from other land uses, as over 70% of global N_2O emissions are associated with agricultural land use (IPCC 2007).

The main benefit of jatropha lies in the very low emissions from cultivation, even if land use change is taken into account. The various co-products further improve the GHG balance, offering a competitive advantage of jatropha compared to other first generation biodiesel crops. Assuming technical suitability of jatropha oil, a market which values GHG-performance may thus lead to higher prices. The GHG reduction compared to diesel has been estimated at 68% under the assumptions used in the EU Renewable Energy Directive (RED) (Dehue and Hettinga 2008; EC 2009).

Looking more broadly at all environmental impacts, biodiesel from jatropha shows both advantages (e.g. saving non-renewable energy) and disadvantages (e.g. acidification and eutrophication) compared to fossil diesel. Therefore, a decision for or against a particular fuel is inevitably subjective in terms of overall lifecycle impacts. Such decisions must be made on the basis of prioritised characteristics (IFEU 2007). If GHG reduction is given the highest priority, then jatropha biodiesel will outperform fossil diesel, unless extensive land use change occurred when the jatropha plantations were set up.

In general, an improvement of the impacts of jatropha requires optimisation of the by-product utilisation. Improved by-product use will in some cases depend on centralised production, where the options increase for different by-products. Therefore, local production and use may not be as effective as central production when it comes to GHG reductions (Tomomatsu and Swallow 2007). However, where jatropha biodiesel is used to offset diesel in electricity production in small generators, additional GHG savings are expected due to the efficiencies of local use, i.e. savings in transport and distribution.

4.6 Conclusions

Jatropha cultivation for biofuels is linked to both small-scale and large-scale production methods and markets. The example of Thomro Biofuels illustrated the potential chain of products and the associated upstream and downstream industries. The fact that these products have markets at various scales, ranging from internal household consumption to national and international markets, shows that cross-scale effects may be important in this industry and may help to cushion it from external shocks that occur in export-focused agro-industries.

The fact that the jatropha industry is open for participation to people from all socio-economic strata, especially in rural communities throughout Zambia, can enhance the ownership and sustainability of the industry. Positive consequences of this ownership include improved household and national economies, food and energy security as well as overall economic development and improved social stability in rural areas. The contribution at national level will depend on how the local markets and small growers are eventually linked to the national biofuels framework, a process that is under development in Zambia.

References

Chungu G (2007) Jatropha curcas in Zambia. Presented at the regional Jatropha practitioners networking workshop on small-scale Jatropha farming, Arusha, Tanzania, 8–11 November 2007

Dehue, Hettinga (2008) GHG performance Jatropha biodiesel. Commissioned by D1 oils, Ref no. PBIONL073010. Ecofys reference, 2 June, Utrecht, The Netherlands

EC (2009) Directive 2009/28/EC of the European Parliament and of the Council of 23 April 2009 on the promotion of the use of energy from renewable sources and amending and subsequently repealing Directives 2001/77/EC and 2003/30/EC

Endelevu Energy (2009) Jatropha reality check – a field assessment of the agronomic and economic viability of Jatropha and other oilseed crops in Kenya. World Agro-forestry Centre and Kenya Forestry Research Institute; GTZ, Germany

IFEU (2007) Screening lifecycle assessment of jatropha biodiesel. Report commissioned by Daimler-Chrysler, IFEU, Heidelburg, 11 December

IPCC (2007) Fourth Assessment Report of the International Panel on Climate Change. www.ipcc.ch

Johnson FX, Matsika E (2006) Bioenergy trade and regional development: the case study of bioethanol in Southern Africa. Energ Sustain Dev X(1):42–54, March 2006

Kakeya M, Sugiyama Y, Oyama S (2006) The citemene system, social leveling mechanism, and agrarian changes in the Bemba villages of northern Zambia: an overview of 23 years of 'fixed-point' research. African Study Monographs 27(1):27–38

Muloshi (2007) Impacts of human activities on the climate of Northern Zambia. AIP Conf Proc 923:296, July 23, 2007

MEWD (2008) National Energy Policy. Published by the Ministry of Energy and Water Development (MEWD), Government of the Republic of Zambia, Lusaka, Zambia

Muok B, Källbäck L (2008) Feasibility study of jatropha curcas as a biofuel feedstock in Kenya. ACTS/PISCES/EXPORTRÅDET, Nairobi

Republic of Zambia (2008) National Energy Policy. Published by the Ministry of Energy and Water Development. May 2008

Tomomatsu Y, Swallow B (2007) Jatropha curcas biodiesel production in Kenya – economics and potential value chain development for smallholder farmers. World Agro-forestry Centre, Nairobi, Kenya. Working Paper No. 154. http://www.worldagro-forestry.org/downloads/publications/PDFs/WP15396.PDF

van Eijck J (2007) Transition towards Jatropha biofuels in Tanzania? An analysis with Strategic Niche Management, vol 3, African Studies Collection. African Studies Centre, Netherlands, ISBN 978.90.5448.072.3

Chapter 5
Economic Evaluation of Sweet Sorghum in Biofuel Production as a Multi-purpose Crop: The Case of Zambia

Clarietta Chagwiza and Gavin Fraser

Abstract Increasing awareness of the environmental damage caused by the existing fossil fuel-based energy system and rising fossil fuel prices explain, in part, the growing interest in renewable energy sources. In March 2008, oil broke through the psychological ceiling of 100 US$ a barrel, and later in early June 2008 rose to 140 US$ on the way to 150 US$. In addition, rising issues on global warming have resulted in the need to consider alternative sources of energy. This also resulted in debates about biofuels in most developing and developed countries. Zambia is faced with an energy crisis from importation of large amounts of crude oil and the high cost of fuel and petroleum products. Sweet sorghum has been flagged as a potential biofuel feedstock in Zambia. This chapter evaluates different varieties of sweet sorghum and identifies production scenarios under which sweet sorghum can be produced in Zambia by the use of Gross Margin Analysis. It also evaluates notable trade-offs in producing sweet sorghum instead of grain sorghum. The results show that identifying high sweet sorghum yielding varieties and optimum production scenarios are important pre-requisites for the successful implementation of the use of sweet sorghum in biofuel production. The results indicate a positive relationship between the yield of sweet sorghum and the production regime.

Keywords Biofuel • Economic • Evaluation • Gross Margin Analysis • Multi-purpose • Socio-economic • Sweet sorghum

---

C. Chagwiza
University of Fort Hare, Private Bag X1314, Alice, 5700, South Africa
e-mail: clarrytee@yahoo.co.uk

G. Fraser (✉)
Rhodes University, PO Box 94, Grahamstown, 6140, South Africa
e-mail: g.fraser@ru.ac.za

R. Janssen and D. Rutz (eds.), *Bioenergy for Sustainable Development in Africa*,
DOI 10.1007/978-94-007-2181-4_5,

5.1 Introduction

In view of the rising crude oil prices, forecasted shortages of fossil fuels, the need for new income and employment opportunities for rural farmers, biofuels have become a topical issue in Zambia. Furthermore, Hernandez and Reddy (2007) point out that environmental pollution associated with fossil oil use has resulted in increased worldwide interest in the production and use of biofuels. Leigh (2008) highlighted that in March 2008 oil broke through the psychological ceiling of 100 US$ a barrel, and later in early June 2008 rose to 140 US$ on the way to 150 US$.

Banda (2007) indicated that Zambia spends in excess of 500 million US$ a year on the importation of crude oil to meet its national demand for fuel, which accounts for 10% of the country's foreign exchange expenditure and 14% of the national energy requirement (Johnson and Rosillo-Calle 2007). This calls for alternative sources of fuels to cut down on the expenses incurred for importing oil from the developed world. O'Keeffe (2007) indicated that 53% of the petroleum imported to Zambia is channelled into the transport sector.

A number of potential crops have been noted, for example maize and sweet sorghum. According to Munyinda (2005), sweet sorghum (*Sorghum bicolor*) has been recognized as a possible alternative or complementary crop to provide raw material to the sugarcane industry in southern Africa. Sweet sorghum is unique in its agronomic characteristics. Hence, Grassi (2001) considered it a crop of universal value, because it can be grown on all continents, in tropical, sub-tropical, temperate regions as well as in poor quality soils and in semi-arid regions.

However, biofuels cause controversy and invoke opposition to their implementation. According to O'Keeffe (2007), serious legitimate social and environmental concerns are raised as to the impact of biofuels, such as fears over deforestation, land consumed by fuel crops to the detriment of food crops, and the quantities of land needed to grow typically low energy density crops, which could lead to the marginalization of small-scale farmers.

As one of the alternative sources of energy and a multi-purpose crop, sweet sorghum needs to be assessed in terms of its socio-economic impacts. This pertains to issues like food security, income generation and general improvements of livelihoods of the people in Zambia. Variety selection is an important decision in sweet sorghum production when feed, fuel, food and fibre needs are to be satisfied (Janssen et al. 2010). This chapter seeks to evaluate the different varieties of sweet sorghum in terms of their fresh stem, sugar and grain yields. It also assesses the socio-economic impacts of sweet sorghum as a multi-purpose crop in Zambia.

5.2 Implications of Using Sweet Sorghum in Biofuel Production

The economic motivation for biofuels is that they are convenient, low cost, domestically produced substitutes for crude oil, a fuel that is getting costlier by the day and is imported from politically volatile regions (Rajagopal and Zilberman 2007). Biofuels have the potential to improve foreign reserves of most countries. By substituting biofuels partially for imported oil, cash-strapped developing countries can invest their scarce capital in their own farms and industries rather than exporting it to wealthier oil-producing nations (ICRISAT 2007).

However, biofuels offer a number of pressing issues that need to be addressed before they are fully embraced and commercialized. Producing biofuels on a large-scale could require huge areas of land. Many countries, especially in Africa, cannot afford to divert land from food production since this may have negative impact on their food security. Food and biomass require the same resources for production, namely land, water and agrochemicals (Arungu-Olende 2007). Sweet sorghum has a strong pro-poor advantage since it has a triple product potential: grain for human consumption, juice for ethanol, and stillage or bagasse for livestock feed or power generation.

5.2.1 *Potential of Sweet Sorghum for Biofuels*

Sweet sorghum is similar to grain sorghum and some other energy crops but features more rapid growth, higher biomass production, wider adaptation, and has great potential for ethanol production. Sweet sorghum being a C_4 species is more water use efficient and can be successfully grown in semi-arid tropics, where other crops such as maize fail to thrive. Woods (2001) described this as an important factor in drought-prone sugar-producing regions of the world. Another important attribute of sweet sorghum is that even the resource-poor farmers can sustain themselves and realize improved income levels through growing this multi-purpose crop.

Munyinda (2005) highlighted that in developing countries like Zambia local biomass production from locally grown crops can cut dependence and cash expenditure on imported fuels, increase community self-reliance, and provide a spur for local job creation and growth. This implies that the cash saved from cutting fuel imports can be used for other social services to improve the livelihoods of the rural poor.

5.2.2 *Criticism and Challenges of Sweet Sorghum in Biofuel Production*

Despite the benefits from biofuel production using sweet sorghum, developing countries face a number of economic, social and environmental challenges. Dufey (2006) highlighted that the cultivation of energy crops on a large-scale will have very little impact on rural labour due to the issue of economies of scale. Hence, the need for these economies of scale can act as a driver for establishing large-scale cultivation of energy crops, thereby crowding out small farmers' cultivation.

Another challenge in using sweet sorghum juice is the harvest time that is limited to 3–4 months per year and the maintenance of juice stability. A number of reports suggest that juice extraction should occur soon after harvest and processing needs to take place immediately (Gnansounou et al. 2005, Kundiyana et al. 2006, as cited by Veal 2007). Additionally, Gnansounou et al. (2005) indicated that delays in juice extraction results in the deterioration of sugar level in sweet sorghum stalks. Munyinda (2005) argued that the crop has a large drawback, as it is bulky to transport and cannot be stored. In fact, processing has to start within 24 hours after harvesting or sugar will be lost.

According to Grassi (2001), one of the most challenging problems to solve for sweet sorghum is to overcome its seasonality problem and the instable characteristics of its fermentable sugars, that require high investments for fast bioethanol production units (~70–90 days).

5.3 Analysis of Findings

Yields (fresh stem and grain) from nine sweet sorghum varieties were assessed under four different production scenarios: (1) small-scale rain-fed, (2) improved rain-fed, (3) improved single-cropping with supplementary irrigation and (4) improved double-cropping with supplementary irrigation. Sweet sorghum varieties investigated include *Sima* (a local variety), *TSI*, *Madhura*, *Praj 1*, *GE 2*, *GE 3*, *Wray*, *Cowley* and *Keller*. Some of the varieties produced substantially higher yields. For example, the variety *Wray* under the improved double cropping with supplementary irrigation has a stem yield of 136.8 t/ha. In addition, the variety *TS1* has a higher grain yield of 14.78 t/ha under the double cropping production scenario.

Table 5.1 presents varieties of sweet sorghum with highest fresh stem and grain yields. On average, *Wray* produced the highest fresh stem yields making this variety suitable for ethanol production. The varieties *Praj-1* and *GE3* also produced high yields of fresh stems. The varieties TS1 and Praj-1 produced the highest grain yields making these varieties the optimum choice for farmers focusing on food security.

5.3.1 Gross Margin Analysis of Sweet Sorghum for Different Scenarios of Production

A gross margin for an enterprise is its financial output minus its allocable variable costs. The gross margin analysis model has been used to estimate the profitability of cultivation of energy crops per hectare based on the assumptions about yield, output price and cost of production. Mathematically, it is expressed as follows:

GROSS MARGIN = Value of Yield − Costs

Where: Value of yield = Price × Output

Costs = All variable costs

Table 5.1 Fresh stem and grain yields of selected sweet sorghum varieties (Source: Chagwiza 2008)

Scenario	Variety	Stem yield (t/ha)	Variety	Grain yield (t/ha)
Small-scale	Praj-1	13.60	Praj-1	1.47
Improved rain-fed	Wray	22.82	Praj-1	3.57
Single-cropping	GE3	78.20	TS1	8.09
Double-cropping	Wray	136.8	TS1	14.78

A simulation of four models of production was employed where gross margins for different scenarios of production were calculated in order to identify which scenario Zambia could follow if sweet sorghum were to contribute significantly to ethanol production as well as food security. The framework used was adapted from the study by Cardno Agrisystems Limited (2007) for various major agricultural commodities in Zambia. Gross margins of each sweet sorghum variety under different production scenarios were calculated. For the sake of this study, the farming systems for the production of sweet sorghum in Zambia were divided according to the type of agricultural management practised. Four types identified were:

- Small-scale rain-fed
- Improved rain-fed
- Improved single cropping with supplementary irrigation
- Improved double cropping with supplementary irrigation

The gross margin calculations have been based on the following assumptions:

- The opportunity cost (i.e. the value forgone by not using the resources in the most profitable alternative way) of a subsistence farmer is negligible. When a farmer decides to hire labour for a task, costs are incurred in the production system. Following this, the labour costs for the small-scale rain-fed scenario is considered zero. Hence;
  - 0% hired is assumed for the small-scale rain-fed production system.
  - 50% hired labour is assumed for improved rain-fed and single crop production system.
  - 100% hired labour is assumed for double crop production.
- An exchange rate of 1 US$ = ZMK 3,360 is assumed.
- Due to the fluctuating nature of commodity prices, all prices used are based on July 2008 rates.
- A minimum wage (i.e. minimum hourly wage necessary for a person to achieve a specific standard of living) of US$ 1 per hour is assumed in calculating the labour costs.
- A cost of 59.50 US$ per 50 kg of both compound "D" and Urea is assumed.
- Price of ethanol is 0.45 US$ per litre. Given that Zambia has not yet sold ethanol from sweet sorghum, the US price of ethanol was used as benchmark.
- All the parameters (yield and output) were calculated on per hectare basis.
- Gross margins do not include transport costs to the processing plant.

Table 5.2 Gross margin for high yielding sweet sorghum varieties (Source: Chagwiza 2008)

Scenario	Total income (US$)	Total costs (US$)	Gross margin (US$)
Small-scale	661	707	(46)
Improved rain-fed	1,486	1,132	353
Single-cropping	2,286	1,312	974
Double-cropping	2,572	1,048	1,524

- A price of 357 US$ per tonne of grain is assumed.
- Price of sweet sorghum stalks is estimated at 10 US$ per tonne.
- Irrigation estimated to be 100 US$ per hectare assuming the same amounts were applied to the single and double cropping production scenarios.

From the calculations carried out in Table 5.2 for different production scenarios, the highest gross margin is realized in scenario 4, "improved double-cropping with supplementary irrigation", followed by scenario 3 "improved single-cropping with supplementary irrigation". A negative gross margin was realized in the "small-scale rain-fed production" scenario (scenario 1). This shows that if sweet sorghum is to contribute significantly to ethanol production, it has to be produced with supplementary irrigation and not with small-scale rain-fed production systems.

An analysis of variation was used to reveal the variation within and between each scenario of production (Chagwiza 2008). The analysis indicates there is a significant influence of production scenario on gross margin. The more advanced the production scenario, the higher the gross margin that can be obtained. The same applies to the total income obtained under each scenario. Therefore, if sweet sorghum is to contribute significantly to biofuel production and a higher gross margin realized, then it should be produced under the double-cropping with supplementary irrigation production scenario.

A *post-hoc* analysis was applied to test the variation in total income and the gross margin in response to the different scenarios of production. Sufficient evidence exists that there is a significant difference in total income under scenario 1 versus scenarios 3 and 4, but insufficient evidence that there is a significant difference on total income between scenarios 1 and 2. There is sufficient evidence to claim that there is a significant difference in gross margin between scenarios 4 and scenarios 1, 2 and 3. This shows that the use of sweet sorghum in biofuel production can be more viable if it is produced at a more advanced management regime, *viz.* scenario 4.

5.4 Conclusions and Recommendations

It is important to note that the sweet sorghum variety *Wray* produced high quantities of fresh stems in all four production scenarios. The higher the stem yield, the higher the amount of juice that can be extracted from a given variety. Hence, a variety that produces large amounts of juice has good economic returns when it comes to ethanol production.

With respect to grain yields, the varieties *TS1* and *Praj-1* showed the best performance. However, the challenge for plant breeders remains to develop varieties producing high grain and fresh stem yields simultaneously.

As evidenced in this research, sweet sorghum has potential as biofuel feedstock. It is important to identify varieties with optimal yields and the type of production regime under which they can best be produced. This chapter has suggested that sweet sorghum cannot be produced under small-scale management regimes due to existing lack of expertise and (capital) resources. Substantially higher yields have been realized under the improved single and double cropping scenarios with supplementary irrigation. Consequently, these production scenarios are economically viable when it comes to bioethanol production since they have indicated positive gross margins.

However, if biofuel production is to ensure that the rural poor capture a large proportion of the benefits, small-scale farmers need to be empowered so that they can also participate fully and make a noticeable contribution. This can be done through the establishment of collectives where a number of small-scale farmers join production activities, hence increasing economies of scale, risk-sharing and reducing individual costs. Biofuels could provide developing countries such as Zambia with a means to invest in their own rural areas instead of exporting their capital to purchase fossil fuel from politically unstable environments.

References

Arungu-Olende S (2007) Article from a paper presented at a Kenya National Academy of Sciences workshop in Nairobi, Kenya in June 2007. http://forests.org/shared/reader/welcome. Accessed 15 Mar 2008

Banda J (2007) Biofuel - experts in Zambia call for caution, The Times of Zambia. Online: http://allafrica.com/stories/200702270140.html. Accessed 27 May 2007

CARDNO AGRISYSTEMS LIMITED (2007) Support to agricultural diversification and food security in western and north-western provinces. A final report of an EU funded project

Chagwiza C (2008) Economic evaluation of sweet sorghum in biofuel production as a multi-purpose crop: case of Zambia. Unpublished M Sc Agric dissertation, University of Fort Hare, Alice

Dufey A (2006) Biofuels production, trade and sustainable development: emerging issues, vol 2, Sustainable markets discussion paper. International Institute for Environment and Development, London

Gnansounou E, Dauriat A, Wyman CE (2005) Refining sweet sorghum to ethanol and sugar: economic trade-offs in the context of North China. Bioresour Technol 96(9):985–1002

Grassi G (2001) Sweet sorghum: one of the best world food-feed-energy crop. http://www.eubia.org/fileadmin/template/main/res/pdf/publications. Accessed 24 Sept 2007

Hernandez EE, Reddy BVS (2007) Sweet sorghum pilot-sites in the Philippines. http://www.mixph.com/2007/08/sweet-sorghum-pilot-sites-in-the-philippines.html. Accessed 17 Nov 2007

ICRISAT (2007) Why biofuels? Pro-poor biofuels outlook for Asia and Africa: ICRISAT's perspective. http://www.icrisat.org/Investors/Biofuel.pdf. Accessed 27 Nov 2007

Janssen R, Rutz D, Braconnier S, Reddy B, Rao S, Schaffert R, Parella R, Zaccharias A, Rettenmaier N, Reinhardt G, Monti A, Amaducci S, Marocco A, Snijman W, Terblanche H, Zavala-Garcia F (2010) Sweet sorghum – an alternative energy crop. Proceedings of the 18th

European biomass conference and exhibition, Lyon, France, 3–7 May 2010. ISBN 978-88-89407-56-5
Johnson FX, Rosillo-Calle F (2007) Biomass, livelihoods and international trade: challenges and opportunities for the EU and Southern Africa. Stockholm Environment Institute, Stockholm
Leigh J (2008) Crude oil price retreat: sunrise or a lull before the storm? Energy Bulletin. http://energybulletin.net/node/46199. Accessed 14 Aug 2008
Kundiyana D, Bellmer D, Huhnke R, Wilkins M (2006) "Sorganol": production of from ethanol sweet sorghum, Paper No. 066070, ASAE Annual Meeting, St Joseph, Michigan
Munyinda K (2005) Evaluation of sweet sorghum as an alternative bioenergy feedstock. Thesis, UNZA, Lusaka
O'Keeffe S (2007) The sustainability of large-scale bioethanol production in Zambia, with a special focus on the potential contribution of sweet sorghum. Unpublished M Sc Thesis, Imperial College, London
Rajagopal D, Zilberman D (2007) Review of environmental, economic and policy aspects of biofuels, No. 4341, Policy research working paper. World Bank, Washington DC
Veal M (2007) On-farm biofuel production from sweet sorghum juice. http://www.southernsare.uga.edu/winnerfiles/07winners/2007onfarmwinners.htm. Accessed 07 Jan 2008
Woods J (2001) The potential for energy production using sweet sorghum in Southern Africa. Energ Sustain Dev 5(1):31–38

Part II
Biomass Technologies and Markets

Chapter 6
Keynote Introduction: Biomass Technologies and Markets in Africa

Dominik Rutz and Rainer Janssen

Abstract Globally, many different biomass technologies for energy production are available. They all have the same general common feedstock, namely biomass which is the organic matter derived from plants and animals, such as agricultural crops, forests, organic wastes, and animal waste. Biomass can be used for different purposes (feed, fibre, food, fuel) including energy production. Bioenergy is the final energy output derived from biomass/feedstock. This includes liquid, gaseous, and solid biofuels which are converted into heat, electricity, light, transportation energy (kinetic energy) and other energy types. In Africa, currently mainly traditional technologies for biomass conversion are used. This includes the use of wood-fuels for cooking. However, also modern technologies are increasingly important in Africa. The present chapter describes different technologies applied or suitable for African framework conditions. It furthermore provides an overview on the existing bioenergy markets in Africa.

Keywords Biofuel technologies • Markets • Value-chain • Feedstock • Process • Energy carrier • Energy service

6.1 Introduction

Globally, many different biomass technologies for energy production are available. They all have the same general common feedstock, namely biomass which is the organic matter derived from plants and animals, such as agricultural crops, forest, organic wastes, and animal waste. Biomass can be used for different purposes

D. Rutz (✉) • R. Janssen
WIP Renewable Energies, Sylvensteinstr. 2, 81369 Munich, Germany
e-mail: dominik.rutz@wip-munich.de; rainer.janssen@wip-munich.de

R. Janssen and D. Rutz (eds.), *Bioenergy for Sustainable Development in Africa*,
DOI 10.1007/978-94-007-2181-4_6, © Springer Science+Business Media B.V. 2012

(feed, fibre, food, fuel) including energy production. Bioenergy is the final energy output derived from biomass/feedstock. This includes liquid, gaseous, and solid biofuels which are converted into heat, electricity, light, transportation energy (kinetic energy) and other energy types.

In Africa, currently mainly traditional technologies for biomass are used. This includes the use of wood-fuels for cooking. However, also modern technologies are increasingly important in Africa. The present chapter describes different technologies applied or suitable for African framework conditions. It furthermore provides an overview on the existing bioenergy markets in Africa.

6.2 Biomass Value Chains and Options in Africa

In order to assess and evaluate bioenergy technologies, the whole value chain of a bioenergy system has to be considered. Typical bioenergy value chains include biomass production, transport, conversion, and consumption. Along the value chain, different impacts can be measured, such as social, economic and environmental impacts. Value chains and their impacts are mainly influenced by the following four characteristics:

- Feedstock type (waste, lignocellulosic material, oily crops, sugar crops, starchy crops, dedicated energy crops, co-products)
- Conversion technology (biological, mechanical, thermal, chemical processes)
- Intermediate energy carrier type (solid, gaseous, liquid energy carrier)
- Energy service (electricity, transport, heat, cooking and lighting)

The options for bioenergy in Africa are manifold since the use of biomass is applicable at different scales, under various climatic conditions and under different framework conditions. Bioenergy options in Africa range from small-scale household applications (e.g. wood-fuels for cooking) to large-scale liquid biofuel production for transport. For all scales, different mechanical, thermo-chemical and biological conversion technologies are available.

Due to the potential of different bioenergy pathways to substitute other energy sources, bioenergy could be a solution to meet all types of energy needs: cooking, lighting, electricity, heat, and transport, depending on the feedstock and on the technology. Bioenergy options for biomass feedstock, technologies and energy needs in Africa are shown in Table 6.1.

Considering the different options listed in Table 6.1, the main criterion for sustainability is related to the feedstock production. Thereby waste material is usually a preferred feedstock source since it avoids land use and other sustainability conflicts, and since it contributes to sustainable waste management practices. Another main criterion to be considered is the efficiency of the systems. Inefficient systems need to be improved. For instance, many African households are still cooking with inefficient three-stone stoves. By improving the stove technology, considerable amounts of wood-fuel could be saved. Furthermore, a general challenge in Africa is

Table 6.1 Traditional and potential bioenergy options for biomass feedstock, technologies and energy needs in Africa

Feedstock	Processes	Energy carrier	Energy service
Non-renewable wood *(wood from land where trees/shrubs do not regrow)*	*Combustion in stoves*[a] *Conversion into charcoal*[a,b,c] *Gasification*[c] *Co-firing in medium-large combustion plants*[c]	*Firewood*[a,b,c] *Charcoal*[a,b,c]	*Cooking*[a] *Electricity*[b,c] *Process heat*[c]
Renewable wood (wood from sustainable managed forests)	Chipping[c,d] Pelletisation[c,d] Combustion in stoves[a] Conversion into charcoal[a,b,c] Gasification[c] Co-firing in medium-large combustion plants[c]	Woodchips[c,d] Pellets[c,d] Firewood[a,b,c] Charcoal[a,b,c] Syngas[c,d]	Cooking[a] Electricity[b,c] Process heat[c,d] Second gen. biofuels[d]
Oil crops (e.g. jatropha, castor, palm oil, ximenia)	Pressing[a,b,c,d] Transesterification for biodiesel production[c,d]	Straight vegetable oil[a,b] Biodiesel[c,d]	Cooking (SVO)[a] Electricity (SVO, BD)[b] Transport (SVO, BD)[c,d]
Waste oil (waste cooking oil, animal fat, fish oil)	Collection[b,c,d] Transesterification for biodiesel production[c,d]	Straight vegetable oil[a,b,d] Biodiesel[c,d]	Electricity (SVO, BD)[b,c] Transport (SVO, BD)[c,d]
Sugar and starch crops (e.g. sugarcane, sweet sorghum, cassava)	Fermentation and destillation[c,d] Geling[c]	Bioethanol[c,d] Ethanol gelfuel[c]	Transport (ethanol)[c,d] Cooking (ethanol and gelfuel)[a]
Dedicated lignocellulosic energy crops (e.g. acacia, dichrostachys, prosopis, miscanthus, typha)	Chipping[c,d] Pelletisation[c,d] Gasification[c,d] Hydrolysis, fermentation, distillation[d] Torrefaction[d]	Woodchips[c,d] Pellets[c,d] Charcoal[c,d] Syngas[c,d] Pyrolysis oil[c,d] Torrefied biomass[d]	Electricity[c,d] Process heat[c,d] Second gen. biofuels[d]
Manure *(manure collected from the field, manure from stables)*	*Dried manure combusted in stoves*[a] Wet manure for anaerobic digestion[a,b,c,d]	*Dried manure for cooking*[a] Biogas[a,b,c,d] Biomethane[c,d]	Cooking[a] Lighting[a] Electricity and heat[b,c] Biomethane in transport[d]
Dry industrial organic waste (e.g. nutshells, bagasse, rice husks)	Conversion into charcoal[c,d] Gasification[c,d] Hydrolysis, fermentation, distillation[d] Co-firing in medium-large combustion plants[c,d]	Dried waste material[c,d] Charcoal[c] Syngas[c,d] Ethanol[c,d] FT diesel[d]	Electricity[c,d] Process heat[c,d] Second gen. biofuels[d]

(continued)

Table 6.1 (continued)

Feedstock	Processes	Energy carrier	Energy service
Wet industrial organic waste (e.g. molasse, palm oil mill effluent – POME)	Fermentation and destillation[c,d] Geling[c] Anaerobic digestion[c,d]	Bioethanol[c,d] Biogas[c,d] Biomethane[d] Ethanol gelfuel[c]	Transport (ethanol)[c,d] Cooking (ethanol and gelfuel)[a] Electricity and heat[b,c] Biomethane in transport[d]
Organic solid household waste (organic waste collected from households)	Anaerobic digestion[c,d] Co-firing in medium-large combustion plants[c,d]	Biogas[c,d] Biomethane[d] Dried waste[c,d]	Electricity and heat from biogas[c,d] Biomethane in transport[d] *Electricity and heat from co-firing*[d]
Food waste (e.g. catering waste from restaurants, expired food from supermarkets, spoiled waste)	Anaerobic digestion[c,d]	Biogas[c,d] Biomethane[d]	Electricity and heat[c,d] Biomethane in transport[d]
Sewage sludge (from wastewater treatment plants)	Anaerobic digestion[c,d]	Biogas[c,d] Biomethane[d]	Electricity and heat[c,d] Biomethane in transport[d]

SVO: straight vegetable oil, BD: biodiesel
Italic: This option should be phased out as soon as possible
[a]Application at household level
[b]Application at village level
[c]Application at medium enterprise level
[d]Application at industrial level

the efficiency of the agricultural sector, independently if bioenergy or other uses of agricultural crops are considered.

6.3 Overview of Energetic Biomass Use in Africa

As it is shown in Table 6.1, the value chains are highly influenced by the energy carriers (e.g. charcoal, pellets, biogas, biofuels) which shall be produced. The energy carrier influences the feedstock selection, the conversion technology and the final energy service. Therefore, the following overview of the energetic biomass use in Africa is presented for a variety of different energy carriers.

6.3.1 Charcoal and Firewood

Wood based fuels or so called wood-fuels include firewood and charcoal. In Sub-Saharan Africa, wood-fuels provide more than 70% of the total energy consumption. The majority of the population depends on wood-fuels because modern energy such as LPG, kerosene and electricity are either unavailable or unaffordable (Legros et al. 2009).

Firewood for cooking is mainly used in rural areas whereas charcoal is mainly used in urban areas. For instance in Tanzania, more than 80% of the urban population depends on charcoal for their daily cooking (Sawe 2009). Charcoal in urban areas is preferred to firewood mainly since it has a higher calorific value per unit weight and thus can be transported more efficiently over longer distances. Furthermore, it needs less storage room, is more stable to store, and cleaner in combustion.

Wood-fuels are used by many African people since they are cheap, available and easy to handle. Thus, they constitute an important energy source for Africa. However, major problems associated with the use of wood-fuels are deforestation, and forest degradation, as well as health and safety issues when used as cooking fuel.

In comparison to the rather simple value chain of firewood production and use, the value chain for charcoal is larger and more complex, comprising a wide range of actors and operators with varying interests and stakes (Vos and Vis 2010).

Charcoal is produced from biomass (usually wood) by the carbonisation process (slow pyrolysis) which is the conversion of biomass in absence of air, under high temperatures and with long reaction times. This is typically done in batches of 1–5 tons in earth, brick or steel kilns. In Africa charcoal production is often illegal and unregulated causing deforestation at large-scale.

6.3.2 Agricultural and Forestry Residues

One of the current success stories of modern bioenergy implementation in Africa is the use of agricultural residues (bagasse) from the sugar sector in Mauritius for the production of electricity and process heat. Bagasse based co-generation is common practice in the Sub-Saharan sugar industry since many years, however Mauritius was the first country in Sub-Saharan Africa to promote the use of highly efficient high pressure co-generation equipment to increase the production of modern bioenergy. Since the 1990s bagasse based electricity is exported to the national grid. In 2004 the installed co-generation capacity in the Mauritian sugar sector was 242 MW with 318 GWh (16.5% of the electricity consumption in Mauritius) produced from bagasse and 407 GWh (21.2%) produced from coal (Deepchand 2005). Today, the sugar industry in Mauritius is self sufficient in energy and contributing about 50% of the national electricity supply through electricity exports to the grid (Karekezi and Kimani 2010). Co-generation in Mauritius is using bagasse during the harvest season (about 6 months) and coal during the rest of the year.

The development of bagasse based co-generation was strongly supported by the government of Mauritius (Karakezi et al. 2008). In 1985, the Sugar Sector Package Deal Act was enacted to encourage the production of bagasse for the generation of electricity. The Sugar Industry Efficiency Act (1988) provided tax incentives for investments in the generation of electricity and encouraged small farmers to provide bagasse for electricity generation. Together with the Bagasse Energy Development Programme (BEDP) of 1991 and the abolishment of sugar export duties in 1994, these measures resulted in a steady growth of bagasse-based electricity production in Mauritius.

In addition to the political support another important prerequisite for the success of the sugar cane bagasse based co-generation programme in Mauritius is its effective revenue sharing measures. Thereby, it is guaranteed that the monetary benefits from the sales of electricity are shared between all stakeholders of the sugar value chain including the poor smallholder sugar cane farmers.

In recent years several Sub-Saharan African countries have implemented initiatives to follow the Mauritian success story in the use of agricultural residues for energy production. In 2007 the regional initiative 'Cogen for Africa' (http://cogen.unep.org), funded by the Global Environmental Facility (GEF), was launched in order to significantly scale up the use of efficient cogeneration systems initially in seven Eastern and Southern African countries (Kenya, Ethiopia, Malawi, Sudan, Uganda, Tanzania and Swaziland). Agro-industries that are actively participating in the project include private-sector-owned sugar companies as well as private sector entities involved in agro-processing industries such as pulp and paper, forest products, palm oil, ground nuts, sisal and rice.

By February 2011, more than 100 investment opportunities for co-generation projects based on agricultural residues have been identified by 'Cogen for Africa'. Furthermore, it is estimated based on current sugar production in Sub-Saharan Africa that bagasse based cogeneration from sugar industries can meet about 5% of the total electricity demand in the region (Karekezi and Kimani 2010). Including the residues of other agro-industries and forestry industries about 10% of the electricity could be generated through co-generation.

Finally, one of the main advantages making co-generation from agricultural and forestry residues a promising option for modern bioenergy production in Sub-Saharan Africa is the fact that well established agro-industries and forestry industries have available financial resources in order to implement efficient bioenergy technologies in the short term. Private investment in bioenergy technologies, however, needs appropriate and supportive legal and regulatory frameworks in place providing suitable incentives as well as economically viable access to electricity grids (including reasonable feed-in tariff structures).

6.3.3 *Torrefied Biomass*

Torrefaction is a thermo-chemical treatment of biomass between 200 and 340°C. Thereby biomass is partly decomposed releasing volatile compounds. Torrefied biomass has approximately 30% more energy content per unit of mass than the biomass feedstock used. Currently, the interest in the use of torrefied biomass especially in large co-combustion power plants is increasing amongst European and North American stakeholders. In Africa, torrefaction could be a suitable technology, but is not yet applied.

6.3.4 *Briquettes, Pellets, and Woodchips*

Besides wood-fuels and torrefied biomass, solid biofuels include briquettes, pellets and woodchips. The production and use of these fuels in Africa is still very limited.

Biomass briquettes are blocks or balls of compacted small biomass particles. Common briquettes are charcoal and biomass briquettes made from sawdust, carbonised sawdust or waste materials (e.g. charcoal residues, sunflower shells, paper). There are various ways of producing briquettes including mechanical treatment (pressure) and mixing different substances (e.g. waxes, water). The latter technology is promoted mainly for small-scale applications in various development projects. Depending on the process and the input material, the quality of briquettes varies largely. The quality influences the use of the briquettes which can be either used at small-scale (for cooking) or at industrial-scale for large co-combustion facilities.

Pellets are generally made from compacted sawdust or other biomass (e.g. straw) that was grinded e.g. in a hammer mill. The pelletizing process is rather industrialised. Pellets are very energy dense with low humidity. Due to their often standardised nature (quality standards exist e.g. in Europe) in size and quality, pellets can be easily traded and marketed. There is an increasing market for pellets in Europe and North America to be used in biomass heating systems at household to medium-size level. In Africa, pellets are currently not used and produced at large-scale. However, due to their characteristics, pellets produced in Africa could be easily exported to international markets (e.g. Europe).

Woodchips are small pieces of wood resulting from cutting or chipping trees, branches, and other woody material. Woodchips can be used to produce heat and electricity. Although the potential for woodchips in Africa is considerable, the use of woodchips is not yet widely applied.

6.3.5 *Biogas and Biomethane*

Biogas is produced by anaerobic digestion (AD) which is a microbiological process of decomposition of organic matter (manure, sewage sludge, wet organic waste, energy crops) by microorganisms in the absence of oxygen (Al Seadi et al. 2008). AD is common to many natural environments and largely applied today to produce biogas in airproof reactor tanks and covered lagoons commonly named digesters. Technologies range in scale from domestic systems for thermal (cooking) energy to multi-megawatt grid-connected combined heat and power generation systems, or even systems that inject biomethane (upgraded biogas to $<95\%$ CH_4 content) into natural gas grids.

A wide range of micro-organisms are involved in the anaerobic process which has two main end products: biogas and digestate. Biogas is a combustible gas consisting of methane, carbon dioxide and small amounts of other gases and trace elements.

The methane fraction of biogas can be burned to produce light, electricity and heat. Digestate is the decomposed substrate, rich in macro- and micro nutrients and therefore suitable to be used as fertiliser.

In several countries in Europe there is currently a boom in agricultural medium sized biogas plants with an average electrical plant capacity of about 400 kW. Currently, in Germany alone, more than 6,000 biogas plants are in operation.

Also in Africa the numbers of biogas installations is increasing, especially for small-scale domestic biogas plants supplying households with energy for lighting and cooking (Austin and Morris 2011). A number of national domestic biogas programmes were set up in Africa each with national targets of over 10,000 domestic systems to be installed in the next 5 years. National programmes in Africa are currently implemented in Rwanda, Tanzania, Kenya, Uganda, Ethiopia, Cameroon, Benin and Burkina Faso (SNV 2010).

One of the main advantages of medium to large-scale biogas production is the potential to use wet organic waste materials which are converted to biogas and digestate which can be used as fertiliser. This presents on the one hand large opportunities for Africa, but on the other hand, this depends on the introduction of suitable waste management practices and logistics.

6.3.6 *Pure Plant Oil and Biodiesel*

A large portion of liquid biofuels can be received from lipid sources such as plant, animal and waste oils (Rutz and Janssen 2008). There exist mainly two types of fuels that are based on lipids: pure plant oil (PPO) and biodiesel obtained from further processing PPO.

Feedstock material in Africa is diverse and includes for example jatropha, palm oil, soy, castor, sunflower, coconut, and many other crops (Walimwipi et al. 2011). Also waste cooking oil and animal fats from the rendering or fish industry could be potentially used. In the PPO and biodiesel value chain, the feedstock selection highly affects the quality and the properties of the fuel which influences the potential use. PPO and biodiesel for transport applications must meet high quality standards in order to ensure smooth engine operations. Several African countries have started to introduce national targets and initiatives for the use of biofuels (both biodiesel and bioethanol) in transport, including for example Angola, Benin, Ethiopia, Ghana, Kenya, Madagascar, Malawi, Mali, Namibia, Nigeria, Senegal, South Africa, Swaziland, Tanzania, Zambia, and Zimbabwe (Walimwipi et al. 2011; Janssen and Rutz 2011).

However, many approaches in Africa do not target the use of PPO and biodiesel in transport, but in small-scale stationary power stations for rural electrification. Often, these projects include multifunctional platforms (MFP) for different services. Small-scale projects usually use pure plant oil directly, whereas large-scale projects aim at further processing of the oil into biodiesel.

Technologies for the pressing, purification and transesterification processes are mature and available for different sizes and purposes. Biodiesel production can range from small-scale, often batch-production, to large-scale industrial facilities.

For example in South Africa, several micro- to small-scale biodiesel plants are operated by small fleet owners or home producers using waste oils.

A relatively new approach is to produce biodiesel from algae at industrial scale. However, technologies are not yet mature and it is unclear if the algae to biodiesel chain will be commercial in the mid- to long-term. It may be also possible that oils from algae are rather used for chemical purposes instead of bioenergy due to their high production prices. Several areas in Africa may be suitable for the production of algae.

6.3.7 *Bioethanol*

The use of bioethanol in the energy sector is widely applied today. The main producers of bioethanol are the USA and Brazil and bioethanol is traded internationally. It is mainly produced by fermentation of sugar and starch. Bioethanol can be produced at different scales; however, most facilities are large-scale.

The main feedstock today are sugarcane, sugar beets, corn, and wheat. In Africa, the main feedstock is sugarcane, but in recent years much research was put on breeding sweet sorghum varieties for bioethanol production (Munyinda et al. 2011; Chagwiza and Fraser 2011; Janssen et al. 2010). The advantage of sweet sorghum is its higher production efficiency compared to sugarcane and its tolerance to difficult growth conditions (e.g. drought, soil acidity). Other potential feedstock sources in Africa are grain sorghum, cassava, and maize. Also molasses from the sugar industry can be used for bioethanol production.

In recent years, many efforts were placed on the production of ethanol from lignocellulosic biomass, so called second generation bioethanol. It is estimated that the technology of second generation bioethanol production is close to commercialisation. Besides waste materials (e.g. bagasse), potential second generation feedstock in Africa could be miscanthus and bamboo (Munyinda et al. 2011).

Munyinda et al. (2011) estimated the bioethanol potential in selected African countries for different feedstocks at 35 EJ, with the Democratic Republic of Congo having the largest potential, followed by Angola, Sudan, Zambia and Tanzania. Bioethanol production in Africa was 637 million l in 2007, with the largest production in South Africa, followed by Egypt and Nigeria.

A number of African countries have introduced initiatives and targets to promote bioethanol development for transport, including Ethiopia, Kenya, Malawi, Nigeria, South Africa, Sudan, and Zimbabwe. A niche market exists for the use of liquid ethanol and ethanol-gel for domestic use in cooking facilities.

6.3.8 *Pyrolysis Oil, Syngas and Other Thermo-Chemical Conversion Products*

Lignocellulosic biomass can be converted by thermo-chemical processes into various valuable intermediate products, including syngas, pyrolysis oil, char, and slurry (mixture of char and pyrolysis oil). The desired product is influenced by the reactor design and several parameters, such as the temperature, reaction time, reaction agent, and pressure. Suitable feedstock are all types of dry lignocellulosic materials such as waste materials (e.g. rice husks, nutshells, straw) and dedicated energy crops such as short rotation shrubs and trees as well as grasses. Potential feedstock in Africa would be acacia, eucalyptus, miscanthus, bamboo, or typha.

Some of the above mentioned intermediate products can be directly converted into energy. For instance, syngas can be used in gas engines and pyrolysis oil in diesel engines to produce electricity. Some pyrolysis plants for electricity production are e.g. installed in West Africa (Novis 2010). Pyrolysis oil, char and slurry can be also gasified in order to obtain syngas.

Syngas, either obtained by direct gasification of biomass or by gasification of pyrolysis oil, char and slurry, can be further processed by e.g. Fischer-Tropsch synthesis to produce Biomass-to-Liquid (BtL) fuels for transport. South Africa has large experience in synthesising syngas from coal and natural gas. However, BtL fuels are not yet produced at commercial scale and it will need several years to make BtL fuels competitive to fossil fuels and other biofuels.

6.4 Conclusion

As in many developing countries, bioenergy in Africa is currently mainly based on traditional use of bioenergy (i.e. wood-fuels used for cooking in simple stoves). Several initiatives in Africa are also promoting the modern conversion of biomass into high value solid, liquid and gaseous energy carriers, but the current market is still comparably small. Currently, the main progress in modern conversion of biomass is achieved in Europe, North America and some countries in Latin America and Asia.

An important challenge is to promote knowledge, technology and experience transfer on modern bioenergy to African countries. Furthermore, the cooperation between African countries has to be encouraged to develop suitable strategies for different technologies and adequate policies. Finally, also experience transfer from African countries to other continents should be supported. For instance the large experience in South Africa in synthesis of syngas could contribute to support the faster commercialisation of BtL fuels. Furthermore, the understanding of local framework conditions and circumstances (climate, policies, soil properties, social aspects) in Africa has to be transferred to foreign investors and technology providers for the successful creation of modern bioenergy markets in Africa.

References

Al Seadi T, Rutz D, Prassl H, Köttner M, Finsterwalder T, Volk S, Janssen R (2008) Biogas handbook. BiG>East Project funded by the Europen Commission (EIE/07/214), University of Southern Denmark Esbjerg, Denmark, p 142, ISBN 978-87-992962-0-0

Austin G, Morris G (2011) Biogas production in Africa. In: Janssen R, Rutz D (eds.) Bioenergy for sustainable development. Springer Verlag, Germany

Chagwiza C, Fraser G (2011) Economic evaluation of sweet sorghum in biofuel production as a multi-purpose crop: the case of Zambia. In: Janssen R, Rutz D (eds.) Bioenergy for sustainable development. Springer Verlag, Germany

Deepchand K (2005) Sugar cane bagasse energy cogeneration – lessons from Mauritius. Paper presented to the parliamentarian forum on energy legislation and sustainable development, Cape Town, South Africa, 5–7 October 2005

Janssen R, Rutz D (2011) Keynote Introduction: overview on bioenergy policies in Africa. In: Janssen R, Rutz D (eds.) Bioenergy for sustainable development. Springer Verlag, Germany

Janssen R, Rutz D, Braconnier S, Reddy B, Rao S, Schaffert R, Parella R, Zaccharias A, Rettenmaier N, Reinhardt G, Monti A, Amaducci S, Marocco A, Snijman W, Terblanche H, Zavala-Garcia F (2010) Sweet sorghum – an alternative energy crop. In: Proceedings of the 18th European Biomass Conference and Exhibition, Lyon, France, 3–7 May 2010, pp 200–206, ISBN 978-88-89407-56-5

Karekezi S, Kimani J (2010) Bioenergy development in sub-Saharan Africa. In: Making It Magazine, 26 March 2010

Karakezi S, Kithyoma W, Muzee K, Oruta A (2008) Scaling up bioenergy in Africa. Presented during the international conference on renewable energy in Africa, Dakar, 16–18 April 2008

Legros G, Havet I, Bruce N, Bonjour S (2009) The energy access situation in developing countries: a review focussed on the least developed countries and sub-Saharan Africa. United Nations Development Programme (UNDP) and World Health Organization 2009

Munyinda K, Yamba FD, Walimwipi H (2011) Bioethanol potential and production in Africa: sweet sorghum as a complementary feedstock. In: Janssen R, Rutz D (eds.) Bioenergy for sustainable development. Springer Verlag, Germany

Novis (2010) http://www.novis.de/?lang_pref=en. Accessed 3 Apr 2011

Rutz D, Janssen R (2008) Biofuel technology handbook, 2nd edn. WIP Renewable Energies, München

Sawe EN (2009) Sustainable charcoal production for poverty reduction in Tanzania. Bioenergy for sustainable development in Africa conference in Brussels, Belgium, 24-25 November 2009. TaTEDO, http://www.compete-bioafrica.net/events/events2/Brussels/S4-3-COMPETE-Conference-Brussels-Nov2009-Sawe.pdf

SNV (2010) Domestic Biogas Newsletter. Issue 3, August 2010

Vos J, Vis M (2010) Making charcoal production in Sub-Sahara Africa sustainable. NL Agency, The Netherlands

Walimwipi H, Yamba FD, Wörgetter M, Rathbauer J (2011) Biodiesel production in Africa. In: Janssen R, Rutz D (eds.) Bioenergy for sustainable development. Springer Verlag, Germany

Chapter 7
Sustainable Charcoal and Firewood Production and Use in Africa

Estomih N. Sawe

Abstract The energy sector in Africa is characterized by low per capita consumption of modern energy in a continent endowed with extensive but under-developed energy resources. The majority of the population depends on wood-fuels (charcoal and firewood) because modern energy such as LPG, kerosene and electricity are either unavailable or unaffordable. Deforestation and forest degradation is a major problem created by the current energy consumption patterns in Africa. The negative environmental impacts of cooking with wood-fuels are increasing in most African countries. Poor forest management and increasing demands are leading to land degradation, which for rural women translates to increasingly long daily treks to gather fuel-wood. On average less than 30% of the Sub-Saharan African population has access to electricity and more than 70% depend on wood-fuels for cooking and heating. The current low access to modern energy services in Africa causes negative impacts on poverty reduction efforts and the attainment of the MDGs. Despite the understanding that wood-fuels are the main energy source for the majority of the African population, improvements in wood-fuel production and use is currently not of high priority in energy development strategies. There are no specific wood-fuel policies, strategies and programmes supported by African governments. However, the use of wood-fuels is critical for meeting basic energy needs of the majority of the population in African countries. Due to this low prioritization of wood-fuels, the sector is poorly funded and lacks good governance. Therefore, securing political commitment for formulating and implementing effective policies and strategies are crucial elements that could ensure sustainable production and use of wood-fuels in Africa.

E.N. Sawe (✉)
TaTEDO, P.O. Box 32794, Mpakani A, Plot No. KJM/MPA/98,
Near Institute of Social Works, Kijitonyama, Dar es Salaam, Tanzania
e-mail: energy@tatedo.org; edirector@tatedo.org

R. Janssen and D. Rutz (eds.), *Bioenergy for Sustainable Development in Africa*,
DOI 10.1007/978-94-007-2181-4_7,

Keywords Sustainable wood-fuels • Modern energy services • Poverty reduction • Wood-fuels production and use • Wood-fuels waste • Climate change • Indoor air pollution • Energy consumption • Policies • Strategies

7.1 Introduction

Wood based fuels (charcoal and firewood) provide more than 70% of the total energy consumption in Sub-Saharan Africa. The majority of the households, institutions, small and medium enterprises will continue to depend on wood based fuels to meet their energy needs for many decades due to lack of affordable alternatives. In particular, the demand for wood-fuels in most of Sub-Saharan African countries continues to grow due to population growth. The present wood-fuel consumption pattern in Africa is clearly unsustainable. The current production and use is based on inefficient technologies and practices. Coupled with most wood-fuel user's inability to access modern energy this leads to uncertainty for future dependence on the already diminishing biomass stock for energy. The situation is further aggravated by the lack of effective policies and strategies to address the growing challenges of wood-fuel production and use in Africa. Policy makers in Africa pay little attention to the ways in which wood-fuels are produced and used. Most governments fail to recognize wood-fuels as the main source of energy for the majority in their countries leaving it to the informal sector. In most countries, the supply and demand of wood-fuels is handled by different ministries. Hence, authority and jurisdiction lack clarity. In many African countries, the forest sector contribution to the national economy is marginal (2–4%). Wood-fuel production, use and marketing are predominantly informal and thus escape official statistics, although estimated revenues from charcoal businesses in most African countries exceed more than 500 million US$ per year. Despite the diminishing wood resource, wood-fuels (especially charcoal) in most countries remain underpriced by more than 30% relative to their economic cost. Production costs are not reflected in the market price of wood-fuels. This undervaluation causes waste and inefficient use. It is also a disincentive for sustainable forest management, and for the adoption of efficient wood-fuel technologies such as improved cook stoves and charcoal production kilns.

7.2 Access to Modern Energy for Sustainable Development in Africa

Poverty reduction and access to modern energy services is crucial for sustainable development and the achievement of the millennium development goals (MDGs) in African countries. Without access to modern energy services, the majority of the African population is deprived of potential income generating opportunities and quality social services. There are more than 800 million people in Africa lacking access to modern energy services. This situation entrenched poverty and enhanced unsustainable use of wood based fuels for cooking and process heat in small, medium and micro-enterprises. The International Energy Agency (IEA) forecasts that the use of wood-fuels will continue to increase in Sub-Saharan Africa due to population growth unless effective policy measures are taken to reverse the situation. The unsustainable use of wood-fuels in Africa is causing deforestation, soil erosion, desertification, increased risk of flooding, biodiversity loss and climate change. The increasing use of wood-fuels has negative effects on human livelihoods, especially that of women and children since cooking on wood-fuels in poorly ventilated houses is a major source of indoor air pollutants which causes acute respiratory disease. Reliance on wood-fuels also entrenches gender disparities as time spent, especially by women, on collecting traditional fuels could be spent on other productive activities and education.

Commercial energy (electricity and petroleum based fuels) accounts for only a fraction of energy consumption in most Sub-Saharan African countries although most investment on energy planning is focused on commercial energy. The expansion of electricity grids is costly and not affordable by the majority of African countries and many of its poor people. Electricity from decentralized sources such as small hydro, solar and wind systems has high initial costs which are unaffordable by the majority of the poor in Africa. The potential for sustainable production and use of wood-fuels through improving efficiency in resource management and the introduction of efficient biomass technologies such as improved charcoal production kilns with efficiency of more than 25% and improved wood-fuel stoves is crucial. Potential wood-fuel alternatives such as LPG, kerosene, biogas need to be carefully assessed, appropriate measures need to be implemented, and policy incentives for scaling up the use of these fuels need to be formulated.

7.3 Sustainable Wood-Fuel Production and Use

The prevailing experience and trend indicate that wood-fuels will remain the major source of energy for most African countries. Other proven renewable energy sources and technologies (such as small hydro, solar, biogas) could significantly reduce the severe energy poverty, provided they are affordable, reliable and financing mechanisms are available. Unfortunately, biomass resources in most African countries are clearly under threat from overexploitation, creating several social, environment and economic challenges. The increasing costs of accessing wood-fuels in terms of walking distance, fuel purchase and decreasing quality of wood-fuels demonstrate the diminishing trend of wood-fuel resources.

Whilst larger efforts are important to increase access of commercial modern energy technologies and services, it is even more crucial to implement effective policies and strategies to ensure sustainable supply of wood-fuels for the majority of the population in Africa who will never have access to electricity and petroleum fuels in the foreseeable future.

African countries need to recognize the important role of wood-fuels in households cooking and provision of process heat in rural industries. The policies and strategies should focus on sustainable wood-fuel production and use through ensuring large-scale uptake and use of efficient wood-fuel technologies and practices. It is unfortunate that a number of African countries continue to insist on unrealistic options that encourage immediate switch to alternative modern energy services which neither the African countries nor their population can presently afford.

The key objective of African governments should be to urgently formulate and implement policies and strategies that clearly recognize wood-fuels as major energy source now and in the foreseeable future. Such initiatives should aim at enhancing capacities of different stakeholders from local to national levels to enable them to contribute effectively in sustainable wood-fuel production and use.

7.3.1 Efficiency Optimization of Wood-Fuel Production and Use

The evaluation of results of wood-fuel programmes implemented by the centre for sustainable modern energy expertise (TaTEDO) in Tanzania has shown that households, SMEs and institutions using improved wood-fuel technologies can significantly reduce costs and hence improve their economic situation. Up to 60% wood and cost savings have been achieved which is 900 US$ per institution per year. Households have been able to save up to 70% on firewood and charcoal use, respectively. In monetary terms this refers to cost reductions of about 80 US$ for firewood users and about 130 US$ for charcoal users per household per year.

Similarly, the average production cost of charcoal using improved basic earth mound kilns was reduced by over 40% from 3 to 1.7 US$ per 30 kg bag. In addition, improved charcoal production kilns increased average productivity of charcoal production from 6.6 to 16.4 bags per week from the same quantity of wood. These experiences indicate that there is tremendous benefit from scaling up and replicating strategies for more efficient wood-fuel production and use in most African countries.

7.3.2 Policy and Institutional Issues

Sustainable wood-fuel production and use cannot succeed without its clear recognition as the main source of energy for both consumptive and productive purposes in most Sub-Saharan African countries. The current trend in many African countries for energy departments focusing on commercial energy at the expense of wood-fuels need to be reconsidered. The present energy policies are inadequate in effectively addressing wood-fuel issues. It is only with official policy recognition of the key role of wood-fuels that the private sector and community initiatives could scale-up capacity building, technology transfer, financing and entrepreneurship on wood-fuels to meet the growing challenges in the sector. It is also important to transfer and share good practices amongst African countries. For successful wood-fuel production, the issue of land tenure needs to be addressed, as in some countries land tenure is a key source of uncertainty for investments in wood-fuel production.

In most African countries there is a clear lack of adequate institutional framework for addressing sustainable wood-fuel production and use. Although there are ministries in charge of forestry and energy, these tend to give low priorities to issues related to wood-fuels. They are in most cases based at national level with no institutions for the development of wood-fuels at local level where they are urgently needed. The few national and local institutions working on wood-fuels tend to undertake activities in an uncoordinated and ad hoc way, in most cases failing to have high impact and often causing duplication of efforts.

A comprehensive bioenergy policy that is well managed and monitored in its implementation will have higher positive impact on sustainable wood-fuel production and use than isolated local level initiatives. The recent initiatives by some African governments to establish rural energy agencies are a welcomed step forwards. It is important that such agencies are adequately empowered and supported to effectively handle wood-fuel issues as priorities. This should include support for formulating and implementing effective legal and regulatory framework and financing mechanism. These agencies need to put greater emphasis on support for the private sector, community participation and other stakeholders in sustainable wood-fuel production, use and marketing.

7.4 Recommendations and Conclusions

Energy policies and strategies of African countries and their development partners should focus on sustainable wood-fuel production and use since this is a priority for the majority of Africa's population and since this has a high potential for the contribution to the MDGs.

Effective wood-fuel policies will ensure sustainability of biomass energy services and contribute to environmental conservation both at local and global level. Value chain efficiency improvements will ensure minimum biomass energy losses with respect to wood-fuel planting, harvesting, processing, use and final disposal.

Wood-fuel waste will be minimized in addition to other social, economic and environmental benefits. It is crucial that wood-fuel technologies and land policies in African countries are reviewed and harmonized, and institutional frameworks and governance structures are put in place. With suitable policies and strategies, and effectively implemented legal and regulation frameworks, wood-fuel production and use in Africa could contribute to sustainable development, poverty reduction and mitigation of climate change.

References

CHAPOSA (2002) Charcoal potential in Southern Africa. Final report Chaposa Research Project. Tanzania

East African Community (2006) Strategy on scaling up access to modern energy services. Arusha, Tanzania

ESMAP (2005) Advancing bioenergy for sustainable development. World Bank, Washington, DC
Malimbwi RE, Zahabu E, Mchome B (2008) Situation Analysis of charcoal sector in Dar es Salaam. Sokoine University of Agriculture, WWF, Dar es Salaam, Tanzania
Sawe EN, Sago S (2009) Charcoal industry in Tanzania with selected TaTEDO experience. Sust Energy Dev Forum Dec 2009(7), TaTEDO
TaTEDO (2009) Integrated improved wood-fuels in Tanzania. Final evaluation report. Submitted by Ringia O, Massawe W, Dar es Salaam, Tanzania, November 2009
United Nations Foundation (2008) Sustainable bioenergy development in UEMOA member countries. Ouagadougu, Mali
UN, Department of Economic and Social Affairs (2007) Small-scale production and use of liquid biofuels in Sub-Sahara Africa: perspective for sustainable development. Background paper no. 2, New York

Chapter 8
Bioethanol Potential and Production in Africa: Sweet Sorghum as a Complementary Feedstock

Kalaluka Munyinda, Francis Davison Yamba, and Hartley Walimwipi

Abstract This chapter presents an overview of bioethanol potential and production in Africa with specific emphasis on sweet sorghum as a complementary feedstock to sugarcane. Bioethanol feedstock types in Africa include sugarcane, sweet sorghum, maize and cassava. The bioethanol potential in selected African countries for different feedstocks is estimated at 35 EJ, with the Democratic Republic of Congo having the largest potential, followed by Angola, Sudan, Zambia and Tanzania. Bioethanol production in Africa was 637 million litres in 2007, with largest production in South Africa, followed by Egypt and Nigeria. Potential ethanol markets in southern African countries based on E5 and E10 are estimated at 0.8 and 1.5 billion litres per year. The biggest challenge for the development of biofuel markets in southern Africa is that although potential markets exist, and conventional technologies are available, feedstock is not available in sufficient quantities. The main feedstock for ethanol production in Africa is sugarcane, but in the recent past the continent has seen an increasing interest in new feedstock types such as sweet sorghum. In this chapter results for yields and sugar brix for eight exotic sweet sorghum varieties (TS1, Madhura, Praj-1, GE2, GE3, Wray, Cowley, and Keller) are presented and compared to a local sweet sorghum variety (Sima). The results show encouraging yields averaging at 30 tons per hectare under rain-fed conditions. Sugar content in Brix% averaged at about 18% for medium maturing varieties (Keller, GE3, and Sima). The highest values of sugar content of sweet sorghum varieties are similar to those of sugarcane (18%).

Keywords Bioethanol • Sweet sorghum • Complimentary feedstock • Brix • Potential

---

K. Munyinda (✉) • F.D. Yamba • H. Walimwipi
Centre for Energy Environment and Engineering Zambia (CEEEZ), 176 Parirenyatwa Road, Suite B, Private Bag E721, Lusaka, Zambia
e-mail: ceeez@coppernet.zm; ceeez@zamnet.zm; munyinda_kalaluka@yahoo.com; ceeez@coppernet.zm; ceeez@zamnet.zm; yambafd@yahoo.com
hartleykabunda@yahoo.com; ceeez@coppernet.zm

R. Janssen and D. Rutz (eds.), *Bioenergy for Sustainable Development in Africa*, DOI 10.1007/978-94-007-2181-4_8,

8.1 Overview of Bioethanol Potential and Production in Africa

8.1.1 Bioethanol Potential

The main resources for bioethanol production in Africa include sugarcane, sweet sorghum, grain sorghum, cassava, and maize as first generation feedstock and miscanthus and bamboo as second generation feedstock. However, there are other indigenous fast growing shrubs and grasses that need to be explored as potential feedstock on the continent (Walimwipi 2008). Key drivers for bioethanol production in Africa are energy security, reduction of oil imports, high oil prices, environmental commitments, rural development opportunities, diversification of agricultural industries and lead/MTBE phase out programmes (Batidzirai 2007).

The most popular feedstock for ethanol production in Africa is sugarcane, but in the recent past the continent has seen an increasing interest in new feedstock types such as sweet sorghum. Maize and cassava are no suitable feedstock for ethanol production, since they may lead to increased competition with food production. The bioethanol potential in selected African countries for different feedstock types is shown in Table 8.1.

The presented resource assessment is based on literature reviews of bioenergy crops that are known to occur naturally or can be grown in African countries. An inventory was set up based on findings from literature. The biofuels potential analysis was based on available arable land and a production mix of various bioenergy crops. The preliminary biofuels potential analysis was scaled down to the realistic assumption of 10% penetration levels of dedicated bioenergy crops into the available arable agricultural land. Further, lower heating values and biofuel yields per hectare for respective crops were utilized to obtain the biofuels potential in PJ (Walimwipi 2008).

8.1.2 Bioethanol Production

Global ethanol production doubled to 46 billion litres between 2000 and 2005. About 13 countries used ethanol fuel in 2003 and at least 30 countries have already or are planning to introduce ethanol fuel programmes (Batidzirai 2007). Table 8.2 shows ethanol production in selected African countries.

A number of African countries have introduced programmes as part of initiatives to promote biofuels development. Some countries have made remarkable progress in terms of policy measures and targets for biofuels. In the following bioethanol programmes and initiatives in selected African countries are briefly presented (Batidzirai 2007).

Zimbabwe has been blending gasoline with ethanol from 1980 to 1992. The ethanol programme was motivated by international sanctions imposed on former Rhodesia, security of supply, foreign currency savings, and low sugar prices. A number

Table 8.1 Bioethanol potential in selected African countries for different feedstock types in PJ

	Sugarcane	Sweet sorghum	Cassava	Maize	Wheat	Total
Angola	1,270	912	612	603	–	3,397
Benin	135	–	65	64	–	264
Botswana	–	189	–	–	–	189
Burkina Faso	336	241	–	–	–	577
Cameroon	433	311	209	209	–	1,162
CAR	918	659	442	–	–	2,019
Chad	716	514	345	–	–	1,575
DR of Congo	2,134	1,533	1,028	1,028	–	5,723
Congo	685	–	64	–	–	749
Côte d'Ivoire	300	–	28	–	–	328
Ethiopia	639	–	–	304	–	943
Gabon	–	–	252	–	–	252
Ghana	187	134	90	89	–	500
Guinea	214	154	103	102	–	573
Kenya	136	98	66	65	–	365
Madagascar	558	400	269	265	–	1,492
Malawi	76	55	37	36	–	204
Mali	412	296	198	196	–	1,102
Mozambique	806	578	388	388	–	2,160
Namibia	193	138	93	92	–	516
Niger	134	96	64	–	–	294
Nigeria	503	361	243	239	–	1,346
South Africa	448	–	–	213	107	768
Sudan	1,476	1,060	–	–	352	2,888
Swaziland	18	–	9	9	–	36
Tanzania	766	550	369	364	–	2,049
Uganda	111		53	53		217
Zambia	913	655	440	434		2442
Zimbabwe	326	234	157	155		872
Total						35,002

Source: Walimwipi (2008)

of factors contributed to the success of the ethanol programme. These include public-private partnerships, use of local materials and labour, well developed agriculture and industry, clear pricing policies, well planned implementation strategies as well as the absence of food-fuel conflicts.

In **Malawi** ethanol-gasoline blending was introduced in 1982, influenced by costly imports and security of supply (regional instability). Factors that fostered success of the programme include clear and consistent policies, incentives and competitive pricing, as well as steady availability of feedstock and irrigation water (Lake Malawi).

Table 8.2 Ethanol production in selected African countries

Country	Annual production (million litres)
Egypt	30
Kenya	15
Mauritius	23
Nigeria	30
South Africa	410
Swaziland	13
Zimbabwe	25
Other Africa	92
Total	638

Source: Batidzirai (2007)

Kenya blended ethanol with gasoline from 1983 to 1993. Due to continuous losses as a result of uncompetitive pricing, poor management, resistance from oil companies, and loan servicing burdens, blending was stopped in 1993 and the ethanol is currently exported.

South Africa has introduced an Industrial Biofuel Strategy in 2007. Mandatory E10 blending legislation is currently pending.

Ethiopia introduced an E5 mandate in Addis in 2007.

Nigeria has entered into a partnership with Brazil to incrproduction in Nigeria. This approach aims at using the Brazilian model to start a bioethanol programme in Nigeria including the Presidential Initiative on cassava. Both countries signed a memorandum of understanding in 2005.

Sudan introduced a bioethanol programme in 2007. The new 10-year sugar strategy includes a 250 million litres ethanol plant in Eljazeera.

8.2 Sweet Sorghum as Complementary Feedstock

Grain sorghum today is the world's fifth largest grain crop in production, with 42 million hectares under cultivation. The leading producers are: USA, Nigeria, India, China, Mexico, Sudan and Argentina. Some of this land in Asia and Africa could be replaced by sweet sorghum varieties, which are capable of producing both grain and energy (bioethanol). Sweet sorghum may give 23% additional returns to farmers compared to grain sorghum (in India) (Kyritsis 2010).

Depending on biofuel policies and especially on mandatory blending quota, significant bioethanol markets could be created in southern Africa. This potential is illustrated in Table 8.3 for the fuel blends E5 (95% gasoline and 5% ethanol) and E10 (90% gasoline and 10% ethanol).

The biggest challenge for the development of biofuel markets in southern Africa is that although potential markets exist and conventional technologies are available, feedstocks are not available in sufficient quantities. For example, in South Africa the ethanol demand for E10 is estimated at 1 billion litres (at 2,000 levels) against a

Table 8.3 Potential ethanol markets in southern African countries based on E5 and E10 (million litres)

	2000		2015	
Country	E5	E10	E5	E10
Malawi	4.81	9.62	6.54	13.07
Mozambique	3.33	6.66	4.71	9.41
South Africa	511.9	1,023.8	691.45	1,382.9
Swaziland	4.64	9.28	6.21	12.42
Zambia	8.89	17.7	19.0	25.31
Zimbabwe	23.46	46.91	28.89	57.78
Total	557.03	1,113.97	756.8	1,500.89

potential supply of 0.37 billion litres from C molasses. Thereby, it is assumed that all sugarcane factories in South Africa convert most of the C molasses into ethanol.

This scenario leaves a deficit of 0.74 billion litres increasing to 1.0 billion litres by the year 2015. Even if existing sugarcane factories are allowed to expand at acceptable levels of 2–3% per year, available molasses will not be sufficient to cover the demand. Ethanol from lignocellulosic sources such as wood, grass and bagasse is promising; however, conversion technologies are currently not competitive, whilst feedstocks are abundant. For these reasons it is advisable to seek for alternative sugar based feedstocks in the short and medium term such as sweet sorghum.

8.3 Sweet Sorghum Characteristics

Sugarcane and sweet sorghum are C4 plants which have high photosynthesis potential and high biomass productivity compared to other crop types (Munyinda 2008). The genus sorghum includes grain sorghum and sweet sorghum noted for their high yields. Sweet sorghum differs from grain sorghum by a few genes which are controlling plant height and the presence of sugar in the stem. Sweet sorghum accumulates sugars in the stem just as sugarcane. The main advantages of sweet sorghum include (1) shorter growing period (100–130 days), (2) low production cost, and (3) relative drought resistance. Table 8.4 presents a comparison of sugarcane, sugarbeet, and sweet sorghum characteristics (Kyritsis 2010).

8.4 Sweet Sorghum Agronomic Performance Field Tests

8.4.1 Methodology

Field tests with sweet sorghum varieties in three agro-ecological regions of Zambia were undertaken with respect to biomass production, sugar content and accumulation, and optimum time of harvest under rain fed conditions. Eight exotic sweet

Table 8.4 Comparison of sugarcane, sugarbeet and sweet sorghum characteristics

Crop duration	Sugarcane	Sugarbeet	Sweet sorghum
Growing season	About 7 months	About 5–6 months	About 4 months
Soil requirements	Only one season	Only one season	One season in temperate and two or three seasons in tropical area
Water management	Grows well in drain soil	Grows well in sandy loam; also tolerates alkalinity	All types of drained soil
Crop management	36,000 m³/h	18,000 m³/h	12,000 m³/h
Yield per hectare	Requires good management	Greater fertilizer requirements; requires moderate management	Little fertilizer required; less pest and disease complex; easy management
Yield per ha	70–80 tons	30–40 tons	54–69 tons
Sugar content on weight basis	10–12%	15–18%	7–12%
Sugar yield	7–8 tons	5–6 tons/ha	6–8 tons
Ethanol production directly from juice	3,000–5,000 l/ha	5,000–6,000 l/ha	3,000 l/ha
Harvesting	Mechanical harvested	Very simple; normally manual	Very simple; both manual and through mechanical harvested

Source: Kyritsis (2010)

sorghum varieties (TS1, Madhura, Praj-1, GE2, GE3, Wray, Cowley, and Keller) were compared to a local sweet sorghum variety (Sima). The second phase of the study involved agronomic performance of sweet sorghum under supplementary irrigation conditions (varieties were planted initially under irrigation and subsequent growth occurred under rainfed conditions).

8.4.2 Results and Analysis

Results of biomass production of sweet sorghum varieties at different growth stages and accumulation of sugar in different varieties are shown in Figs. 8.1 and 8.2, respectively, under rain fed conditions. The growth stage T1 refers to the booting stage of sweet sorghum varieties (i.e. opening of inflorescence), whereas T2 refers to the soft dough stage (i.e. the stage of the grain before hardening).

The stem yields presented in Fig. 8.2 are averages across different environments influenced by soil type and climate. The results at T2 show that yields tend to be highest for the variety TS1 (in average 32 t/ha). Wray had similar yields as Praj-1, Madhura, GE2, and GE3, namely 21.2 Mt/ha, whereas Cowley, Madhura, Keller,

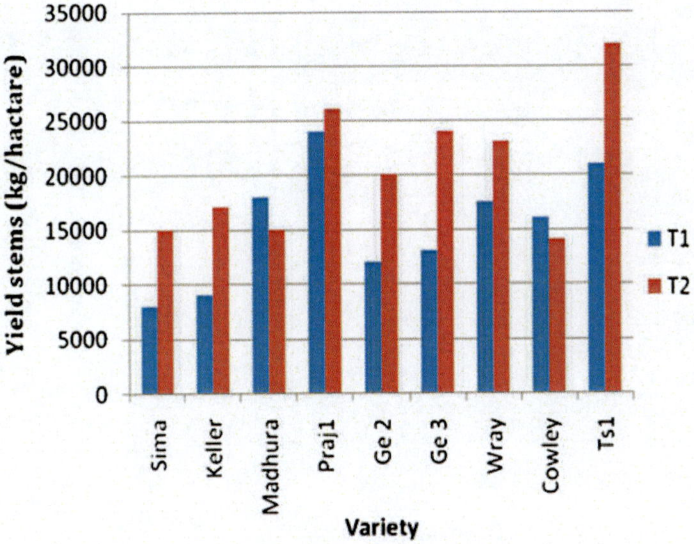

Fig. 8.1 Harvest of sweet sorghum at different growth stages

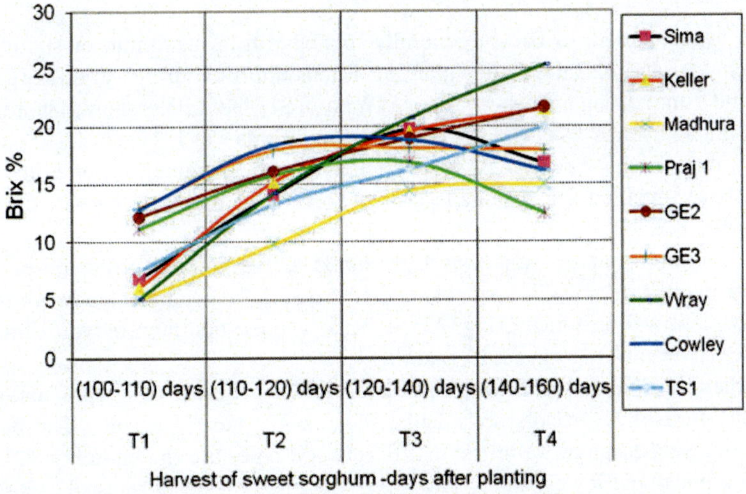

Fig. 8.2 Accumulation of sugar in different varieties of sweet sorghum at UNZA Farm

and Sima had the lowest yields of 13.8 t/ha. The yields were highest in Phaeozems, whereas yields were lowest in Ferralsols (i.e. acidic soils with high aluminium saturation). The stem yields of sweet sorghum varieties under rainfed conditions were lower than those for sugarcane (75 t/ha). The yields are, however, higher under supplementary irrigation conditions averaging at 70 t/ha for single cropping.

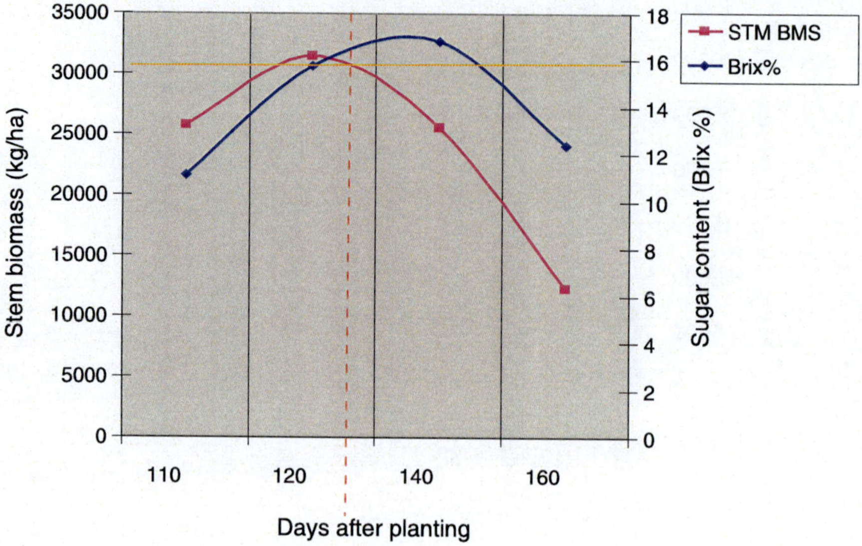

Fig. 8.3 Optimum harvest time of sweet sorghum

The sugar content in Brix% generally increases to a maximum of about 18% between 110 and 120 days after planting for medium maturing varieties (Keller, GE3, and Sima). Long maturing varieties (Wray, TS1, and Keller) showed increasing sugar content even at T4 (140 days after planting). Sima was lowest in sugar content with a value of 15%. The highest sugar content of sweet sorghum varieties are similar to those obtained for sugarcane (18%). Results of optimum harvest time are shown Fig. 8.3.

The optimum harvest time was determined at the point of inflexion of stem yields, i.e. when stem yields started to decrease due to the drying of stems whilst the sugar content was still increasing. Across varieties, the optimum harvest time was observed at around 122 days after planting.

Furthermore, stem yields were evaluated under different soil types including Acrisols, Lixisols, Vertisols, Ferralsols, Phaeozems and Cambisols. The performance of sweet sorghum varieties on different soil types are shown in Fig. 8.4.

The highest and lowest stem yields were obtained on Phaeozems (32 t/ha) and Ferrolsols (9 t/ha) across all varieties. Table 8.5 shows yield increase of sweet sorghum varieties on Phaeozems and Lixisols compared to Ferralsols and Vertisols.

The results show the adaptation (i.e. differential response) of sweet sorghum varieties on soil types. The lowest yields for Madhura, GE2, Praj-1, TS1, Cowley, Keller, Wray and Sima are obtained on Ferralsols. On the other hand, GE3 and Wray also perform poorly on Vertisols. For all varieties, maximum yields were obtained on Phaeozems. Although the lowest yield of Wray was obtained on Ferralsols, it was higher than for other varieties on the same soils. This shows that Wray is more adapted to acidic soils.

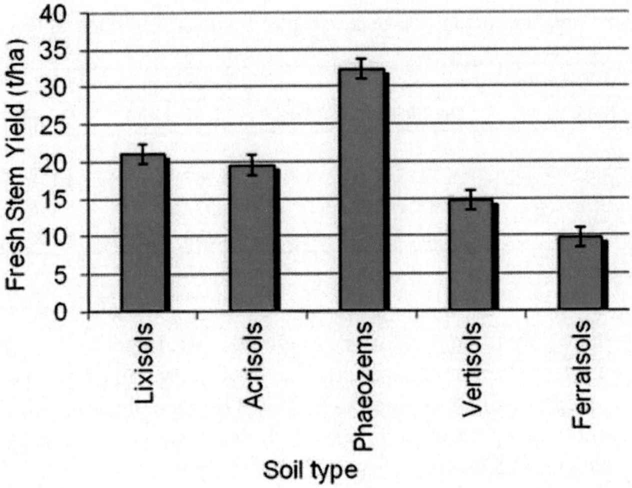

Fig. 8.4 Performance of sweet sorghum varieties on different soil types

Table 8.5 Yield increase of sweet sorghum varieties on Phaeozems and Lixisols compared to Ferralsols and Vertisols (kg/ha)

Variety	Minimum yield	Soil type	Maximum yield	Soil type	Yield increase
Madhura	4,816	Ferralsols	32,370	Phaeozems	85.1
GE2	7,056	Ferralsols	43,030	Phaeozems	83.6
Praj-1	6,545	Ferralsols	38,720	Phaeozems	83.1
TS1	5,223	Ferralsols	29,690	Lixisols	82.4
GE3	9,495	Vertisols	34,590	Phaeozems	72.5
Cowley	10,190	Ferralsols	32,370	Phaeozems	68.5
Keller	11,900	Ferralsols	33,430	Phaeozems	64.4
Wray	12,140	Ferralsols	28,670	Phaeozems	57.7
Sima	9,293	Ferralsols	20,960	Phaeozems	55.7
GE3	9,744	Vertisols	43,030	Phaeozems	77.4
Wray	22,340	Vertisols	28,670	Phaeozems	22.1

In order to improve soil and crop management practices for sweet sorghum, varieties will require the following measures: (1) addition of macro- and microelements through fertilizer application, (2) application of supplementary irrigation, (3) timely planting and control of weeds and pests, (4) implementation of cultural practices that increase soil organic matter content (such practices also improve the stability of highly erodible soils like Acrisols, Alisols and Arenosols), and (5) adjusted lime application to neutralize soil aluminum and increase availability of phosphate and other essential elements to produce high stem yields which is especially relevant for acidic soils.

Table 8.6 Sweet sorghum production under supplementary irrigation conditions with double cropping

	Millable stalk (t/ha)		Ethanol production (litres/ha)	
Variety	Single crop	Double (ratoon) crop	Single crop	Double (ratoon) crop
GE3	82.5	135.3	3,926	6,439
Cowley	71.4	94.2	3,167	4,178
Wray	70.2	143.8	3,571	7,316
Madhura	41.8	85.2	1,230	2,506

Further research was undertaken to assess the performance of sweet sorghum varieties under supplementary irrigation with double cropping (i.e. harvesting the first crop under supplementary irrigation followed by second harvesting). Results of stem yields and corresponding ethanol production for selected sweet sorghum varieties are shown in Table 8.6.

Much higher production of sweet sorghum, and therefore higher ethanol production, was obtained for cultivation under supplementary irrigation with double cropping. Stem yield was highest for Wray with similar yields for the first and the ratoon crop. It was observed that Wray was able to withstand rationing compared to other varieties. GE3 and Wray had higher ethanol production under double cropping compared to sugarcane (typically 5,000 l/ha).

8.5 Conclusion

The bioethanol potential in selected African countries for different feedstocks is estimated at 35 EJ, with the Democratic Republic of Congo having the largest potential, followed by Angola, Sudan, Zambia and Tanzania. Bioethanol production in Africa was 637 million litres in 2007, with the largest production in South Africa, followed by Egypt and Nigeria. Potential ethanol markets based on E5 and E10 in southern African countries are estimated at 0.8 and 1.5 billion litres per year.

The main feedstock for ethanol production in Africa is sugarcane, but in the recent past the continent has seen an increasing interest in new feedstock types such as sweet sorghum. Results for yields and sugar brix for eight exotic sweet sorghum varieties (TS1, Madhura, Praj-1, GE2, GE3, Wray, Cowley, and Keller) are presented and compared to a local sweet sorghum variety (Sima).

Highest stem yieldy are observed for the variety TS1 (32 t/ha), followed by Praj-1, Madhura, GE2, and GE3 (21.2 t/ha), whereas Cowley, Madhura, Keller, and Sima had the lowest yield of 13.8 t/ha. The stem yields of sweet sorghum varieties under rainfed conditions were lower than those for sugarcane (75 t/ha). However, the yields are higher under supplementary irrigation conditions averaging at 70 t/ha for single cropping. The sugar content in Brix% is about 18% for medium maturing varieties (Keller, GE3, and Sima). The highest values of sugar content of sweet sorghum varieties are similar to those obtained for sugarcane (18%).

In view of the encouraging results of yields and sugar content of sweet sorghum varieties, it is recommended to further improve the yields of sweet sorghum varieties thorough the following measures: (1) improved crop and soil management on currently available sweet sorghum varieties, (2) appropriate agronomic packages for sustained agricultural production, (3) research on (low cost) pest control, especially stem borers, (4) evaluation of promising bio-control measures such as the use of nematodes, and (5) control of diseases such as anthracnose, bacterial stripe, blight, grey leaf spot, sorghum rust, sooty stripe and sheath blight. In the longer term, the focus of programmes should be on crop improvements. Local and exotic sweet sorghum germplasm should be selected to improve yields. The material screened for tolerance to pests and diseases and crop improvements should be conducted with marker assisted selection. The use of molecular markers will enable rapid selection of sweet sorghum lines with desirable traits. This will enable the crop improvement programme on sweet sorghum to bc more efficient and cut down on cost and time required to develop new varieties of sweet sorghum. The new varieties of sweet sorghum will thus be available to growers in the shortest possible time. To achieve higher stem yields, sweet sorghum should be grown under rain-fed and supplementary irrigation with single or double cropping schemes, and research should be conducted in an interdisciplinary framework involving all important stakeholders.

References

Batidzirai B (2007) Bioethanol technologies in Africa. UNIDO/AU/Brazil first high-level biofuels seminar in Africa (30 July–1 August 2007) Addis Ababa, Ethiopia

Kyritsis S (2010) The emerging importance of the new feedstocks for bioethanol production. Agricultural University of Athens, Crops2industry workshop, Winschoten, 12 Mar 2010

Munyinda K (2008) Evaluation of sorghum production and potential for bioethanol production in Zambia. Department of Energy Conference on Bioenergy

Walimwipi H (2008) Biofuels development in Africa. ICS UNIDO in collaboration with Deutsches BiomasseForschungsZentrum gemeinnützige GmbH

Chapter 9
Biodiesel Production in Africa

Hartley Walimwipi, Francis Davison Yamba, Manfred Wörgetter, Josef Rathbauer, and Dina Bacovsky

Abstract This chapter provides a synopsis of biodiesel resources in African countries, production pathways and current biodiesel production technologies. It presents an overview of different feedstock types converted into biodiesel at different scales as well as options of biodiesel production technologies such as single and multi-feedstock technologies and small-scale technologies according to current practice in the industry. Finally, the chapter identifies appropriate biodiesel production pathways, technologies, biodiesel markets and biofuel policy targets in several African countries.

Keywords Biodiesel • Trans-esterification • Esterification • Production pathway • Production technology • Markets • Feedstock • First generation biodiesel • Next generation biodiesel • FAME • FAEE

H. Walimwipi (✉)
Centre for Energy, Environment and Engineering Zambia, Parirenyatwa road, 176, Private Bag E 721, Lusaka, Zambia
e-mail: hartleykabunda@yahoo.co.uk; ceeez@coppernet.zm

F.D. Yamba
Centre for Energy Environment and Engineering Zambia (CEEEZ), 176 Parirenyatwa Road, Suite B, Private Bag E721, Lusaka, Zambia
e-mail: ceeez@coppernet.zm; ceeez@zamnet.zm; yambafd@yahoo.com

M. Wörgetter • J. Rathbauer
Francisco Josephinum - Federal Secondary School and Research Institute for Agriculture, Agricultural Engineering and Food Technology, BLT – Biomass, Logistics, Technology, Rottenhauserstrasse 1, Wieselburg, AT 3250, Austria
e-mail: manfred.woergetter@fjblt.bmlfuw.gv.at; josef.rathbauer@fjblt.bmlfuw.gv.at

D. Bacovsky
BIOENERGY 2020+, Gewerbepark Haag 3, 3250, Wieselburg-Land, Austria
e-mail: dina.bacovsky@bioenergy2020.eu

R. Janssen and D. Rutz (eds.), *Bioenergy for Sustainable Development in Africa*,
DOI 10.1007/978-94-007-2181-4_9,

9.1 Introduction

Depending on the biomass feedstock and the type of technology employed in the production, biodiesel can be named either first generation or second generation biodiesel. Biodiesel produced from food crops (oil crops) and other food based feedstock (e.g. waste oil, animal waste) is often referred to as first generation biodiesel. First generation biodiesel today has a considerable market share and its production technologies are well established. Second generation biodiesel is produced form lingo-cellulosic biomass.

In relation to the conversion technology, the *bio-chemical conversion* pathway is referred to first generation and the *thermo-chemical* pathway to the second generation production process (Fig. 9.1).

Chemically, first generation biodiesel is equivalent to fatty acid methyl esters or ethyl esters, produced from triacylglycerols via trans-esterification or fatty acids via esterification. Fatty acid methyl esters (FAME) today are the most commonly used biodiesel type, whereas fatty acid ethyl esters (FAEE) so far have been only produced at laboratory or pilot scale (Bacovsky et al. 2007).

There are many options to use different biomass feedstock types for pure plant oil (PPO) and biodiesel production. Besides dedicated oilseed crops such as rapeseed and soybean, microalgae, animal fats and waste oil provide viable feedstock opportunities for fuel production. However, these last three feedstock types are not yet used on a large-scale (Rutz and Janssen 2008).

The most common bioenergy crop for biodiesel production in Sub-Saharan Africa is jatropha, mainly because it is non edible, drought tolerant and suitable for cultivation in almost all countries. Other potential feedstocks include coconut, oil palm, sunflower, soybean, animal fat, and castor oil.

Second generation biodiesel can be produced from a wider range of feedstocks, which are represented mainly by non-food crops such as lignocellulosic materials. Mainly thermo-chemical processes are employed in converting such biomass feedstock into biodiesel (Arumugam et al. 2007).

9.2 Second Generation Biodiesel Production Technologies

Through thermo-chemical processes (use of high temperatures to pyrolyse or gasify biomass) lignocellulosic biomass can be converted to a raw gas or oil. The resulting gas is then treated and conditioned into synthesis gas (syngas), consisting mainly of carbon monoxide and hydrogen. This gas can further be processed into different types of liquid and gaseous fuels via different fuel syntheses. Fuels from this route are called 'synthetic biofuels' (Arumugam et al. 2007).

The most promising liquid synthetic biofuel currently is BtL fuel (Biomass-to-Liquid) produced with the Fischer-Tropsch (FT) process. Gaseous synthetic biofuels are e.g. dimethylether (DME) and Bio-SNG (Synthetic Natural Gas).

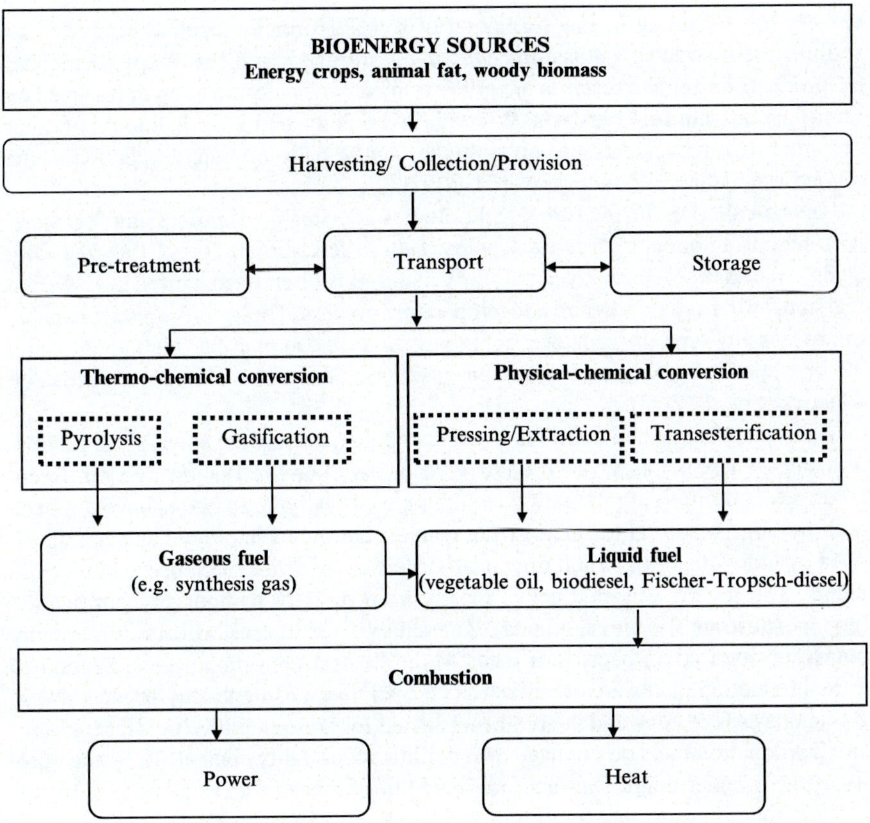

Fig. 9.1 Biodiesel production pathways

Bio-oil obtained from biomass via pyrolysis or hydrothermal treatment can be converted into high quality liquid fuels by deoxygenation (e.g. HTU diesel). On the other hand, bio-chemical conversion involves pressing and/or extraction of oil from oil crops followed by the transesterification process (Arumugam et al. 2007).

9.3 First Generation Biodiesel Production Technologies

There are different possibilities to classify different biodiesel production technologies, namely according to the type of catalyst (homogenously or heterogeneously catalyzed processes), according to reaction conditions (low and high temperature and pressure reactions), or between continuous or batch operation. On the other hand, it is also possible to classify according to the type of feedstock. The so-called single feedstock technologies use half or fully refined vegetable oils like rapeseed, soybean, sunflower, etc. With these technologies the content of free fatty acids should be very low, resulting in low formation of soaps. Normally alkaline catalysts like sodium methoxide or potassium hydroxide are used, and the soaps formed as by-products during the reaction are either removed by water washing or recycled by esterification with acid catalysts. With this technology also a small amount of other feedstock like recycled frying oil or higher acidic palm oil can be blended to the refined vegetable oils (Bacovsky et al. 2007).

The so-called multi-feedstock technologies are capable of processing feedstock with higher amounts of free fatty acids. Here, pre-esterification of the free fatty acids is necessary. Alternatively, all fatty material is directly converted to FAME in one step during a high pressure and temperature process. These processes are capable to process any type of feedstock, including acid oils, animal fat, high acidic palm oil or even fatty acids, and they can easily be adapted to change of feedstock (Bacovsky et al. 2007).

Apart from single and multi feedstock technologies there are small-scale production units. These plants have a production capacity of up to 5,000 t/a, using different feedstock and different production technologies. Mostly these plants have not been built by large biodiesel technology companies, but the technology has been developed by individual groups and organizations based on own experience and developments. The glycerol by-product is mostly used directly without any purification (e.g. as substrate for biogas plants). The catalyst for transesterification is mainly potassium hydroxide, because it leads to the highest conversion rates. Several of these production plants are organized as co-operatives, using locally produced vegetable oils as feedstock and the biodiesel as fuel for agricultural vehicles. Most very small production units do not have own facilities for quality control. Thus, the quality of the product might vary and not meet the European fuel standard EN 14214, representing a serious risk for diesel engines (Bacovsky et al. 2007).

9.3.1 *Catalysts for Transesterification and Esterification Reactions*

9.3.1.1 Homogeneous Catalysts

Alkaline or basic catalysis is by far the most commonly used reaction type for biodiesel production. The main advantage of this form of catalysis over acid-catalyzed transesterification is high conversion under mild conditions in comparatively short reaction times (Bacovsky et al. 2007).

Moreover, alkaline catalysts are less corrosive to industrial equipment, and thus enable the use of less expensive carbon-steel reactor material. The main drawback of the technology is the sensitivity of alkaline catalysts to free fatty acids contained in the feedstock material. Therefore, alkali-catalyzed transesterifications optimally work with high-quality, low-acidic vegetable oils, which are however more expensive than waste oils. If low-cost materials, such as waste fats with a high amount of free fatty acids, are processed by alkaline catalysis, deacidification or pre-esterification steps are required. Acid catalysis offers the advantage of also esterifying free fatty acids contained in the fats and oils and is therefore especially suited for the transesterification of highly acidic fatty materials (Bacovsky et al. 2007).

However, acid-catalyzed transesterification is usually far slower than alkali-catalyzed reactions and requires higher temperatures and pressures as well as higher amounts of alcohol. The typical reaction conditions for homogeneous acid-catalyzed methanolysis are temperatures of up to 100°C and pressures of up to 5 bars. A further disadvantage of acid catalysis, probably prompted by the higher reaction temperatures, is an increased formation of unwanted secondary products, such as dialkylethers or glycerol ethers (Bacovsky et al. 2007).

The major disadvantage of homogeneous catalysts is that they cannot be reused. Moreover, catalyst residues have to be removed from the ester product, usually necessitating several washing steps which increase production costs.

9.3.1.2 Heterogeneous Catalysis

Traditional heterogeneous catalysis offer a series of advantages, such as easy separation, re-usable pure glycerol and no side products (salts) (Mittelbach 2005). There have been various attempts aimed at simplifying product purification by applying heterogeneous catalysts, which can be recovered by decantation or filtration or are alternatively used in a fixed-bed catalyst arrangement. The most frequently cited heterogeneous alkaline catalysts are carbonates and oxides of alkali metals and alkaline earth metals (Bacovsky et al. 2007).

9.3.1.3 Enzymes as Catalysts

In addition to the inorganic or metallo-organic catalysts presented so far, also the use of lipases from various microorganisms has become a topic in biodiesel production. Lipases are enzymes which catalyze both the hydrolytic cleavage and the synthesis of ester bonds in glycerol esters (Bacovsky et al. 2007).

As compared to other catalyst types, biocatalysts have several advantages. They enable conversion under mild temperature, pressure, and pH-conditions. Neither the ester product nor the glycerol phase has to be purified from basic catalyst residues or soaps. Therefore, phase separation is easier, high-quality glycerol can be sold as a by-product, and environmental problems due to alkaline wastewater are eliminated. Moreover, both the transesterification of triglycerides and the esterification of free fatty acids occur in one process step (Bacovsky et al. 2007).

However, lipase-catalyzed transesterifications also entail a series of drawbacks. As compared to conventional alkaline catalysis, reaction efficiency tends to be poor, so that biocatalysis usually necessitates far longer reaction times and higher catalyst concentrations. The main hurdle to the application of lipases in industrial biodiesel production is their high price, especially if they are used in the form of highly-purified, extra cellular enzyme preparations, which cannot be recovered from the reaction products (Bacovsky et al. 2007).

9.4 Biodiesel Production Technologies for Sub-Saharan Africa

From the overview of biodiesel production technologies presented above, several first generation biodiesel production technologies can be employed in Africa. These technologies include single feedstock, multi-feedstock and small-scale production units. In terms of catalyst application, homogeneous catalysts are favourable for many African countries due to their low cost. In addition, homogenous alkaline catalysts are cheap and utilization as fertilizer is possible.

Biodiesel production technologies today use homogeneous alkaline catalysts such as alkali alkoxides and hydroxides. New trends in biodiesel production include heterogeneous catalysts, enzymes and supercritical alcohols (Mittelbach 2005). The choice of catalyst thereby requires a fine balance amongst feedstock type, quantity of free-fatty acids in the feedstock, re-usability, cost, reaction time, alcohol consumption, and reaction temperature and pressure.

High investment costs of the use of heterogeneous catalysts make such catalyst non competitive from an African point of view and thus inappropriate for application in most African countries. Similarly, due to high associated costs enzymes and supercritical alcohols are currently inappropriate for application in most African countries.

Biodiesel production with thermo-chemical processes by converting lignocellulosic biomass to oil is currently neither competitive in Africa nor in other parts of the world due to high cost of production. In future, however Africa can benefit greatly from this technology considering its large potential for cellulosic feedstock production.

9.5 Feedstock for Biodiesel in Africa

Africa has a wide ranging variety of crops for bioenergy purposes owing to its suitable climatic and soil conditions. The large potential of currently underutilized arable land places Africa in a strategic position as a continent with enormous potential for biofuels production. Regions with significant potential are in tropical wet and in tropical wet and dry areas. Also, regions in semi-arid areas have the potential to produce significant amounts of bioenergy (particularly drought resistant crops) (Walimwipi 2008).

The most favourite bioenergy crop for biodiesel production in Sub-Sahara Africa is jatropha, mainly because it is non edible, drought tolerant and can grow

Table 9.1 Oil crops distribution in Africa

	West Africa	East Africa	North Africa	Southern Africa	Central Africa
Coconut (*Cocos nucifera*)	X	X	X	X	
Jatropha (*Jatropha curcas*)	X	X		X	X
Castor oil (*Ricinus communis*)	X	X		X	X
Olive (*Olea europaea subsp. cuspidata*)	X	X	X	X	X
Linaceae (Linseed)	X	X		X	
Sunflower (*Helianthus annuus*)		X		X	
Niger Seed (*Guizotia abyssinica*)		X		X	X
Soy bean (*Glycine max*)	X	X		X	X
Oil palm (*Elaeis guineensis*)	X	X		X	X
Abyssinian mustard (*Crambe hispanica*)		X			X
Safflower (*Carthamus tinctorius*)	X	X		X	
Green thorn (*Balanites maughamii*)		X		X	
Groundnut (*Arachis hypogaea*)	X	X		X	X
Mkange (*Allanblackia stuhmannii*)		X			
Allanblackia floribunda	X				X
Allanblackia parviflora	X				

Source: FAO (2008), Nguyen (2006), Raymond et al. (2007), WCMC/INBAR (2004), www.eoearth.org, EcoWorld (2008), (www.israel21c.org), (Biofuels Market Africa), (African Biodiversity Network)

Notes: (1) X –implies regions where a particular crop is grown or occurs naturally (2) These crops may not grow in all countries in a particular region

in almost all countries. Other potential feedstocks include coconut, oil palm, sunflower, soybean, animal fat, and castor oil. Table 9.1 provides a distribution of oil crops across the African continent according to regions where a particular crop grows naturally or can be cultivated. It must be pointed out, however, that an indication of crop suitability in a particular region (e.g. olives in southern and central Africa) does not imply that it grows in all countries in the region (Walimwipi 2008).

9.6 Biodiesel Market in Africa

Markets as key components for biofuels development are influenced by national fuel consumption of gasoline and diesel, potential blending ratios and the availability of feedstocks. Driving forces for markets include high oil prices, policy decisions on biofuel blending, local employment and poverty (Yamba 2007).

Biofuel production affects many different markets (through direct and indirect market interactions), including markets for inputs (e.g. land and water) as well as markets for agricultural products and biofuel co-products (e.g. food and animal feed) (Walimwipi 2008).

Growth of global demand for biofuels has so far resulted in large increases in the scale of production of ethanol and FAME biodiesel. One indicator of the magnitude of this increase is its impact on prices in large, established agricultural commodity markets.

With respect to biodiesel in Africa, jatropha emerges as a favourite biodiesel feedstock which has generated a lot of interest across the continent. In the following a brief summary of the current status of jatropha biodiesel in Africa is presented. In northern Africa there are very little jatropha related activities due to the extremely arid climatic conditions. However, several pilot projects using sewage water for year-round irrigation are tested in Egypt. In West Africa, Mali and Cape Verde Islands have long-traditions in jatropha cultivation. The focus in Mali lies on the use of PPO for village energy supply. Large-scale projects are currently under preparation in several West African countries, such as Ghana, Nigeria and Cameroon.

In East Africa the largest project developments are reported in Tanzania, followed by Ethiopia. Jatropha-related activities have started at small-scale also in Kenya and Uganda and are likely to rise dynamically. Today, significant investments in cultivating jatropha as an energy crop take place in Africa. However, there are strong regional disparities. Apart from Botswana, Angola and South Africa (due to the prohibition of commercial jatropha plantations), ambitious commercial operations are currently developed throughout southern Africa (Walimwipi 2008).

9.7 Targets for Biofuels in Africa

Until today, many African countries have started to embrace biofuels and a number of countries have made remarkable progress in terms of policy measures and targets for biofuels. Several African countries including Ethiopia, Ghana, Kenya, Malawi, Nigeria, Senegal, South Africa, and Zimbabwe plan to expand biofuels production and use (Walimwipi 2008). A number of initiatives are being employed in African countries to develop biofuel industries. The following are examples of initiatives currently implemented in African countries.

Botswana recently launched a Biomass Energy Strategy (BEST) with support from the German Development Cooperation (GTZ) and the Partnership Dialogue Facility of the EU Energy Initiative (EUEI PDF). The objective of BEST is to ensure that biomass energy is produced, supplied and used in a socially, economically and environmentally sustainable manner. Within BEST, studies are performed to identify opportunities for interventions on woody biomass, wet biomass, energy crops and residues (COMPETE 2009).

The government of **Tanzania** established a Biofuels Task Force under the lead of the Ministry of Energy and Minerals with the participation of a variety of government ministries and institutions. This Task Force elaborated draft biofuel guidelines which aim at guiding the sector until policies, legislation and an institutional framework are in place. These guidelines focus on ensuring socio-economic sustainability of bioenergy development, the avoidance of food-fuel conflicts, and sufficient value creation for the local rural population. Different land acquisition and tenure systems were introduced for bioenergy projects including shorter leasing periods of 5–25 years, the possibility to use land as equity, and mandatory villager shares in projects. Furthermore, investors will be required to use part of the land allocated for the production of food crops (COMPETE 2009).

In **Kenya** initiatives of bioenergy policy implementation include the development of strategies in the areas of biodiesel, bioethanol, cogeneration and wood-fuel. In the field of biodiesel a National Biofuels Committee was set up in 2006 with the objective to develop a strategy for 2008–2012. The motivation for the biodiesel strategy was to increase security of energy supply by reducing dependence on imported fossil fuels, to diversify rural energy sources by supplementing with biodiesel, to contribute to poverty alleviation through diversification of income sources, and to address global warming (COMPETE 2009).

In **Niger**, various efforts have been undertaken to address issues related to biofuels which include the Poverty Reduction Strategy (SRP), the National Renewable Energy Strategy (SENR) and the Reference Program of Energy Access (PRASE). In the field of biofuels preliminary studies have identified three potential feedstock sources, namely jatropha, neem, and moringa. Furthermore, successful experiments have been performed to test the use of neem oil in engines (COMPETE 2009).

African countries which have set up bioenergy policies and/or targets include: (1) Zimbabwe has set blending targets of 8–13% ethanol in gasoline, (2) blending ratios of E10, and B5–B20 are proposed in Zambia, (3) South Africa has introduced a penetration target of 2% for liquid road transportation fuels by 2013 with E8 and B2 as blending ratios, as well as fuel levy exemptions, and incentives for producers on underutilized land, (4) targets of 10% ethanol in gasoline (E10) and 5% biodiesel (B5) in fossil diesel are set in Mozambique, (5) Botswana has a proposed blending ratio of E5 and either B5 or B10, (6) the biodiesel strategy in Kenya includes blending ratios of 5% (B5) by 2012 and 10% (B10) by 2020 to ensure a national market for biodiesel (Walimwipi 2008).

9.8 Conclusions

Africa has a large variety of suitable crops for biodiesel production due to its suitable climatic and soil conditions. The large potential of currently underutilized arable land places Africa in a strategic position as a continent with enormous opportunities for biofuels production (both bioethanol and biodiesel). The emerging biofuels industry in Africa has seen a strong interest amongst African countries with several initiatives to expand the production and use of biofuels. The technology for biodiesel production from first generation feedstock is available at reasonable cost.

Africa has a huge market potential for biodiesel both internally and externally. A number of countries have already established targets and blending ratios for biodiesel and other countries are expected to follow. These targets in African countries together with targets set in the EU and elsewhere provide good market opportunities for biodiesel produced in Africa.

References

Arumugam S, Zinoviev S, Foransiero P, Stanislav M, Müller-Langer F, Kaltschmitt M, Vogel A, Thraen D (2007) Bio-fuels technology status and future trends, technology assessment and decision support tools

Bacovsky D, Körbitz W, Mittelbach M, Wörgetter M (2007) Biodiesel production: technologies and European providers. IEA Task 39 Report T39-B6, 104 p

Biofuels Markets Africa – the 2nd annual Pan-African meeting place for the biofuels industry, accelerating sustainable growth in Africa

COMPETE (2009) Bioenergy policy implementation in Africa. COMPETE Policy Conference, Lusaka, 26–28 May 2009

Eco World (2008) http://www.ecoworld.com/home/articles2.cfm?tid=367. Accessed 2008

FAO (2008): Food and Agriculture Organisation,http://www.fao.org. Accessed 2008

Mittelbach M (2005) Biodiesel: production technologies and perspectives. Working Group Renewable Energy Resources

Nguyen (2006) A simple rule for bioenergy conversion plant size optimisation: Bioethanol from sugar cane and sweety sorghum

Raymond et al (2007) Biofuels: ACP's response to fossil fuel dependence The role of science, technology and innovation in supporting sustainable biofuel policies

Rutz D, Janssen R (2008) Biofuel technology handbook. 2nd Version. WIP Renewable Energies, München

Walimwipi H (2008) Biofuels development in Africa. for ICS UNIDO in collaboration with Deutsches BiomasseForschungsZentrum gemeinnützige GmbH

WCMC/INBAR (2004) Bamboo biodieversity, Africa, Madagascar and the Americas Nadia Bystriakova, Valerie Kapos, Igor LysenkoUNEP

Yamba FD (2007) Biofuels framework for Zambia. For Government Republic of Zambia Ministry of Energy and Water Development

Chapter 10
Biogas Production in Africa

Greg Austin and Glynn Morris

Abstract Biogas is derived from anaerobic digestion of biomass, which can be broadly grouped into municipal sewage and solid waste (including food wastes), livestock and agroprocessing residues, and energy crops. Technologies range in scale from domestic systems for thermal (cooking) energy to multi-megawatt grid-connected combined heat and power generation systems, or even systems that inject biomethane into natural gas grids. Biogas technology provides distinctive advantages compared to other renewable energy technologies: it combines energy (gas) storage with generation; it provides valuable co-products such as nutrient-rich bioslurry; and it can easily co-digest a range of feedstocks thus providing an integrated waste management service. At the same time, the subtleties associated with the technology arise from a unique combination of factors including that: the feedstock source is often a waste or problem product, and hence its use for energetic utilization resolves waste management problems; the nutrients made more available through the anaerobic digestion process are extremely valuable (typically, the nutrients are at least five times as valuable as the biogas generated); and co-digestion of different substrates in one system improves returns on investment. The trend now is that of increasing numbers of biogas installations across Africa. This is largely apparent in the domestic energy sector, which has in recent years seen the start of a number of national domestic biogas programmes each with national targets of over 10,000 domestic systems to be installed in the next 5 years. National programmes in Africa are currently implemented in Rwanda, Tanzania, Kenya, Uganda, Ethiopia, Cameroon, Benin and Burkina Faso. In many African countries, the sewerage infrastructure and waste management within the urban and peri-urban perimeters is often non-existent, or failing. Many major cities have an immediate opportunity through

G. Austin (✉) • G. Morris
AGAMA Energy, The Green Building, 9b Bell Crescent Close, Westlake Business Park, 7945, Westlake, South Africa
e-mail: greg.austin@agama.co.za; glynn.morris@agama.co.za

R. Janssen and D. Rutz (eds.), *Bioenergy for Sustainable Development in Africa*,
DOI 10.1007/978-94-007-2181-4_10,

biogas technology implementation to reduce the serious, negative impact of the status quo on the aquatic environment, on human health and hygiene, and on pests and methane emissions from rotting organic solid waste, be it at the roadsides or at dump sites. The biogas solutions are additional and complementary to the existing infrastructure. The benefits of the technology, compared with other renewable energy technologies (RETs) are often subtle and hence more difficult to measure and realize, yet with the correct policies and financial tariffs/incentives in place, the use of the technology is most certainly viable. The levels and types of service offering and the associated business models are key elements to deepening the access to the technology across Africa.

Keywords Biogas • Anaerobic digestion • Bioslurry • Renewable energy • Energy services • Integrated waste management • Co-digestion

10.1 Introduction

Biogas derived from anaerobic digestion of non-lignocellulosic biomass is possible in three broadly defined feedstock categories: municipal sewage and solid waste (including food wastes), livestock and agroprocessing residues, and energy crops (Al Seadi et al. 2008). More recent technology developments also allow for anaerobic digestion of lignocellulosic feedstocks as well. The benefits of anaerobic digestion, compared with other renewable energy technologies (RETs) are often subtle and hence more difficult to measure and realize; yet with the correct support frameworks in place, the use of the technology is certainly viable and the uptake is increasing.

Hence, the trend in Africa now is that of increasing numbers of biogas installations. This is largely apparent in the domestic energy sector, which has in recent years seen the start of a number of national domestic biogas programmes each with national targets of some 10,000 domestic systems to be installed in the next 5 years. At present the countries with national domestic biogas programmes are Rwanda, Kenya, Tanzania, Ethiopia, Uganda, Cameroon, Benin and Burkina Faso. In the period 2007 to July 2010, the total number of domestic biogas systems installed totalled 1,665 under the formal programmes running in the aforementioned countries (SNV 2010), with additional digesters having been built prior to and outside of the national programmatic frameworks. These programmes incorporate multilateral, co-ordinated financing in order to leverage national investments, and programme-level technical, marketing and financial elements to transform the market uptake. Subsidies are invariably necessary and carbon finance is clearly an attractive option to supplement the funding. Increasingly the programmatic or ‘nationally appropriate mitigation activities’ (NAMA’s) approaches lend themselves to aggregating many thousands of domestic digesters into large individual carbon projects to raise additional funding to offset capital costs.

In many African countries, the sewerage infrastructure and waste management within the urban and peri-urban perimeters is often non-existent, or failing.

Many major cities have an immediate opportunity to reduce the serious, negative impact of the status quo on the aquatic environment, on human health and hygiene, and on pests and methane emissions from rotting organic solid waste, be it at the roadsides or at dump sites.

In summary, service offerings of waste treatment, energy provision, water recycling and nutrient capture for agriculture are valuable inputs to societies and economies in Africa. These services are critically important requirements for enhancing (or maintaining) the quality of life of people and communities as well as for productive activities – and are key underlying requirements for meeting the Millennium Development Goals. A significant and important proportion of the demand for these services can be provided from systems based on biogas digester technologies. This chapter thus explores the opportunities for biogas systems to provide these services based on experience in Africa.

10.2 Services Offerings

The key service offerings, or requirements, as experienced by customers that are made available from the operation of biogas systems include:

- waste and waste water treatment and management – the treatment by means of anaerobic digestion of organic wastes from a wide range of sources including human and animal wastes, food wastes and agricultural wastes
- energy supply (including energy storage) – the production of biogas for thermal energy services such as cooking, baking, water heating, industrial process heating or electricity generation
- water re-use/recycling – the integrated utilization of water for sanitation in a manner in which the water can be treated for re-use on site
- nutrient capture for food production – the use of the nutrient rich digestate from the output of a digester for food production in food gardens or agriculture

Of the four service offerings mentioned, the focus traditionally has been on the use of biogas as a fuel or energy substitute to replace either fossil fuels, or more often, woody biomass that is extracted on a non-renewable basis and which therefore leads to deforestation, erosion and degradation of soil quality. However, by extension, and over time, these additional co-benefits can be and are readily attained.

The traditional application of domestic biogas technology for energy provision is typically for homes with four or more cattle supplying all the cooking requirements of the household without any additional inputs of biodegradable resources to the digester (Austin and Blignaut 2008). This is based on the widely practised rural South African livestock management system, where cattle feed widely in communal areas during the day and are only penned (corralled) at night. Hence to achieve the minimum amount of twenty kilograms of cattle manure, four head of cattle are required.

As a side-effect, the limiting access of the technology to households with four or more head of cattle excludes the poorest in society, whilst benefiting the better off (on a rural relative scale) in the community. Technology extension leads to smaller, cheaper digesters being made more accessible to those with three, then two cows etc., as has been the case in Nepal for example. However, an alternative service delivery method is possible that can yet lead to more equitable access to the benefits of biogas – in the form of energy, or nutrients, or even just employment – across a community, in the case of a minority of that community holding a majority of the cattle.

10.3 Biogas Technology

Whilst manure of any kind can be digested to produce biogas, albeit with varying amounts of biogas output per kilogram of feedstock, the focus here on cattle manure is based on the strong ownership of cattle in rural African communities, as well as the cultural relationship between cattle ownership and status and wealth in a given community.

Domestic biogas digester systems require a certain amount of liquid feedstock per day for the fermentation processes. In the ideal situation all the urine from the cattle would be captured together with the solids, which would then be fed into the digester with the correct ratio of liquids to solids for the anaerobic process, which is in the order of one component solids to one component liquids. In practice, the urine is seldom captured and therefore in the four-cow scenario where 20 kg of solids are collected, access to at least 20 l of water per day (corresponding to two buckets) is essential. It is important to note that access to fresh water for this purpose is not a limiting factor for the uptake of the technology, since the water used in the digester can just as readily be used or 'grey' water from within the household. Nevertheless, a distance of not more that 1 km from the household is felt to be the maximum distance people should walk to get water, in order for water access not to be a limiting factor for domestic biogas technology uptake.

There are three main philosophies employed in the design of domestic biogas digesters, namely the floating drum digester, the fixed dome digester, and the flexible balloon digester. These configurations are shown in Figs. 10.1–10.3.

The microbiological processes taking place in these digesters are essentially the same. Due to the generally warm climate in Africa, at most locations ambient temperature is sufficient to maintain the fermentation process and no artificial heating is required. In practice, biogas installations in Africa are based on psychrophilic (<20°C) or mesophilic (30–42°C) anaerobic digestion. Since the digestion temperature is related to the alternating ambient temperature, slight process variations and hence biogas production rates occur. However, generally the processes can be kept stable. In case of microbiological breakdown, digesters can be usually emptied and re-filled without major problems due to the small-scale of the digesters.

The cheapest design to implement is the **flexible balloon** digester (Fig. 10.1). This design is also technically very simple, but not very durable. This digester is placed in a trench in the ground and consists of a plastic or rubber digester bag, in

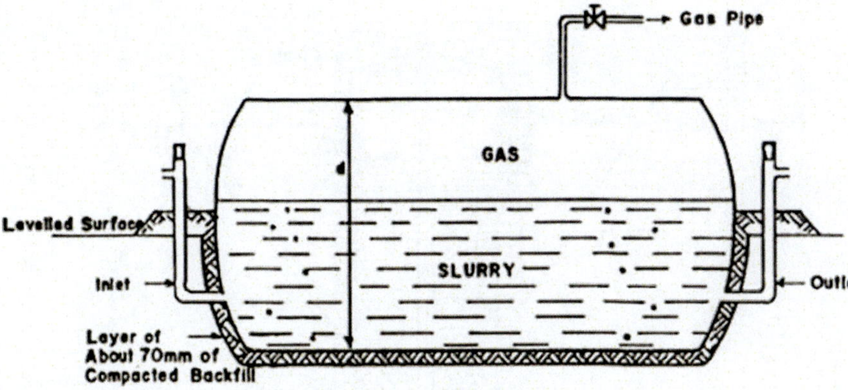

Fig. 10.1 Schematic of a flexible balloon digester (FAO 1996)

the upper part of which the gas is stored. The inlet and outlet are attached direct to the skin of the balloon. When the gas space is full, the plant works like a fixed-dome plant, in that the balloon is not inflated since it is not very elastic.

The fermentation slurry is agitated slightly by the movement of the balloon skin, which is favourable to the digestion process. Even difficult feed materials, such as water hyacinths, can be used in a balloon digester.

Advantages include low cost, ease of transportation, shallow construction (important if the water table is high), high digester temperatures, uncomplicated cleaning, emptying and maintenance. Disadvantages include a short lifetime (about 5 years), easy damage risk, no real local employment creation, and little scope for self-help (Sasse 1988). Very limited gas storage is another limitation of this design.

This design is not suitable for a programmatic approach to digester installations where a financing mechanism (including subsidies linked to quality assurance measures, as well as long-term production of voluntary or certified emissions reductions) and therefore long-term functionality, is required. Balloon plants can be recommended wherever the balloon skin is not likely to be damaged and where the temperature is relatively constant high.

The **floating drum digester** (Fig. 10.2) consists of a digester and a moving gasholder. The gasholder floats either directly on the fermentation slurry or in a water jacket of its own. The gas collects in the gas drum, which thereby rises. If gas is drawn off, it falls again. The gas drum is prevented from tilting by a guide frame (Sasse 1988). This design offers the most consistent pressure of the different designs, since the biogas produced is kept under pressure by the weight of the gas drum itself.

Advantages of this design include that it has a simple operation, a constant gas pressure, and that the volume of stored gas is visible directly (by the level of the gas drum). Disadvantages include the high construction cost (up to 50% greater than that of a fixed dome digester), use of many steel parts liable to corrosion, resulting in a short life (up to 15 years; in tropical coastal regions about 5 years for the drum), and regular maintenance costs due to painting.

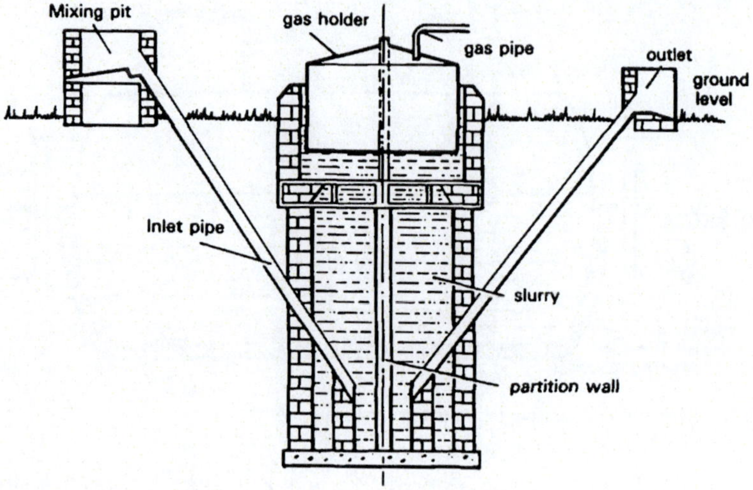

Fig. 10.2 Schematic of a floating drum digester (Frankel 1986)

The **fixed dome digester** (Insert Fig. 10.3) consists of an enclosed volume with a fixed, non-movable gas space. The gas is stored in the upper part of the digester. When gas production commences, the slurry is displaced into the compensating tank (Sasse 1988). Gas pressure increases with the volume of gas stored, up to a maximum predetermined gas pressure. The gas produced is kept under pressure by the water level differential between the digester vessel/gas holder and the compensation/removal tank and hence if there is little gas in the holder, the gas pressure is low.

Advantages include a relatively low construction cost (about two-thirds of the cost of the floating drum design), no moving parts, no rusting steel parts, hence long life (20 years or more), underground construction, affording protection from winter cold and saving space, and local employment in its construction.

Disadvantages include that these digesters are often not gastight (porosity and cracks), and gas pressure fluctuates substantially. The gas tightness is an issue that pertains only to the built systems, and not prefabricated systems.

The fixed dome and floating drum designs take approximately the same length of time to construct (up to three weeks) whilst the flexible tube would take only a few days at the most. Both the floating drum and fixed dome designs have been increasingly implemented over the past 10 years using prefabricated or partially prefabricated concepts, usually manufactured from plastic or glass fibre.

From the trend over the past 15 years it is clear that the fixed dome design is generally accepted to be the most viable design approach that allows for an affordable yet reliable solution for the domestic market. This is supported by the current switch across India from the floating drum design – their preferred technology during the latter half of the twentieth century – to the fixed dome design.

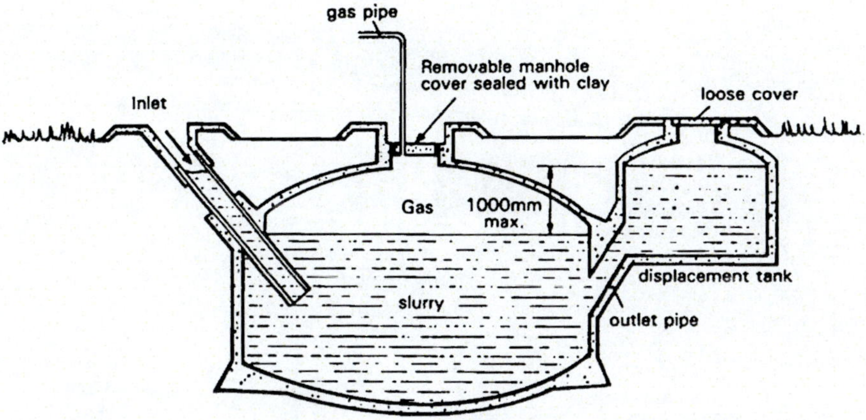

Fig. 10.3 Schematic of a fixed dome digester (Frankel 1986)

10.4 Scale and Integration

A biogas installation is not just as a renewable fuel supply, but rather an integrated technology whereby various feedstocks and waste water are co-digested, and valuable additional benefits can be derived from the same fixed infrastructure investment. Comparing the primary outputs of a biogas system – biogas and bioslurry - it is interesting to note that the value of the nutrients available in the bioslurry leaving the digester is five to six times that of the biogas. The 'trick' is to plan and implement the system in such a way that both these – and other – value opportunities can be realized. This therefore is not just a question of energy supply and demand, but more about the social, financial and economic context for a given project. Therefore, from a national programme perspective a significant focus and budget should be allocated to capacity building around, for example, bioslurry utilization.

As for all forms of RETs, site-specific factors have to be taken into account. The subtleties associated with the use of biogas technology arise from a range of unique characteristics: the energy source is often a waste or problem product, and hence their use for energetic utilization resolves waste management problems; the nutrients made more available through the anaerobic digestion process are extremely valuable; and co-digestion of different substrates in one system leads to improved returns on investment.

Because of the subtleties referred to above, the application of the biogas technology must increasingly target the integration of different inputs and outputs to and from the technology itself, in order to fully realize the value inherent in the system. A generalized model of this is outlined in Fig. 10.4.

There are two distinct primary advantages that biogas technology has over other RETs: energy storage, and scalability. For renewable energies generally, the issue of intermittency with the associated need to match energy supply with user demand, either remains unsolved or only resolvable with much greater cost to the owner. In biogas, and

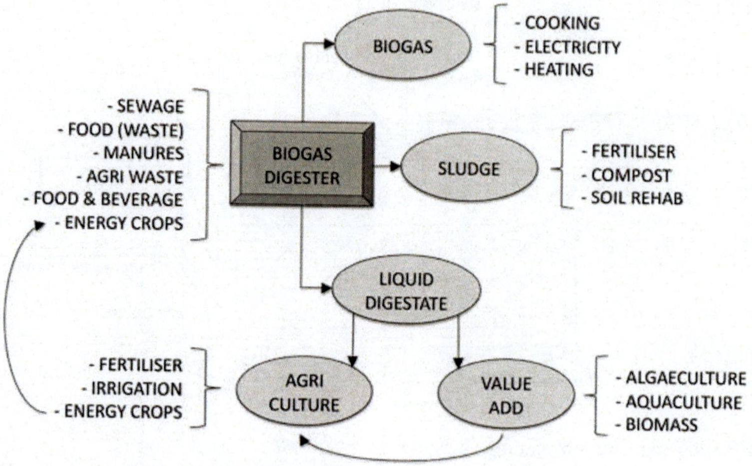

Fig. 10.4 Generalized integrated biogas model

particularly with the floating drum or fixed dome designs, the technology has in-built gas storage. This storage is typically sized to accommodate up to fifteen hours of gas production, to allow for a minimum of twice-daily meal preparation, on the basis of small biogas amounts being produced continuously and stored over the fifteen-hour period.

The anaerobic digestion technology is tremendously adaptable with respect to its scale: plants range in scale from domestic systems for thermal (cooking) energy to multi-megawatt grid-connected combined heat and power (CHP) systems. As the confidence and experience with biogas grows each year, more innovation is evident in the application of the technology. Thus for example the ARTI biogas digester developed in India and recently introduced to East Africa is promising. It digests household food waste, thus providing a small solution to an (urban) waste management and energy supply problem (ARTI 2011). This technology is prefabricated from plastic and based on the floating drum design; whilst a plastic fixed dome prefabricated biogas digester has been introduced into Africa by AGAMA Biogas (AGAMA Biogas 2010).

10.5 Modes of Service Provision

Experiences with different modes of service provisions show that the costs of services are significantly influenced by operational and business models for the particular service delivery. The advent of reliable and more affordable domestic scale biogas digesters and the potential for large-scale service provision from distributed infrastructure, in the form of many small digesters, further suggests the merit of addressing the question of the comparative advantages of different modes of biogas technology service provision in Africa.

The different modes of service provision are essentially categorized as ownership-based and fee-for-service modes. In the former, the assets which provide the services are ultimately owned by the individual customers who derive the benefit of the services from the assets whereas, in the latter case, the assets which provide the services remain with a service provider or utility. Clearly, in the case of larger or community-scale infrastructure the customer ownership mode of service provision is not applicable except through some form of shared ownership or shareholding in the entity which owns the more centralized biogas digester asset.

In an African context there is an important need to link the fee-for-service provision – which might include biogas or other forms of energy, nutrients, water, etc. – with the needs of the community, to cement both the owner and user commitments to the on-going operational needs of the facility. The resulting transaction must provide the required service at an affordable level for the customers, and with sufficient margin to ensure a sustainable income for the owners (which may in turn be the customers themselves, through a co-operative ownership model).

As an example, for a low-income, dense urban area characterized by poor or non-existent sanitation and ablution facilities, the typical facility might include toilets, hot water linked to showers and clothes washing facilities, and a biogas fuel component. For a very low and affordable (0.02 US$–0.03 US$ per use), some 2,000 residents make use of the facility on a daily basis, with this income (1) repaying the capital cost and finance cost of the facility, (2) employing sufficient personnel to ensure a clean and secure environment for the users and (3) making provision for maintenance and replacement costs.

The biogas component creates an additional value-add, since the fuel and nutrient components can be sold on or used by the owner-operator for additional activities (e.g. fuel for a restaurant facility and food gardens). Sufficient transactional linkages can be created for the community to enjoy the facility long into the future. Versions of this example are already being implemented in some Indian and Indonesian cities, and should be piloted in African cities.

10.6 Practical Applications and Experience in Africa

Biogas digester technology is well established in parts of the world, with a history dating back as far as the tenth century B.C, when it is recorded as having being used in Assyria (mostly Irak and Syria today). After early establishment in China, biogas digesters remained in use in China during the past several centuries to the present day. Mass dissemination of biogas technology was first initiated in China around 1958 (Gregory 2010). Subsequently, the technology has been implemented on a large-scale in China, India, and Nepal, countries that now have considerable expertise in biogas digester technology programme implementation. As an example, there are currently in the order of 30–40 million domestic scale digesters in China and over 220,000 in Nepal, and a number of other Asian countries where biogas digesters have been constructed in notable numbers.

Table 10.1 Domestic biogas installations in selected countries

Country	Programme started in	2010 (January to July)	Cumulative up to July 2010
Asia			
Nepal	1992	14,514	219,117
Vietnam	2003	14,447	90,278
Bangladesh	2006	3,242	13,261
Cambodia	2006	1,850	8,207
Lao PDR	2006	425	1,454
Indonesia	2009	214	264
Pakistan	2009	203	303
Africa			
Rwanda	2007	259	693
Tanzania	2009	254	360
Kenya	2009	170	173
Ethiopia	2008	131	259
Uganda	2009	94	134
Cameroon	2009	10	33
Benin	2010	8	8
Burkina Faso	2009	4	5
Total		35,780	334,549

Biogas experiences in Africa have been on a far smaller scale and have often been disappointing at the household level. The capital cost, maintenance, and management support required have been higher than expected. However, this somewhat bleak picture has improved through the recent initiation of the Africa Biogas Partnership Program (ABPP) which has been developed based on experiences over some 20 years in Asia. The programme aims at constructing 70,000 biogas plants in Ethiopia, Kenya, Tanzania, Uganda, Cameroon, Benin and Burkina Faso providing about half a million people access to a sustainable source of energy by the year 2013. The current status of the rollout of the programme in these African countries, as well as in Asia, is summarized in Table 10.1 (SNV 2010).

This programmatic approach relies on the successful establishment of market-oriented programmes, and appropriate financial structuring including subsidies.

10.7 Practical Applications and Experience in South Africa

In many respects South Africa's use of biogas technology to address rural energy poverty and access remains behind many of its northern neighbours. There are many factors behind this, but at the same time there has been good practical experience and the stage is set to increase the use of this technology in the rural parts of the country.

To date there are approximately 100 small-scale digesters in use across the country, with approximately 30 of these in rural households, and two trial projects on small to large farms. Some digesters of the flexible balloon design have been installed without too much success, and the bulk of the working systems are of the fixed dome design. About half of these are prefabricated, for the good reasons of assurance around gas tightness, speed of installation, and ease of installation.

The projects have started to attract the attention of a range of stakeholders, and include the National Treasury who has included four types of biogas applications in their Working for Energy Programme. The Programme is located within the national Department of Energy, and implemented through the South African National Energy Development Institute (SANEDI). The Programme is seeking to maximize job creation opportunities in the renewable energy and energy efficiency sectors, with the four biogas-focussed areas being household, municipal solid waste, municipal waste water and agricultural residues.

A South African feasibility study was undertaken in 2008 that assessed the technical, financial and economic viability of a rural household biogas programme for South Africa (Austin and Blignaut 2008). Some of the key findings from the study are that:

- There are (conservatively) 310,000 households (9.5% of South Africa's rural households) showing technical viability to participate in a rural biogas programme (these are the households that inter alia have four cows or more, do not have access to grid electricity, and are situated within a 1 km radius of water). This is confirmation of the technical potential and scale of the programme.
- Using the most conservative assumptions throughout, and with the capital subsidy for a biogas programme at 30%, the programme would have significant financial and economic benefits: 15% and 67% financial and economic internal rates of return (IRR) respectively, averaged across the six provinces investigated. These numbers confirm the financial and economic viability of such a programme.
- Implementing the programme presents an opportunity to co-ordinate various rural developmental programmes under one banner, and as a result to harmonize different public funding streams in capital and operational subsidies.

Under the Working for Energy Programme it is anticipated that a detailed programme implementation plan will be developed, co-ordinated by the Department of Energy. In the meantime research and development work is underway on a broad front in South Africa. One important example is a five-year research project being led by the University of KwaZulu-Natal assessing the impacts of installing rainwater harvesting and biogas systems on grasslands and cattle management, as well as the social dimensions pertaining to biogas energy and bioslurry utilization (Austin et al. 2010).

10.8 Conclusions

Biogas technology has an important role to play in the waste management, renewable energy, water and nutrient (food security) sectors in Africa. Multiple economic, social and environmental benefits that are realized from the introduction of domestic biogas digesters provide a powerful motivation for supportive measures to increase the uptake of the technology on the continent. It is evident that support in the form of national programmes – which provide marketing, technical, research and development, co-ordination and financial support frameworks and which have recently being initiated in some African countries – are achieving success. Hence, with the correct policies and financial tariffs/incentives in place, the use of the technology is certainly providing real benefits and the uptake is increasing.

Good opportunities exist for widening the application of and mode of service delivery for biogas technology. On the one hand programmatic approaches are necessary to bridge the financial gap between digester affordability and capital costs, whilst there are many opportunities to pilot fee-for-service business models that narrow this gap and link more closely the technology benefits with a household or community's needs. This requires re-evaluating models of distributed asset ownership, financing mechanisms, and the like, which in turn necessitates development of resource-economic models which include all the costs and benefits of a fee-for-service model. These would include local economic development opportunities, reduced environmental impacts, improved livelihoods, improved health, and so forth.

Today, technical standards and codes for domestic biogas digesters for both installation and maintenance, as well as testing methodologies are lacking. Development of such standards is necessary to support a more rigorous and profession approach to provision of the technology, where issues relating to return on investment are clearly linked to gas tightness and other performance issues. Experience in Asia has identified common problems which can be addressed through the development and application of technical standards and codes of practice.

Prefabricated technologies are playing a major role in accelerating the deployment of domestic digesters as well as providing improved quality assurance (as more quality checks are possible within a factory environment). These technologies are now being introduced with good success into African countries.

References

AGAMA Biogas (2010) http://www.biogaspro.com. Accessed 3 Jan 2011

Al Seadi T, Rutz D, Prassl H, Köttner M, Finsterwalder T, Volk S, Janssen R (2008) Biogas handbook. BiG>East project funded by the European Commission (EIE/07/214), University of Southern Denmark Esbjerg, Denmark, p142, ISBN: 978-87-992962-0-0

ARTI (2011) http://arti-africa.org/. Accessed 3 Jan 2011

Austin G, Blignaut J (2008) South African national rural domestic biogas feasibility assessment. Ministry of Development Co-operation, the Netherlands

Austin G, Everson C, Everson T, Gambiza J, Salomon M, Tau M and Van Ierland J (2010) Improving the livestock carrying capacity with rainwater harvesting and conservation on grasslands for extensive and/or intensive livestock production and biogas generation from manure in rural areas of South Africa. Water Research Commission, project number K5/1955

FAO (1996) Biogas technology – a training manual for extension. Support for Development of National Biogas Programme (FAO/TCP/NEP/4451-T). Food and Agriculture Organisation of the United Nations

Frankel PL (1986) Water lifting devices. http://www.fao.org/docrep/010/ah810e/AH810E13.htm. Accessed 8 Dec 2010

Gregory R (2010) China – biogas. http://www.ecotippingpoints.org/our-stories/indepth/china-biogas.html. Accessed 1 Dec 2010

Sasse L (1988) Biogas plants. A publication of the Deutsches Zentrum für Entwicklungs technologien – GATE, a division of the Deutsche Gesellschaft für Technische Zusammenarbeit (GTZ) GmbH

SNV (2010) Domestic Biogas Newsletter. Issue 3, Aug 2010

Chapter 11
New Conversion Technologies for Liquid Biofuels Production in Africa

Bothwell Batidzirai, Edward M.W. Smeets, and André P.C. Faaij

Abstract On the longer term, the production of second generation biofuels from lignocellulosic biomass is expected to become economically competitive with gasoline and diesel. A pre-requisite is that several technological hurdles will be overcome and that a large, stable supply of lignocellulosic biomass will be guaranteed. Studies have shown that Sub-Saharan Africa has the potential to contribute significantly to the global supply of biomass energy derived from lignocellulosic resources. Due to the high investment costs of establishing large-scale second generation biofuel processing plants, the production and export of pre-treated biomass (e.g. pellets) to industrialised countries is a potentially interesting short-term option. An illustrated example is the production of lignocellulosic biomass in Mozambique which has the capacity to annually contribute up to 2% of global energy supply in the form of bioenergy under a strict sustainability framework. However, rationalisation in agriculture will be essential for realising this potential and efficient logistics are needed to ensure competitive biomass supply.

Keywords Biomass • Bioenergy • Africa • Mozambique • Second generation biofuels • Lignocellulosic • Fischer-Tropsch • Logistics

B. Batidzirai (✉) • E.M.W. Smeets • A.P.C. Faaij
Copernicus Institute, Utrecht University, Heidelberglaan 2, 3584 CS, Utrecht, The Netherlands
e-mail: B.Batidzirai@uu.nl; esmeets@hotmail.com; a.p.c.faaij@chem.uu.nl

R. Janssen and D. Rutz (eds.), *Bioenergy for Sustainable Development in Africa*,
DOI 10.1007/978-94-007-2181-4_11, © Springer Science+Business Media B.V. 2012

11.1 From First to Second Generation Biofuels

Future transport energy systems in a climate constrained world are anticipated to rely increasingly on biofuels, such as fuel ethanol, biodiesel, biogas or hydrogen. Current estimates by the International Energy Agency (IEA) project the contribution of biofuels to 9% of the total transport fuel demand in 2030 (IEA 2009a). Other scenarios expect the contribution of biofuels to increase to 26% of total transportation fuels by 2050 (IEA 2008a).

Current global biofuel production is over 80 billion litres equivalent to 1.5% of global transport fuel consumption, and this represents an increase of more than 400% since 2000. Over 99% of all currently produced biofuels are classified as “first generation” (i.e. fuels produced primarily from cereals, grains, sugar crops and oil seeds) (IEA 2008b). About 85% of global first generation biofuel production is bioethanol. First generation bioethanol is produced by fermenting plant sugars of feedstocks such as starchy and sugar crops. Biodiesel on the other hand is derived mainly from oil-seed crops (e.g. palm oil, rapeseed or soybean), animal fats or waste greases. Bioethanol and biodiesel can be used as transportation fuels either as neat fuels or blended with petroleum fuels. Technologies for the production of first generation biofuels are mature and projects are undertaken on commercial basis. In contrast, “second generation” biofuel technologies are still at pilot and demonstration stages.

“Second generation” biofuels are produced from lignocellulosic feedstocks such as agricultural and forest residues, as well as purpose-grown energy crops such as vegetative grasses and short rotation forests (SRF). These feedstocks largely consist of cellulose, hemicellulose and lignin. Conversion to bioethanol fuel is via complex hydrolysis of the cellulose and hemicellulose to sugar, after which fermentation of sugar is performed. These feedstocks can also be converted to fuel via gasification or pyrolysis to produce synthetic diesel, bio-oil and other fuels. To be competitive with fossil fuels, there is need to overcome several technical challenges – the focus of current R&D (Rutz et al. 2009).

11.1.1 Drawbacks of First Generation Biofuels

In spite of their early hype, production of most first generation biofuels is facing heavy criticism regarding socio-economic and environmental impacts. There are concerns about impacts on food security and agricultural commodity prices, as well as worries about GHG emissions associated with land use changes.

One of the key drivers for biofuel production is their vaunted ability to reduce GHG emissions. However, life cycle assessments indicate that current biofuels provide only limited GHG reduction benefits (with the exception of sugarcane ethanol). As shown in Fig. 11.1, there is a large emission range for each of the different biofuel options, indicating wide uncertainty of actual GHG emissions of biofuels. Furthermore, the presented LCA studies still exclude indirect land use change impacts.

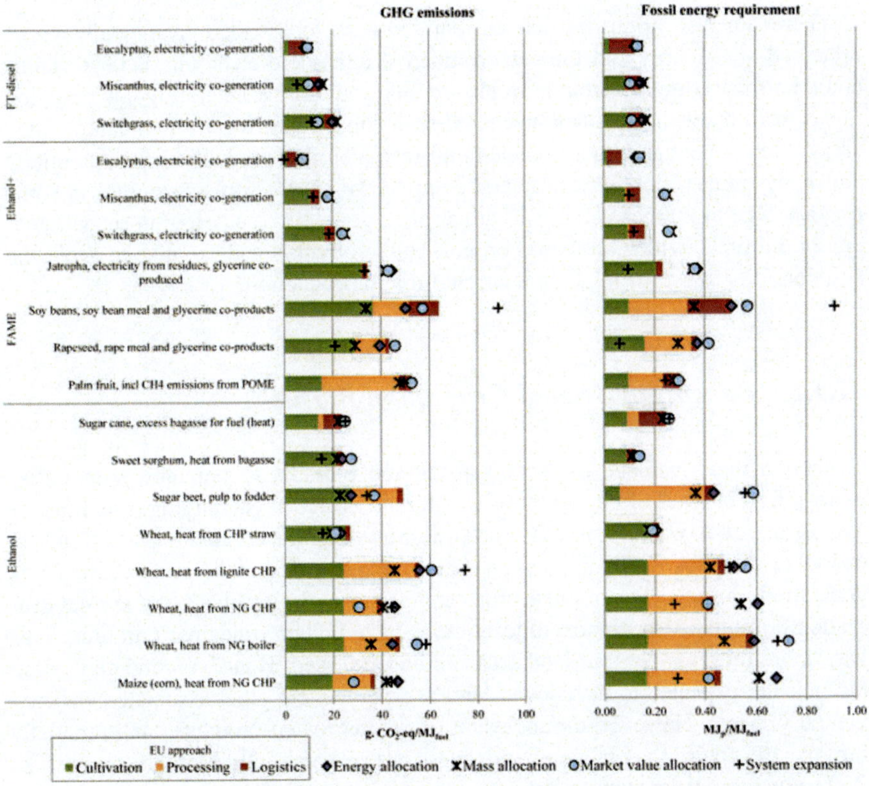

Fig. 11.1 Comparison of life cycle GHG emissions for biofuels and petroleum fuels (Source: Hoefnagels et al. 2010)

Current biofuel crops are not efficient energy producers and require vast surfaces of arable land that will not be available for other purposes such as food production, thus causing potentially adverse consequences for food security. This could lead to competition for scarce productive land resources and social conflicts especially between large corporate entities and local communities. By drawing on food crop resources such as corn, first generation biofuels are said to undermine food security and contributed to recent increases in world prices for food and animal feed. Although biofuels are not the only factor contributing to food shortages and high prices, competition with food remains an issue as long as the bulk of first generation biofuels are produced from food crops.

Production costs of current biofuels are high and they have become an expensive option for energy security and GHG mitigation. The IEA estimates that GHG mitigation from biodiesel and corn ethanol (whilst varying by country and pathway), mostly exceed 250 US$ per t CO_2 avoided (IEA 2010). As these technologies are mature, there is little scope for improving the mitigation costs compared to other options.

There are also concerns that current biomass feedstock may not always be produced sustainably and thus environmental and socio-economic benefit claims could be misleading. Current biofuels are reported to be accelerating deforestation (apart from other indirect land use effects) in some countries. In high nature value areas (such as peatlands and tropical rainforests), biofuel production can potentially adversely impact on biodiversity and compete for scarce water resources in some regions. Due to all these concerns, attention is gradually focussing on second generation biofuels, which are based on non-edible biomass and promise to avoid the sustainability concerns related to current biofuel production.

11.1.2 Benefits of Second Generation Biofuels

Generally, the advantage of second generation biofuels is their ability to utilise many different types of lignocellulosic materials as feedstock, ranging from forestry and agricultural waste materials such as straw, bio-wastes and wood offcuts to energy crops. Whilst only parts of an energy crop are used in the production of first generation biofuels, the complete crop can be used in the production of second generation biofuels. Also, lignocellulosic crops have higher yields per unit area. This way, more biomass per unit land area can be harnessed. Hence, second generation biofuels are expected to have lower land use change and other associated environmental impacts. However, the inclusion of credits for co-products can potentially improve the environmental competitiveness of first generation biofuels.

The successful development of commercially viable biofuel programmes depends on the continuous supply of appropriate feedstocks. There is limited scope for rapidly expanding the supply base of first generation feedstocks. In contrast, lignocellulosic biomass is the most abundant biological material on earth, and its use would significantly expand the volume and choice of feedstocks for second generation biofuel production.

Dedicated cellulosic energy crops have high potential as feedstocks for second generation biofuels. These crops include short-rotation woody crops such as eucalyptus or grasses such as miscanthus and switchgrass. Apart from providing better returns in terms of biomass production per hectare, some fast-growing perennials can grow on degraded or marginal soils where food crop production is not optimal. Thus, competition for land with food and feed production is reduced and degraded land can potentially be rehabilitated.

As shown in Fig. 11.1, second generation biofuels could dramatically reduce lifecycle GHG emissions compared to petroleum fuels and first generation biofuels. This stems from both the higher energy yields per hectare and energy savings in the conversion process. Second generation feedstock production based on perennial crops provide year-round soil cover and require less soil preparations than annual crops. This assists in erosion control and increases water-retention capacity. The soil carbon stock can also be improved through both roots and leaf litter. However, ultimately the environmental impact of lignocellulosic biofuels depends on the conversion route, the feedstock and site-specific conditions. Similar sustainability issues also apply for second generation biofuels and supply chains need to be carefully analysed.

Furthermore, the use of agricultural residues in biofuel production could benefit farmers by adding value to agricultural waste, which mostly has no commercial value in many African countries. There is also potential to create jobs along the value-chain of second generation biofuels which could boost rural income and development.

However, second generation technologies are still under development, although commercialisation is anticipated in the next decade. During this period, it is expected that first generation technologies will continue to play a role as the industry makes a transition to a more sustainable future.

11.2 Overview of State-of-the-Art Second Generation Biofuels Technologies

There are two main promising routes used to process biofuels from lignocellulosic feedstock: bio-chemical and thermo-chemical. In the bio-chemical route, enzymes and other micro-organisms are used to convert cellulose and hemicellulose components of the feedstocks to sugars prior to their fermentation to produce ethanol. The remaining lignin can be used to generate power or steam. The thermo-chemical pathway (commonly referred to as Biomass-to-Liquids (BtL) technology) employs gasification or pyrolysis to produce a synthesis gas or syngas from which a wide range of long carbon chain biofuels, such as synthetic diesel or aviation fuel, can be derived (see Fig. 11.2).

11.2.1 Syngas Based Biofuels

The advantage of the BtL route to liquid transport fuels lies in the ability to use almost any type of biomass, with little pre-treatment other than moisture control. This vastly widens the feedstock base and improves the GHG balance. Thermo-chemical conversion of biomass to biofuels generally involves higher temperatures and pressure than those needed for the biochemical route. It is based on either gasification or pyrolysis. Biomass feedstock is pre-treated to required specifications before being fed into a gasifier. The syngas produced is a mixture of mainly CO and H_2, and some CH_4. Further treatment involves cleaning to remove tars, particulates and gaseous contaminants before being fed to a Fischer Tropsch (FT) reactor where syngas interaction with catalysts results in the production of different types of fuels. Most prominent fuels derived include FT liquids and dimethyl-ether (DME). Methanol and hydrogen can also be produced. FT fuels are suitable for use in diesel engines whilst DME can substitute LPG (Liquefied Petroleum Gas). The FT process is an established technology and is already applied on a large-scale to produce liquid fuels from coal or natural gas.

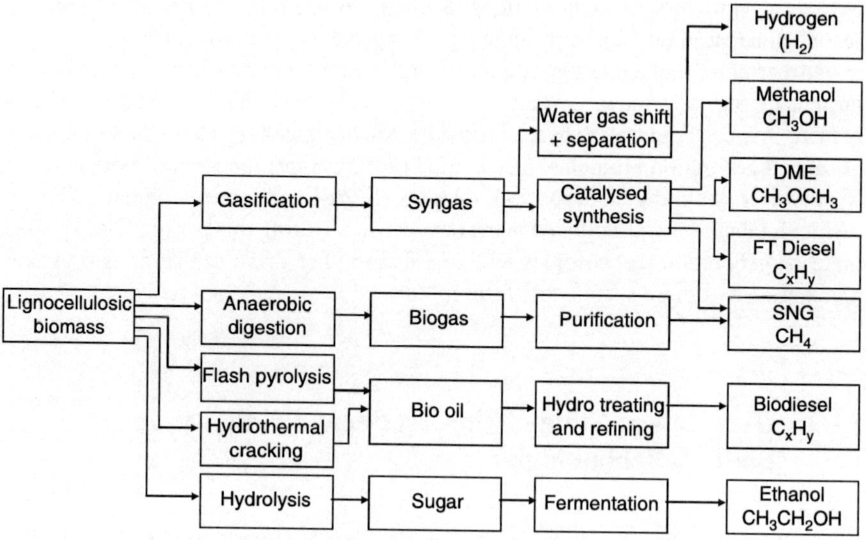

Fig. 11.2 Overview of second generation biofuel production pathways

Developed in the 1920s in Germany, it was used by both Germany and Japan during World War II and later by South Africa and to a lesser extent in the United States. The FT process is a catalysed chemical reaction in which CO and H_2 are converted into liquid hydrocarbons of various forms. Depending on the catalyst, a different fuel is produced. Generally, iron and cobalt catalysts are used in the process.

BtL fuels can also be produced via direct gasification of biomass, as well as via 'fast' or 'flash' pyrolysis. These are thermo-chemical processes in which organic materials are rapidly heated in the absence of oxygen to produce char, tar and organic vapours depending on reaction parameters. Following cooling, bio-oil (or pyrolysis oil) can be separated from the char. Bio-oil can be upgraded to a fuel for use in diesel engines. Fluid bed and ablative reactors are the two principal technologies now available for flash pyrolysis. A related thermo-chemical process is called hydrothermal upgrading (HTU). During HTU, biomass is thermo-chemically converted to a liquid called 'bio-crude'. Bio-crude can be further processed in a traditional oil refinery to various fuels.

Advantages of BtL fuels are their high quality as ultra clean transportation fuel with very low sulphur content and aromatic compounds. FT-diesel can be directly used in vehicles and existing infrastructures without any adaptation.

11.2.2 BtL Development Challenges

Although coal gasification technologies have been commercially available for decades, there is little commercial experience of integrating biomass gasification with FT processes.

Key challenges relate to the gasifier designs, syngas quality, product selectivity in chemical synthesis, process integration and demonstration at industrial scale. In addition, there is need to optimise plant configurations to ensure technical and economic viability of BtL technology (FAO 2008; Rutz et al. 2009). The syngas technology is very capital intensive and large-scale conversion facilities (in the Gigawatt thermal range) are required for cost effective operations. Because of such scale requirements sourcing adequate biomass feedstock to meet demand at viable scale is a challenge. Supply chain logistics need to be optimised as the scale of operation may be limited by the distance over which biomass can be transported to a processing BtL plant at an economic price. Operational and maintenance costs of BtL plants are also high.

11.2.3 Ethanol Production from Lignocellulosic Biomass

The production process of lignocellulosic biomass to ethanol consists of several stages, including biomass pre-treatment, hydrolysis, fermentation and distillation. Chemical and physical pre-treatment breaks down cell structures and separates the lignin from cellulose and hemicellulose and thereby facilitates the hydrolysis (saccharification). Acid or enzymatic hydrolysis converts the cellulose and hemicellulose into fermentable monomeric and oligomeric sugars, with enzymatic hydrolysis using cellulases and hemicellulases being the preferred route. The lignin residue can be used for generation of electricity. The sugars are fermented to ethanol, which is then purified and dehydrated.

11.2.4 Lignocellulosic Ethanol Development Challenges

Lignocellulosic ethanol is still at the demonstration stage and several technical challenges need to be overcome to make it competitive. Key challenges include the development of low cost enzyme production, the development of robust microbes, the demonstration of processes at industrial scale and the potential for integration of process with production of other fuels and fuel additives. Pre-treatment is considered the most costly step for this technology affecting downstream stages. Two trends are possible for future processes, a continuing consolidation of hydrolysis-fermentation reactions in fewer reactor vessels and with fewer micro organisms, or an optimisation of separate reactions. The second biggest challenge is optimising ethanol fermenting micro-organisms that can tolerate adverse process conditions, and convert all sugars including pentoses like xylose and arabinose. Further optimisation can be achieved through ‘consolidated bioprocessing’ i.e. by producing all required enzymes within the reactor vessel, thus using the same ‘microbial community’ to produce both the enzymes that break down cellulose into sugars and those that ferment the sugars to ethanol.

11.2.5 Developments in Advanced Biofuel R&D and Economics

Most second generation biofuel technology R&D is currently taking place in developed countries (e.g. US and Europe) and in some emerging economies (e.g. China and Brazil). Significant progress is being made in R&D and demonstration, and it is likely that commercial scale plants will be deployed over the next decade. A host of technical improvements, cost reductions and environmental concerns have to be addressed to make the technologies competitive. Figure 11.3 shows the global cost supply curve for BtL diesel under the four IPCC SRES (Intergovernmental Panel on Climate Change - Special Reports on Emission Scenarios) scenarios[1] projected up to 2050, assuming that the feedstock is based on energy crops produced on abandoned agricultural land and rest land.

11.3 Development of Advanced Biofuels in Africa

Recent studies show that Africa has a high potential for the production of dedicated bioenergy crops (at low cost) suitable as second generation biofuel feedstocks. Smeets et al. (2007) estimated potential surplus land of up to 0.7 Gha in Africa that could be used to grow energy crops with annual yields of up to 317 EJ, assuming

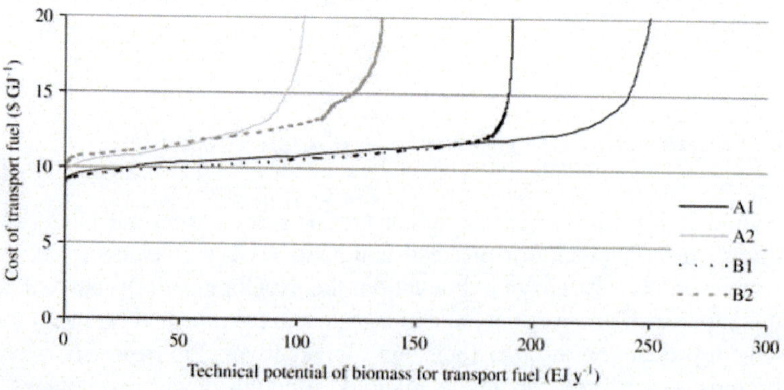

Fig. 11.3 Cost supply curve of BtL diesel under four IPCC SRES scenarios (Source: Hoogwijk et al. 2009)

[1] The four IPCC SRES storylines (A1, A2, B1 and B2) describe different probable global social, economic, technological, environmental, and policy development pathways. They are constructed along two dimensions, i.e. the degree of orientation on material versus social and ecological values (A – B), and the degree of globalisation versus regionalisation (1–2), i.e. A1 is so-called "Global Economy" scenario, A2 is "Continental Markets", B1 "Global Co-operation", and B2 "Regional Communities" scenarios (see Nakicenovic (2000) for further details).

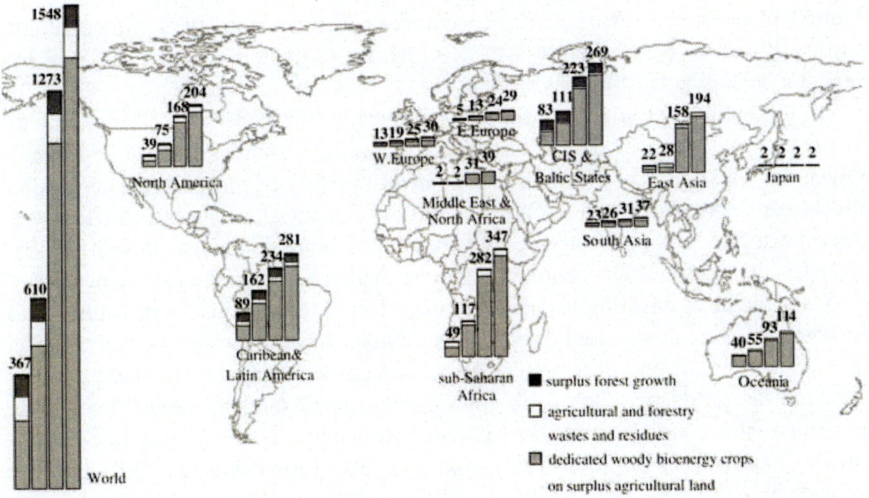

Fig. 11.4 Projected technical bioenergy production potential in 2050 (EJ yr^{-1}) assuming four different agricultural management systems (the *rightmost bar* represents a very high level of agricultural technology) (Source: Smeets et al. 2007)

improvements in agriculture productivity of up to 8 times[2] (Fig. 11.4). Using scenario analysis, Hoogwijk et al. (2004) also projected biomass production potential of up to 134 EJ per year for Africa, using abandoned agricultural land and non-productive land. Wicke et al. (2010) estimated the annual technical bioenergy potential for semi-arid and arid regions of eight Sub-Saharan African countries to be 300 PJ for cassava based ethanol, 600 PJ for jatropha based biodiesel and 4 EJ for fuel-wood from SRF.

Recent analysis in Hoogwijk et al. (2009) show that Africa may become an important low cost supplier of lignocellulosic biomass produced on abandoned land and non-productive land (excluding bio-reserves, forest, agricultural and urban areas and assuming the demand for food, fodder and forestry products is met). The study identifies East and West Africa as largest regions with lowest-cost potential (up to 92% of global potential below 1 US\$ GJ^{-1}). The regions also have significant potential to produce biomass at costs below 2 US\$ GJ^{-1}, which costs are competitive compared to fossil fuels. Recent analysis by van Eijck et al. (2010) have also shown that it is possible to produce low cost eucalyptus pellets (at 2.7–3.3 US\$ GJ^{-1}) which is competitive at current market prices with conventional pellets (at 5 US\$ GJ^{-1}). At these cost levels, large-scale biofuel production is expected to become competitive with fossil fuels, provided technological advancements are achieved.

Despite Africa's theoretical bioenergy production potential, realising this potential remains a daunting challenge. High investment costs of second generation

[2] In comparison, total land surface area of SSA is about 2.3 Gha (Bruinsma 2003), and global oil consumption in 2007 was 148 EJ of which 90 EJ was for transportation (IEA 2009b).

biofuel plants are a major barrier to commercial realisation of this technology, especially in Africa. Currently, capital costs are estimated at about 300 M€ for medium scale plants (400 MWth input) to 750 M€ for large-scale plants (2 GWth input) (Hamelinck et al. 2005). Similar to power sector projects, such investments in Africa usually rely on external finance as domestic resources are usually limited. Given the risks and because of these high investment costs, most second generation plants are in the short-term likely to be built in industrialised countries. However, Africa can play an important role as a supplier of biomass feedstock to the global industry, well before the technologies can be implemented on the continent.

A few detailed case studies have been conducted to evaluate the possibilities of biomass energy exports from Africa. These studies have identified Mozambique as a potential biomass energy supplier. Mozambique is considered a promising country for biomass production due to relative abundance of land resources, favourable environmental conditions and low population density. Biomass resource assessments (Batidzirai et al. 2006; Hoogwijk et al. 2004; Smeets et al. 2007) indicate considerable potential within the sub-region. The following section presents some key results of an assessment of the potential, economics and logistics options to produce biofuels in Mozambique for the export market using Rotterdam Harbour as a final destination.

11.4 Biomass Production Potential from Dedicated Energy Crops in Mozambique

Mozambique is a sparsely populated country, with land area covering 80 Mha and an estimated population of 21.7 million. Population density is about 26 pers/km^2, the most densely populated areas are along the coast and in the agriculturally productive central provinces (INE 2009). Although Mozambique has achieved an impressive economic growth in recent years, it remains one of the poorest countries in the world, with an average per capita annual income of 370 US$ in 2008 (World 2009).

Delivery of biomass from Mozambique to international markets can be in the form of pellets, bio-oils or fuels. Biomass supply logistics depends mainly on the biomass distribution density, operational scales, operating windows and relative distances to supply destinations. Logistic operations include biomass harvesting, storage, drying, transportation, sizing, and conversion. Detailed formulation of the logistic concept can be found in Batidzirai et al. (2006).

In the analysis it was assumed that conversion of FT fuels is via direct pressurised oxygen blown Circulating Fluidised Bed (CFB) gasification technology fed with wood chips. A simple comparison was also made of FT fuel production via pressurised oxygen blown Entrained Flow (EF) gasification using pyrolysis oil and pulverised pellets as input. CFB gasification has the disadvantage that it has limited maximum capacity of about 400 MWth for a single unit whereas EF gasification units in the 1,000 MWth range are possible. This allows greater scaling benefits for EF gasification (Hamelinck et al. 2005).

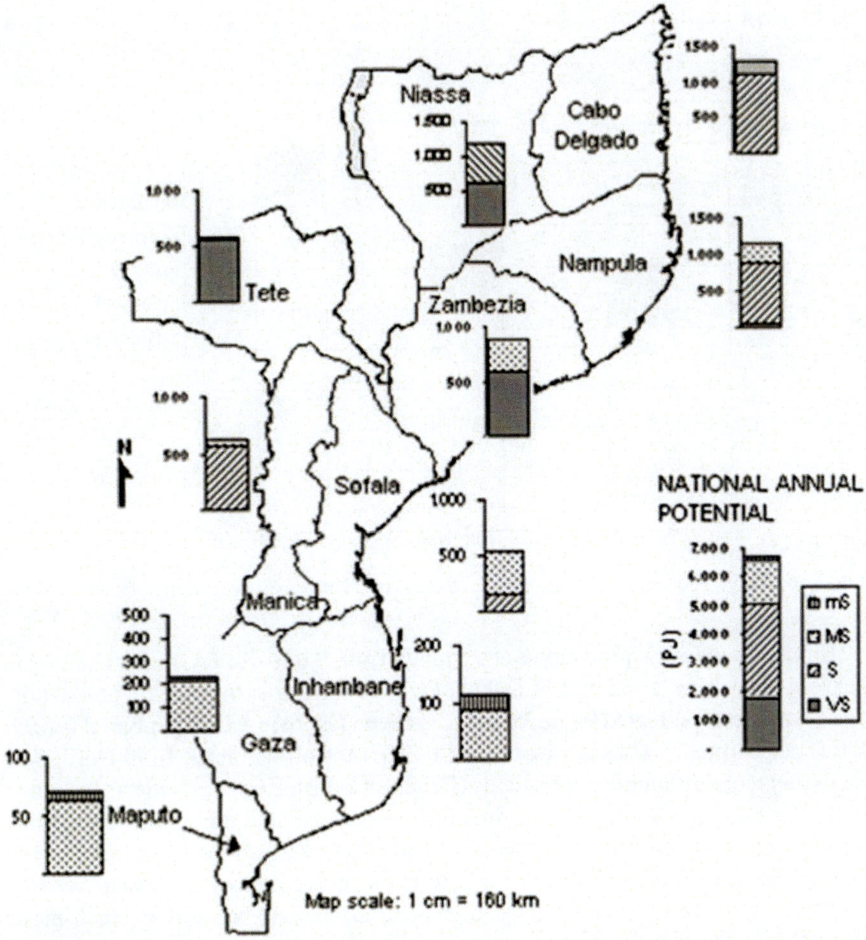

Fig. 11.5 Regional biomass annual production potential in Mozambique (in PJ HHV projected for 2015) (Source: Batidzirai et al. 2006)

Mozambique has the potential to produce up to 6.7 EJ of biomass energy annually using surplus land with moderate agricultural technological inputs. As shown in Fig. 11.5, the northern parts of the country have the highest potential compared to the central and southern regions. This potential is assessed based on estimated availability of surplus land and the yield level, whilst observing key sustainability criteria (mainly ability to meet food demand and no deforestation). Currently, only 5 million hectares (Mha) of the 48 Mha of agricultural land are effectively used for food production (FAOSTAT 2009). Up to 45 Mha of agricultural land could (in theory) be made available for bioenergy crop production in Mozambique by 2015, depending on the level of advancement of agricultural technology and the animal production system. Only 10% of the surplus areas are assigned to eucalyptus plantations in estimating the biomass potential.

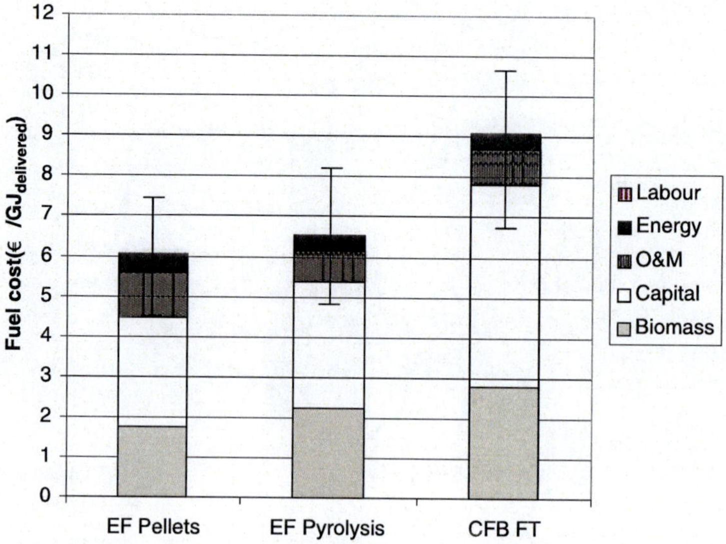

Fig. 11.6 Range of costs for BtL delivered at Rotterdam Harbour (Source: Batidzirai et al. 2006)

Biomass feedstock production costs range from 0.6 to 1.2 €/GJ plus additional harvesting costs of between 0.11 and 0.2 €/GJ. In comparison, crude oil production costs have been estimated in the Middle East to be less than 5 US$ per barrel (~0.71 €/GJ), and up to 15 US$ per barrel (~2.1 €/GJ) for oil extracted from North Sea fields and more difficult Russian fields (Fullerton 2003). However, biomass production costs have minor influence on final fuel costs and inland transport costs as well as pre-treatment costs are more decisive. Also, it was determined that early conversion of biomass to either pellets or pyrolysis oils at local facilities reduces overall logistics costs, indicating that truck transportation of raw biomass is much more important in the overall costs than the benefits of economies of scale achieved at centralised facilities. The opposite is true for Fischer Tropsch conversion.

Pellets are delivered at lower costs (2.6 €/GJ) than pyrolysis oils (3.2 €/GJ) or FT fuels (6.8 €/GJ). This shows that the higher transportation costs due to higher volumes of pellets are compensated by the high conversion cost for FT fuels (2.2 €/GJ) and higher handling costs for pyrolysis oil. In comparison, the production cost of petrol and diesel from crude oil is the range of about 9.2–10.9 €/GJ (Gonsalves 2006). Conversion of pellets and pyrolysis oils to FT fuels via EF gasification at the importing labour results in fuel costs of 4.5 and 4.8 €/GJ (Fig. 11.6). The most attractive supply chain of bioenergy in terms of delivered fuel cost and energy requirements was found to be the local conversion of biomass to pellets with further conversion to FT fuels using the entrained flow gasification method, whereas in terms of CO_2 emissions, the pyrolysis chain is more attractive.

Whilst the best logistics of supplying bioenergy from Mozambique have been found to be local production of pellets and pyrolysis oils in Sofala region and eventual conversion to FT fuels in Rotterdam, this is largely because local synthesis of

FT fuels will demand large amounts of biomass feedstock, which may not be achievable in the short-term. However, in the long-term, it may be possible to establish large-scale FT fuel synthesis plants in Mozambique which may be supplied by biomass from one or several regions. This will require improvement in (especially rail) infrastructure. In addition, there will be need to build more experience and biomass production capacity in the country. For the short to medium term, pellets are recommended as an intermediate biofuel for international trade because the delivered fuel cost is cheaper and energetic losses are lower than either pyrolysis oil or FT fuel (produced at source).

11.5 Conclusion

Second generation biofuels are not yet produced commercially, although a number of pilot and demonstration plants are underway mainly in North America, Europe and a few emerging countries. A number of technological and cost barriers need to be overcome for the successful commercial deployment of second generation biofuel technologies. Africa has potential to play an important role in the production and supply of modern biomass energy, given its competitive advantage. Mozambique is one of the countries in the region which is strategically positioned to produce biomass energy for export to the international market as well as meet some of its own internal energy needs. In line with the Mozambican government's agricultural and economic development policies, sustainable cultivation of energy crops has the potential to open up rural areas, to create employment, as well as to improve degraded land and infrastructure whilst generating much needed foreign currency.

References

Batidzirai B, Faaij A, Smeets E (2006) Biomass and bioenergy supply from Mozambique. Energy Sust Dev X(1):54–81

Bruinsma J (2003) World agriculture: towards 2015/2030. An FAO perspective. Food and Agriculture Organization. Earthscan, London

FAO (2008) The state of food and agriculture. Biofuels: prospects, risks and opportunities. Food and Agriculture Organization of the United Nations, Rome

FAOSTAT (2009) ProdSTAT. Food and Agriculture Organization of the United Nations

Fullerton RL (2003) Blood for oil? United States Air Force Academy. http://www.globalsecurity.org/military/library/report/2003/main0903.htm. Accessed 15 July 2010

Gonsalves JB (2006) An assessment of the biofuels industry in India. UNCTAD, Geneva

Hamelinck CN, van Hooijdonk G, Faaij APC (2005) Ethanol from lignocellulosic biomass: techno-economic performance in short, middle and long-term. Biomass Bioenergy 28(4):384–410

Hoefnagels R, Smeets E, Faaij A (2010) Greenhouse gas footprints of different biofuel production systems. Renew Sust Energy Rev 14:1661–1694. doi:10.1016/j.rser.2010.02.014

Hoogwijk M, Faaij A, van den Broek R, Berndes G, Gielen D, Turkenburg W (2004) Exploration of the ranges of the global potential of biomass for energy. Biomass Bioenergy 25(2):119–133

Hoogwijk M, Faaij A, de Vries B, Turkenburg W (2009) Exploration of regional and global cost–supply curves of biomass energy from short-rotation crops at abandoned cropland and rest land under four IPCC SRES land-use scenarios. Biomass Bioenergy 33:26–43

IEA (2008a) From 1st to second-generation biofuel technologies: an overview of current industry and RTD activities. OECD/IEA, Paris

IEA (2008b) Energy technology perspectives 2008: scenarios and strategies to 2050. OECD/IEA, Paris

IEA (2009a) World energy outlook 2009. IEA, Paris

IEA (2009b) Key world energy statistics: 2009. OECD/IEA, Paris

IEA (2010) Sustainable production of second -generation biofuels: potential and perspectives in major economies and developing countries. Information Paper, Paris

INE (2009) Instituto Nacional de Estatística. Second general population and housing census, 2007, Maputo. http://www.ine.gov.mz/Ingles/publicacoes/censos dir/ recenseamento_geral/. Accessed 9 Mar 2010

Nakicenovic N, Alcamo J, Davis G, de Vries B, Fenhann J, Gaffin S et al (2000) IPCC special report on emissions scenarios. Cambridge University Press, Cambridge, UK and New York, USA. 599 pp

Rutz D, Hiegl W, Janssen R (2009) Biomass-to-liquid production on Latin America: technical opportunities and research needs. WIP Renewable Energies; Report of the EU Project BioTop (FP7); Contract no.: 213320

Smeets EMW, Faaij APC, Lewandowski IM, Turkenburg WC (2007) A bottom-up assessment and review of global bioenergy potentials to 2050. Prog Energy Combust Sci 33(1):56–106

van Eijck J, Smeets E, Faaij A (2010) The economic performance of Jatropha, Cassava and Eucalyptus production systems for energy in an East African smallholder setting. Submitted for publication to Agricultural Systems

Wicke B, Smeets E, Watson H, Faaij A (2010) The current bioenergy production potential of semi-arid and arid regions in sub-Saharan Africa. Biomass Bioenergy 35(7):2773–2786

World Bank (2009) World development indicators database, April 2009. http://web.worldbank.org/. Accessed 9 Mar 2010

Chapter 12
Modern Use of Solid Biomass in Africa: Prospects for Utilization of Agro-Waste Resources in Nigeria

Kemjika Benaiah Ajoku

Abstract Agro and forest resources are important energy sources in sub-Saharan Africa. However, Nigeria largely depends on conventional energy sources such as electricity and fossil energy carriers such as crude oil and natural gas. Nevertheless, Nigeria is blessed with abundant bioenergy resources presenting significant potential for developing sustainable energy supply to the rural areas and for mitigating the demand-supply gap in the country. Two key agro-waste resources are rice husks and wood wastes were examined. They present enormous potentials for bioenergy development in the rural areas of Nigeria, but awareness on these potentials has to be raised. Furthermore, appropriate regulatory frameworks are needed to use this potential. Therefore, theoretical and empirical analysis of the potentials of modern use of solid biomass in Africa was carried out and is described in this chapter.

Keywords Agro-waste • Biomass • Centralized energy generation • Rice husks • Wood-waste • Wood industry • Rice mill

12.1 Introduction

Biomass is one of the major sources of renewable energy globally. It supplies about one third of energy requirements in developing countries of which the level of utilization varies from country to country. According to FAO (1999) biomass contributes to about 14% of the world's energy demand (55 EJ or 25 million barrels oil equivalent). Bioenergy technologies based on sustainable biomass supply lead to net CO_2 emission reductions if used to substitute fossil fuels (Coelho and Walter 2003; Karekezi et al. 2008).

K.B. Ajoku (✉)
Raw Materials Research and Development Council, Abuja, Nigeria
e-mail: ajokukemji@yahoo.co.uk

R. Janssen and D. Rutz (eds.), *Bioenergy for Sustainable Development in Africa*,
DOI 10.1007/978-94-007-2181-4_12,

Bioresources are potentially the world's largest and sustainable source of fuel and chemicals. Residues represent a large potential which is mainly under-utilized at present. The energy content of potentially harvestable residues is estimated at about 93 EJ/year worldwide (FAO 1999). It is evident that the problem of biomass utilization is not availability, but relates to the sustainable management and delivery of energy to those who need it at acceptable cost.

There is increasing evidence that considerable potential exists for modernization of biomass fuels to produce convenient energy carriers such as electricity, gases and transportation fuels whilst continuing to provide traditional uses of biomass. The modernization of biomass and the necessary industrial investment is already happening in many countries. However, future use of bioenergy must be strongly linked to high energy efficiency and environmentally sustainable production and use (Woods and Hall 1994).

Traditional biomass has been estimated to account for more than 17% of total primary energy consumption (Inter Academy Council 2010). Traditional uses of biomass are estimated to account for more than 90% of the biomass contribution to global energy supply, most of which occurs outside the formal market economy and predominantly in developing countries. On the other hand, modern uses of biomass to generate electricity and heat or as transportation fuels are estimated to account for less than 10% of total biomass energy consumption worldwide.

The change from traditional biomass utilization to more modern applications is challenging for most developing countries, especially for countries of sub-Saharan Africa.

This chapter examines the potentials of this transition to modern application of biomass, especially for electricity generation. The impact of this transition on sustainable industrial and overall economic development of Africa is assessed. Specific emphasis is placed on the prospects for the use of agro-wastes for energy generation in Nigeria, especially for addressing household energy needs.

12.2 Overview of Biomass Utilization in Africa

Biomass and bioenergy are essential components of the energy mix in most African countries. Biomass is a renewable energy resource derived from agricultural wastes, residues, forestry resources as well as municipal and industrial wastes.

Agro and forest biomass currently play a vital role in energy generation in Africa, especially in sub-Saharan Africa, and mainly in rural areas where about 75–95% of the population depend on fuel-wood as primary energy source (Brenes 2006). The reliance on bioenergy in Africa is significantly associated with lack of conventional energy sources such as electricity and petroleum products. Modern biomass utilization has the potential to provide improved energy services to rural Africa based on the resources and agricultural residues.

Wastes (such as bagasse) generated from sugar mills in a number of eastern and southern African countries have proven to be a commercial resource for electricity generation. Kerekezi and Kithyoma (2005) reported that up to 16 sub-Saharan

African countries could meet significant proportions of their current electricity consumption from bagasse-based cogeneration in the sugar industry. There is no doubt that the potential for electricity generation from bagasse is high since cogeneration equipment is almost always an integral component of modern sugar factories.

In Kenya for instance, it is reported that about 60% of bagasse produced by six sugar factories in the western part of the country is used as boiler fuel for steam generation and electricity (Karekezi and Kithyoma 2005). The estimated cogeneration potential in Tanzania is 157 GWh, of which only 9% has been developed even though the country has an installed capacity of about 26 MW from sugar and forest industries (Karekezi et al. 2008). Similarly, in Mauritius close to 30% of the country's electricity is generated from the sugar industry using mainly bagasse. Bagasse-based cogeneration development in Mauritius has delivered a number of benefits including reduced dependence on imported oil, diversification in electricity generation and improved efficiency in the power sector in general (Karekezi and Kithyoma 2005).

12.3 Biomass Technologies

Different technologies enable the conversion of biomass into various energy carriers including liquid and gaseous fuels as well as end-energy such as electricity and heat. Technologies such as modern wood fired plants, biomass co-firing, gasification, and co-generation are commercially available and have been increasingly applied worldwide. Contrary to the developments in the industrialized world, there has been persistent reliance on small-scale traditional biomass conversion technologies and limited use of modern technologies in Africa (Onyekwelu and Akindele 2006).

Traditional biomass technologies emphasize the inefficient conversion of wood resources into charcoal. Efforts to improve and modernize small-scale biomass energy production constitute an important component of national energy strategies in many African countries of sub-Saharan Africa. Given the enormous quantity of wastes from wood processing in sub-Saharan Africa, the briquetting industry has a promising future (Onyekwelu and Akindele 2006).

Modern biomass technologies such as co-generation have the potential to provide improved energy services for the rural population in Africa based on available biomass resources and agricultural residues. These resources provide the feedstock for sustained use of combined heat and power generation in rural areas of Africa (AFREPREN/FWD 2006). The availability of economically competitive biomass power in Africa has the potential to provide cleaner, more efficient energy services to support local development, promote environmental protection, provide improved domestic fuels and improve rural livelihoods (Karekezi et al. 2008). Furthermore, the efficient use of improved biomass feedstock for local co-production of heat, electricity, and transportation fuel have profound impact on the ability of rural populations to access modern, cleaner forms of energy (Inter Academy Council 2010). In fact, the widespread use of biomass based combined heat and power generation in rural Africa can help overcome multiple socio-economic and environmental constraints that hinder development (Karekezi et al. 2008).

12.4 Energy Potential of Agro-Wastes in Nigeria

Conventional energy in Nigeria consists of crude oil and natural gas products. Nigeria also relies on hydropower for electricity production, but its availability is currently limited due to technical problems and ecological constraints.

Over the years, the main attention of the Government of Nigeria was on fossil fuels. It therefore implemented energy programmes that target fossil fuels with limited attention to renewable energy resource development and utilization. There is no doubt that the country has very high renewable energy resources besides its oil, gas and coal reserves. With high level of economic activities and increasing rate of urbanization, there has been a sharp increase in energy consumption in the country.

Power demand in both residential and industrial sectors has increased significantly in recent times. Industries have not been able to meet their energy needs for competitive manufacturing production. Most of them have invested considerably in alternative energies which has affected the costs of production.

Nigeria has an installed capacity of 4,663 MW with a maximum demand of about 3,500 MW. The electrification rates of urban and rural areas are estimated at 84.3% and 27.9%, respectively (World Bank 2007).

Nigeria has not yet explored and exploited alternative energy resources such as biomass energy in its energy conservation programme. Biomass energy has, however, been identified as alternative energy source at least to meet the energy needs of the rural and urban poor. The country's biomass energy resources are estimated at 144 million tons per year (Adegbulugbe and Adenikinju 2010). It is equally estimated that the country consumes about 43.4 billion kg of fuel-wood annually with an average daily consumption of about 0.5–1.0 kg of dry fuel-wood per person.

Biomass is a primary source of energy for about 98 million Nigerians (70% of the population) living in rural areas. Even in urban centres, especially amongst the urban poor, considerable amounts of biomass are utilized as energy source. Biomass therefore provides vital and affordable energy for cooking and heating. It has been observed that although the widespread use of traditional and inefficient biomass energy in Nigeria is linked to indoor air pollution as well as to land degradation, deforestation and subsequent soil-erosion, biomass-based economic activities still create significant numbers of jobs and income for the rural poor.

The share of biomass energy in total energy consumption has been associated with poverty especially amongst developing countries. It has been reported that about 92% of the population of Nigeria live below 2 US$ per day and depend on traditional biomass consumption (UNDP 2003).

There is no doubt that biomass has considerable potential to become more important in total energy consumption and its increased utilization could have significant impact (both positive and negative) on agriculture and the rural poor. The efficient exploitation of existing agricultural wastes presents significant potential for bioenergy development. In Nigeria, some of the common agricultural resources suitable for bioenergy development include for example sugarcane bagasse, rice husks, wood waste, sawdust, oil palm waste, maize cobs, and cotton stalks.

These wastes are produced during agro-processing and are often not returned to the field with some exceptions, such as for palm wastes. Nevertheless, the use of agro-waste for energy generation could potentially be a source of revenue for the poor whilst mitigating negative environmental impacts of improper waste disposal.

This chapter shows the bioenergy resource potentials, especially the exploitation of existing agricultural wastes and suggests priority steps for developing bioenergy processes that are beneficial for the Nigerian population.

12.4.1 Rice Production in Nigeria

Nigeria is West Africa's largest producer of rice with an average production of 3.2 million tons of paddy rice annually (i.e. about 2 million tons of milled rice) in the last 7 years. According to the West Africa Rice Development Association (WARDA), it is estimated that Nigeria accounted for about 44% of the total West African rice output and 57% of the total rice producing area (WARDA 1996).

With a population of over 150 million people, Nigeria is also one of the largest importers of rice in the world. The annual demand for rice in the country is estimated at 5 million tons. Rice production in Nigeria has increased over the years and the principal factors driving the increased production are population growth and urbanization.

Rice cultivation extends from Nigeria's northern to southern zones with the major producing areas located in Enugu, Ebonyi, Cross River States in the eastern part of the country and Benue, Borno, Kaduna, Kano, Niger and Taraba States in the northern part of the country. Niger State is the largest rice producer in Nigeria followed by Benue State. Generally, there is low yield per hectare which is associated with poor production systems, aging farming population, and low competitiveness with imported rice. Nigerian rice production systems are characterized as follows:

- Rainfed upland rice on plateau and slopes
- Rainfed lowland rice in valley and flood plains
- Irrigated rice with relatively good water control
- Deep water, floating rice along river beds and banks
- Rice in mangrove swamps

The rainfed lowland rice production system occupies the largest share of national rice growing area (47%) with an average yield of 2.2 tons/ha and an estimated production share of 53%. The irrigated rice production system has an estimated share of national growing area of 17% with a yield of 3.5 tons/ha and a production share of 27%. This is followed by the rainfed upland rice production system with about 30% share of national growing area, 1.7 tons/ha average yield and 17% share of rice production.

Rice in Nigeria is usually produced by small-scale farmers, with an average farm size of 1–2 ha and very low yields. The average rice yield in Nigeria is estimated at 1.85 tons/ha. The national production of rice by smallholder farmers is presented in Fig. 12.1.

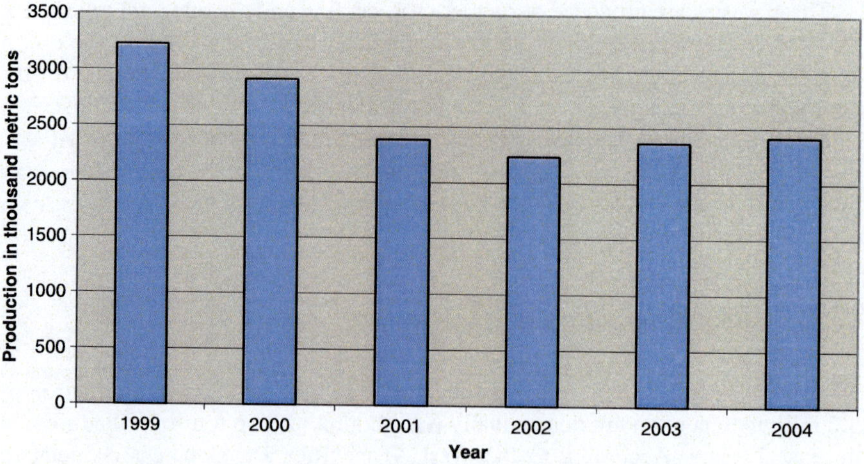

Fig. 12.1 Production of rice by smallholder farmers in Nigeria (in thousand metric tons) (Source: Adapted from Data by Obassi and Ajoku 2007)

Nigeria generates significant amounts of wastes from rice milling enterprises and clusters. The states identified with considerable level of organized clusters include Ebonyi (Abakaliki), Benue (Otukpo, Adikpo, Gboko, Vandekiya), Nasarawa (Lafia), Niger (Bidda, Mokwa) and Kano (Kura). In other states where processing facilities are lacking, rice is bought and transported to neighbouring states for further processing. The evolution of rice clusters resulted from dedicated support in some states including provision of land and processing facilities. Cooperatives were formed to use joint processing and marketing facilities (Obassi and Ajoku 2007).

It is estimated from a field survey that about 664,000 tons of rice husks are generated from about 3.32 million tons of paddy rice per year. These figures are also shown by FAO data in Table 12.1.

A considerable quantity of rice husk heaps are scattered in rice producing and processing localities and clusters across the country. This means that rice husks are not used and that they constitute environmental nuisances (Obassi and Ajoku 2007). The number of rice mills and the quantity of rice husk residues generated in the identified states are presented in Table 12.2.

The wastes derived from processing of rice could serve as energy resource material to generate heat and electricity for domestic use. Improving the efficiency of energy production from these wastes could deliver significant economic benefits to the people and associated industries.

12.4.2 *Potential for Energy Production from Rice Husks*

There is large potential for energy from rice husks in Nigeria. However, as shown above, there are only very limited efforts to exploit this potential. Attempts were

Table 12.1 World paddy rice and potential rice husk production

Country	Paddy rice production in 2002 (t)	Percentage of total paddy rice production (%)	Husk produced (20% of Total) (t)	Potential ash production (18% of husk) (t)
China	177,589,000	30.7	35,517,800	6,393,204
India	123,000,000	21.2	24,600,000	4,428,000
Indonesia	48,654,048	8.4	9,730,810	1,751,546
Bangladesh	39,000,000	6.7	7,800,000	1,404,000
Viet. Nam	31,319,000	5.4	6,263,800	1,127,484
Thailand	27,000,000	4.7	5,400,000	972,000
Myanmar	21,200,000	3.7	4,240,000	763,200
Philippines	12,684,800	2.2	2,536,960	456,653
Japan	11,264,000	1.9	2,252,800	405,504
Brazil	10,489,400	1.8	2,097,800	377,618
USA	9,616,750	1.7	1,923,350	346,203
Korea	7,429,000	1.3	1,485,800	267,444
Pakistan	5,776,000	1.0	1,155,200	207,936
Egypt	5,700,000	1	1,140,000	205,200
Nepal	4,750,000	0.8	950,000	171,000
Cambodia	4,099,016	0.7	819,803	147,565
Nigeria	3,367,000	0.6	673,400	121,212
Sri Lanka	2,794,000	0.5	558,800	100,584
Colombia	2,353,440	0.4	470,688	84,724
Laos	2,300,000	0.4	460,000	82,800
Rest of the World	29,091,358	5.0	5,818,272	1,047,289
Total (World)	579,476,722	100	115,895,344	20,861,162

Source: FAO Statistics (2002)

Table 12.2 Distribution of rice mill clusters and quantity of residues

S/No	States surveyed	Number of clusters	Number of mills	Quantity of rice husk/ residue generated (t/year)
1.0	Benue	5	1,102	39,000
2.0	Ebonyi	8	1,303	54,500
3.0	Kano	1	120	15,000
4.0	Nasarawa	2	222	18,000
5.0	Cross River	1	25	6,000
6.0	Niger	6	73	20,400
7.0	Adamawa	3	35	12,000
8.0	Kwara	4	10	3,000
9.0	Edo	3	37	1,500

Source: Field Survey (Obassi and Ajoku 2007)

made in some locations to utilize husks for the production of briquettes, however with very limited success due to lack of know-how.

Nevertheless, rice husks and other agricultural residues could assist the country in addressing its rural energy needs.

Table 12.3 Energy potential of major rice processing clusters in Nigeria

S/No	State	Cluster name	Daily quantity of husk generated (t)	Yearly quantity of husk generated (t)	Energy potential per year (GJ)
1	Benue	(i) Otukpo Rice Mills Cluster	128	38,400	542,970
		(ii) Makurdi Rice Mills cluster	6.75	2,025	28,640
		(iii) Kwande United Rice Mills, Adikpo	7.2	2,160	30,540
2	Ebonyi	(i) Abakaliki Rice Mills	150	45,000	636,300
		(ii) Onueke Rice Mils	7.3	2,190	31,000
		(iii) Afikpo cluster	3.75	1,125	15,910
		(iv) Ivo LGA clusters	4.8	1,440	20,360
3	Nasarawa	Lafia Rice Mills cluster	33.2	9,960	140,830
4	Niger	(i) Mokwaka Rice Mills,	7.5	2,250	31,820
		(ii) Minna-Kpakungu Clusters	1.35	405	5,730
		(iii) Lemu Market Area, Gboko LGA(not really Cluster)	5.7	1,710	24,180
5	Cross River	Ogoja Central Rice Mills cluster	20	6,000	84,840
6.	Kwara State	(i) Ojagboro Rice mills cluster	1.05	315	4,450
		(ii) Patigi Women Processors	1.5	450	6,360
		(iii) Lafiagi Women Group	2.0	600	8,480
7	Edo State	Agenegbode Rice Mills	3	900	12,730
		Agbede Rice Mills	1.5	450	6,360

Source: Field Survey (Obassi and Ajoku 2007)

The estimation was based on the following data and assumptions: Annual national paddy rice production = 3.32 million metric tons; Percentage of husk by weight of paddy = 20%; Rice husk energy value (heat value) = 14.140 kJ/kg or 0.01414 TJ; Number of working/production days in a year = 300 days

One of the major rice growing and processing states, Ebonyi is currently establishing an energy generating plant using rice husks with assistance of UNIDO. The energy potential of major rice processing clusters in Nigeria is presented in Table 12.3.

The findings revealed that the clusters in Ebonyi and Benue states have great potential for the establishment of biomass energy projects. Three approaches for energy generation from the rice husks could be adopted.

1. **Centralized energy generation from centralized rice mills**

   Centralized energy production from centralized rice mills needs a regulatory framework to be implemented. Furthermore, an appropriate incentive framework ensuring significant benefits to farmers, industries and relevant stakeholders must be provided.

2. **Centralized energy generation from decentralized rice mills**

   In order to guarantee effective utilization of agro-wastes generated at the farm or processing level under a decentralized milling system, an efficient system for collection, transportation, storage handling and fuel preparation is required in order to supply a centralized energy plant. There is a need to put in place a cost effective waste collection system. For rice husks, centralized energy generation systems can be only feasible if the wastes generated are within a particular radius around the cogeneration plant.

3. **Decentralized energy generation from decentralized rice mills**

   Small-scale rice mills have the option to produce decentralized energy for their own consumption. This requires effective support in establishing an appropriate technology that can guarantee quality energy supply services.

The efficient exploitation of existing agricultural wastes such as rice husks presents significant potential for developing bioenergy without unduly disrupting existing agricultural practices and food production. It will equally serve as a source of revenue for the poor.

There are prospects to establish a biomass energy production facility using rice husks as feedstock in both Ebonyi and Benue States. To ensure sustainability in terms of supply of agro-wastes, there is need to improve the current efficiency of existing rice mills.

For optimal and sustainable energy generation from rice husks, joint efforts should be made to revitalize large-scale rice production and processing facilities in the country and to promote cluster formation where processing technology is lacking. The waste generated from increasing rice production could be converted to energy to serve the rural communities where the rice is farmed. The energy generation from wastes could serve as an incentive to improve rice farming and processing facilities, if good energy pricing is provided by the government.

Obassi and Ajoku (2007) noted that the existing rice milling facilities have the capacity to produce about 350,000 tons of rice husks per year. The installed electricity capacity could be 57 MW and at the same time 608 GWh of steam could be produced per year. It is expected that combined heat and power facilities will supply the energy requirements of modern rice mills and the excess will be made available for local use. It is believed that a small to medium sized power plant ranging from 4 MW to 50 MW of nominal capacity should be promoted to utilize the abundant agricultural wastes through application in combined heat and power (CHP) plants.

Furthermore, the government of Nigeria should regulate or prohibit the current practice of dumping rice husks and burning heaps of rice husks which constitutes serious environmental hazards. Efforts should be made through advocacy and technical support to encourage better practices to harness husk heaps for renewable energy production.

12.5 Utilization of Forestry Waste for Energy Production in Nigeria

12.5.1 Wood Processing in Nigeria

The wood processing industry in Nigeria is characterized by formal and informal sectors. Whilst the large-scale public owned enterprises constitute the formal sector, the small and medium enterprises owned by the private entrepreneurs constitute the informal sector. The small and medium enterprises, comprising significant number of wood-working enterprises operating in primary processing units, are scattered all over the country.

The industrial sub-sector for wood and wood products is engaged in primary wood processing activities in the area of mechanical transformation of wood including saw milling, veneer, particle board manufacturing, pulp and paper, and match splint-making. The secondary wood processing industry comprises small-scale enterprises for furniture making, doors, parquet, mouldings, etc. The tertiary wood industries include those in the production of domestic products (such as fuel-wood, charcoal, etc.), industrial products (construction materials, round and sawn wood) and specialty products (mortars and pestles and carvings). Fuel-wood and charcoal are the basic domestic energy sources apart from electricity and petroleum products.

The wood processing industry generates significant volumes of wood wastes, especially sawdust. Saw milling and ply milling generates about 80% of the total wood wastes in the country. It is estimated that there are about 2,800 sawmills in the country with an installed capacity of 5,842,000 m^3 and production output of about 2,711,000 m^3 (Abdullahi and Ajoku 2001). Sawmills are responsible for processing about 80% of the round wood in the country. The states in Nigeria that are richly endowed with round wood resources and account for a large distribution of sawmills include Lagos, Ogun, Oyo, Osun, Ondo, Edo and Delta (Fig. 12.2).

There are about nine large and medium-scale plywood mills in Nigeria:

- African Timber and Plywood (AT&P), Sapele, Delta State
- Piedmont Plywood, Ologbo, Edo State
- Premier Timbers, Bolorunduro, Ondo State
- Delta Plywood, Burutu, Delta, State
- Seromwood Industry Ltd, Calabar, Cross River State
- Omowoods, Ijebu-Ode, Osun State
- Calabar Veneer and Plywood, Ltd, Cross River State
- Nigerian-Romanian Wood Industries Ltd, Ondo State
- Epe Plywood Company Ltd, Epe, Lagos State

The state owned plywood mills have been privatized. The plywood mills with a total installed capacity of 126,000 m^3 currently operate only at 40–45% capacity. Several plywood mills have been shut down due to operational, financial and

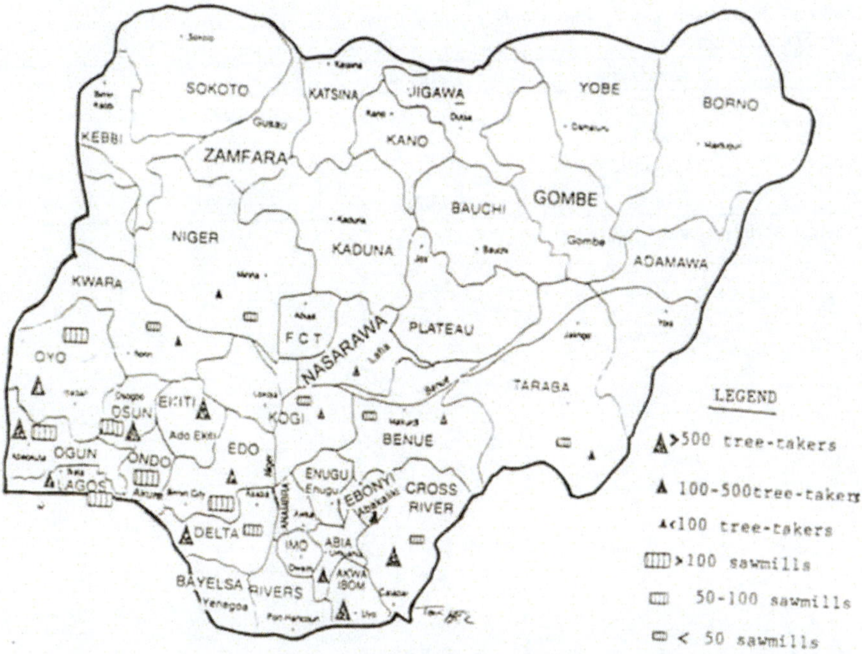

Fig. 12.2 Intensities of saw milling and tree cutting (Source: Abdullahi and Ajoku 2001)

management problems. The few mills still operating do so occasionally especially when they have large contracts to supply plywood. It is projected that the national requirements for plywood and veneer logs will increase in 2015 to about 453,000 m^3 and to about 858,000–1,359,000 m^3 respectively. However, the wood availability is unlikely to meet this demand as there are indications that wood availability will decrease to about 119,000 m^3 by 2015 from 170,000 m^3 recorded in 1990.

There are two principal categories of forest resources in Nigeria, namely the forests of the woodland and savanna regions which are the primary sources of fuel-wood and poles, whilst the rainforests of the southern humid zone supply domestic timber and lumber, with fuel-wood as by-products. Deforestation has steadily reduced the forest potential of the country. The deforestation rate has exceeded 3.6% in the recent past. There are about 1,160 forest reserves with a total area of about 10.8 million ha, whilst forest plantations are about 269,537 ha (Abdullahi and Ajoku 2001).

Fuel-wood accounts for over 90% of the total wood produced, whilst saw logs, poles, pulpwood and veneer logs account for about 4%, 2%, 0.3% and 0.4%, respectively. The demand and supply of wood types such as fuel-wood, saw logs and veneer logs for the period 2007–2010 is projected to decrease significantly (Table 12.4).

Table 12.4 Demand, supply and balance of wood types in Nigeria (2007–2010) (in thousand m^3 roundwood volume)

S/No	Wood type	Demand	Supply	Balance
1	Fuel-wood	88,138	63,099	–25,039
2	Sawn logs	10,205	2,480	–7,725
3	Veneer logs	1,078	114	–964
	Total	99,421	65,693	–33,728

Source: Abdullahi and Ajoku (2001)

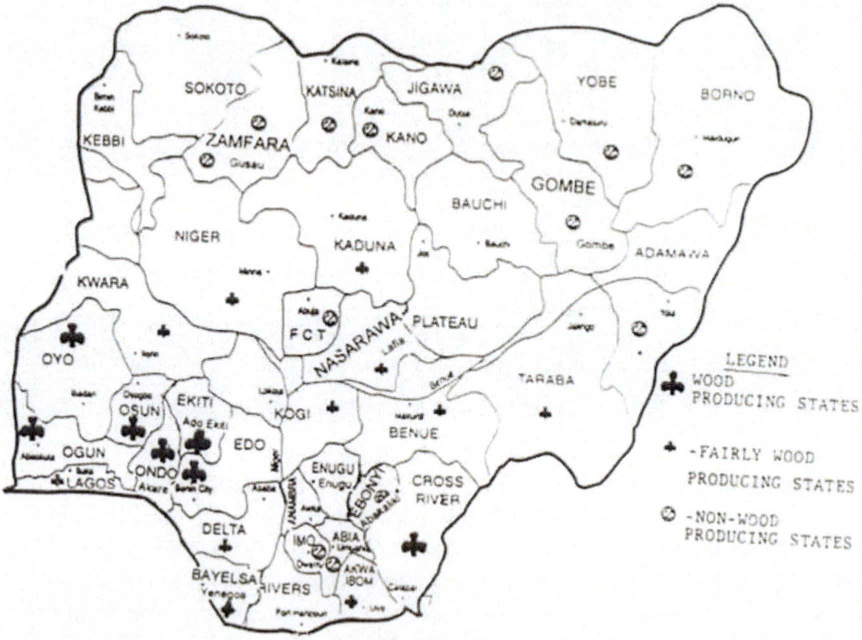

Fig. 12.3 Wood and non-wood processing states in Nigeria (Source: Abdullahi and Ajoku 2001)

It is projected that a deficit of 33.7 million m^3 of fuel-wood, saw logs and veneer logs will be recorded by 2010. The wood and non-wood producing states are illustrated in Fig. 12.3.

Saw logs production comes from the forest reserves in the states of Anambra, Edo, Delta, Kogi, Cross River, Ogun, Ondo, Ekiti, Osun and Oyo States. These states account for about 55% of saw logs. Seventy percent of the veneer log supply is sustained from the forest reserves in Cross River, Edo, Delta, Kogi and Ondo States.

Significant wood waste potential exists from the activities in various segments of the wood industry. However, the decreasing capacity utilization in the processing industry coupled with the shortfall in supply of round wood implies a future reduction

Table 12.5 Energy potential of wastes generated by clusters

S/No	Cluster location	Annual residues generated (t)	Energy potential (GJ)
1	Lagos	137,805	1,757,022
2	Oyo (Ibadan)	34,669	442,035
3	Osun (Oshogbo)	8,358	106,570
4	Ogun (Abeokuta)	4,974	63,429
5	Ondo (Akure)	31,047	395,850
6	Ekiti (Ado-Ekiti)	3,840	48,989
7	Edo (Benin)	2,585	32,966
8	Delta (Asaba/Sapele)	10,806	137,778

Source: Field Survey (Idi and Ajoku 2007)
Note: 1 MWh = 3.6 GJ

in the amount of wood wastes. Nevertheless, the existing processing industries, especially the small and medium wood processing enterprises, are generating significant volumes of wood wastes.

12.5.2 Energy Potential from Wood Waste

The major source of wood waste (especially sawdust) in Nigeria comes from sawmills. Large volumes of sawdust are generated in the states with high wood resources for saw logs and veneer logs. The states Lagos, Oyo, Osun, Ogun, Ondo, Ekiti and Delta have wood processing clusters with large potential to generate significant volumes of wood wastes or saw dust. The analysis of the saw mill clusters showing the annual wood wastes generated and energy potentials is presented in Table 12.5.

It is estimated that about 137,805 tons of sawdust are generated from Lagos State clusters annually. Business activities in this cluster include saw milling, planing, carpentry and furniture making as well as specialty products like photograph frames, mortars and pestles, carvings, etc. The clusters source their round wood from neighbouring states including Edo State. The peak operating season is from November to April, with an off season from June to August. The saw dust generated is dumped by the seaside and at times incinerated.

In Oyo State, it is estimated that about 34,669 tons of wood wastes are generated annually. There are two major clusters which include the Bodija Plank Market and the Alomaja, Saw Millers Association. The two clusters generate over 140 tons of waste daily and have the capacity to generate about 32,935 tons of sawdust annually. The Forestry Research Institute of Nigeria in Ibadan generates about 7.4 tons daily, even though it hardly operates continuously.

The above two locations (Lagos and Oyo) have large wood waste potential from existing sawmill clusters. However, smaller waste generating clusters include those found in Oshogbo, Osun State; Abeokuta, Ogun State; Akure, Ondo State; Sapele, Delta State; Ado-Ekiti, Ekiti State and Benin, Edo State.

In some sawmills, the waste generated is sold to entrepreneurs involved in the production of briquettes. In Sapele for instance, the sawdust is used in the conversion process of wood wastes to charcoal for markets in urban and semi-urban centres. This informal activity in the utilization of wastes presents constraints to energy projects relying on the use of wood wastes, unless the expected benefits outweighs the current economic benefits associated with the current uses. However, in areas with large waste production, there is a potential for conversion of wood wastes into energy, especially in Lagos and Ondo States.

Nigeria has substantial quantities of forest residues from the timber industry that could be utilized to generate renewable energy. Residues resulting from wood harvesting are currently often not utilized. However, they generate significant volumes of dry biomass. Some of these residues are currently utilized inefficiently for heating and cooking, but not for electricity.

Efficient wood-fired power plants should have a capacity of 25–40 MW capable of supplying electricity for 24 h per day as part of the base load system. To produce one MWh of electricity in a 30 MW power plant requires approximately 0.7 dry tons of wood waste (or 1.4 green tons of wood waste). According to the National Association of Forest Industries of Australia, this would be an annual supply of 320–360,000 tons of wood waste (NAFI 2005). A facility of this size can produce enough electricity to supply 30,000 homes in Nigeria assuming an average residential energy consumption of 7.4 MWh per year.

A considerable number of factors determine whether a renewable energy project is commercially viable:

- choice of conversion technology
- source of wood waste, moisture content and energy density
- transport distances for the wood waste
- plant scale
- economic value of co-products
- alternative disposal cost for the wood waste
- availability of subsidies and incentives for the energy products

It is imperative that the choice of conversion technology will depend on the costs of capital, the available volumes and costs of wood waste, access to energy markets, and the prices for renewable energy products.

12.6 Conclusion

The potential for renewable energy production from rice husks and wood wastes in Nigeria is significant. Lagos State was found to have the highest stock of sawdust generated by sawmills whilst Ebonyi and Benue states have the highest resources of rice husk waste. Sawmills and rice mills are scattered all over the country, but are well established especially in states with large wood resources and rice production. Several factors were identified that influence the development of wood waste and rice husks as energy resources:

- Availability of suitable technology for electricity production from wood wastes and rice husks
- Logistical challenges for the transportation of feedstock to the conversion plants
- Cost of infrastructure

Experience has helped to understand the production chain (from resources to markets) regarding the opportunities for producing electricity from biomass. However, a key need identified is the requirement for an agro-waste bioenergy roadmap that will link producers, processors, renewable energy technologies and renewable energy markets. Specific measures must be taken to create awareness and understanding of the options for using agro-wastes to produce renewable energy. The following recommendations are made:

- There is need to create awareness on the potential of renewable energy (RE) in Nigeria using agro-wastes as energy source in meeting rural energy needs.
- Agro processing industries need to be informed and sensitized on the markets currently available for RE production from agro-waste resources.
- Nigeria should seek for partnerships and investment opportunities to promote technologies for agro-waste based electricity generation.
- The government should put in place appropriate regulatory frameworks that promote the utilization of agro-waste as energy resource.
- In view of the large volume of agro-wastes generated in the country, efforts should be made to promote renewable energy projects.
- State governments should be advised to establish policy frameworks that encourage sawmill operators as well as rice millers to form clusters in order to take advantage of generating income from agro-waste energy projects.
- State governments should explore the prospects of generating electricity under their Independent Power Plants (IPPs) programme from agro-wastes to supplement domestic energy needs, especially in rural areas and semi-urban centres.

The transition from traditional use of biomass to modern use provides very large opportunities, but also challenges to African countries. There is no doubt that biomass residues, especially agro-wastes suitable for co-generation, can be found in abundance in most sub-Saharan African countries. Using these residues, allows agro-industries to generate power and heat from renewable energy resources in an environmentally sustainable way.

It also provides opportunities for increasing rural electrification, as well as reducing transmission and distribution losses. Furthermore, it reduces government involvement in electricity generation as well as provides additional income to rural farmers.

References

Abdullahi A, Ajoku KB (2001) Capacity building for sustainable industrial development: Nigerian perspective. Raw Materials research and development Council, Abuja

Adegbulugbe AO, Adenikinju A (2010) Country chapter: Nigeria. In Renewable energies in West Africa. Deutsche Gesellschaft fur Technische Zusammenarbeit (GTZ) GmbH, Germany

AFREPREN/FWD (2006) Potential for cogeneration in Africa (unpublished report), AFREPREN/FWD, Nairobi

Brenes MD (2006) Biomass and bioenergy new research. Nova, New York

Coelho ST, Walter SA (2003) Indigenous technologies for sustainable development. (Chapter 4). Brazil-Country study for sustainable development. Study under development funded by IAE

FAO (1999) Multifunctional character of agriculture and land: the energy function. Background paper 2: bioenergy. conference on the multifunctional character of agriculture and land. FAO/the Netherlands

FAO (2002) Statistics. Food and Agricultural Organization, Rome

Idi A, Ajoku KB (2007) Energy resource assessment of agricultural (Wood) wastes in Nigeria. Country study report under UNIDO funded project

Inter Academy Council (2010) Biomass. Biomass-Inter Academy Council. http://www.interacademycouncil.net/CMS/Reports/11840/11928/11943.asx?

Karekezi S, Kithyoma W (2005) Sustainable energy in Africa: cogeneration and geothermal in east and horn of Africa-Status and prospects. Nairobi, AFREPREN/FWD and Heinrich Boll Foundation Regional office for East and Horn of Africa (HBF)

Karekezi S, Kithyoma W, Oruta A (2008) Scaling-up bioenergy in Africa. Background paper for international conference on scaling-up renewables in Africa, organized by UNIDO from 16–18 Apr, Dakar

NAFI (2005) Cost consideration in using wood waste to produce renewable energy. www.nafi.com.au/bioenergy

National Association of Forest Industries (2005) Report 4 – Converting wood waste into renewable energy, National Association of Forest Industries, Canberra

Obassi E, Ajoku KB (2007) Energy resource assessment of agricultural (Rice husk) wastes in Nigeria. Country Study Report under UNIDO funded project

Onyekwelu JC, Akindele SO (2006) Biomass and bioenergy research in Tropical Africa: state of the art challenges and future direction. In: Brenes M (ed) Biomass and bioenergy new research. Nova Science Publishers, Inc, New York, USA

UNDP (2003) Human development report on Nigeria, United Nations Development Program (UNDP), Nigeria

WARDA (1996) Rice trends in sub-Saharan Africa: a synthesis of statistics on rice production, trade and consumption. Sayce Publishing, London

Woods J, Hall DO (1994) Bioenergy for development: technical and environmental dimensions, FAO Environment and Energy Paper 13, FAO, Rome

World Bank (2007) Nigeria competitiveness and growth: country economic memorandum. Main Report, No. 36483-NG, Nigeria

Chapter 13
International Trade of Biofuels: Current Trends and the Potential Role of Africa

Arnaldo Walter and Anna Segerstedt

Abstract Many countries have turned their attention to the production and consumption of biofuels over the last years. This interest is due to factors like the need to mitigate greenhouse gas (GHG) emissions, counteract surging and increasingly volatile oil prices, search for a more diverse energy matrix and the development of the agricultural sector. For developing countries, biofuels are also seen as an opportunity to reduce oil dependency, minimize foreign debt, improve living conditions in rural areas (including the access to basic energy services) and enhance economic activities. In particular, this is the case of many African countries. Despite the growing interest, large-scale biofuels production and consumption are concentrated to few countries/regions, such as the US and Brazil, in the case of ethanol, as well as European Union (EU), the US and Argentina, in the case of biodiesel. Due to the size of their potential markets, the US and EU will have a crucial impact on trade and may open for more export opportunities for producers in developing countries. However, currently both the US and EU have trade regimes based on tariffs that offset the comparative advantages of some producer countries. In addition, doubts have been raised regarding the real benefits of biofuels. Some criticize their efficiency in mitigating GHG emissions whilst others raise doubts about the pressures on food supply or the non-existence of socio-economic benefits for the population directly affected by biofuels production. Indeed, the effective sustainability of biofuels has been recognized as an essential aspect for their consolidation in the international markets and both in Europe and the US sustainability criteria have been defined. This chapter assesses the international trade of biofuels with particular focus on African countries, which have

A. Walter (✉)
University of Campinas (Unicamp) and Brazilian Bioethanol Science
and Technology Laboratory (CTBE), Mendeleyev, 200, Campinas 13083-860, Brazil
e-mail: awalter@fem.unicamp.br

A. Segerstedt
Institut für Umweltökonomik und Welthandel, Leibniz Universität Hannover,
Königsworther Platz 1, 30167 Hannover, Germany
e-mail: segerstedt@iuw.uni-hannover.de

R. Janssen and D. Rutz (eds.), *Bioenergy for Sustainable Development in Africa*,
DOI 10.1007/978-94-007-2181-4_13, © Springer Science+Business Media B.V. 2012

a reasonable potential for biofuels production, but may face difficulties in building up the required infrastructure and fulfil the sustainability criteria imposed by the main consumer markets. The main conclusion is that African engagement in an international trade of biofuels is far from being reality, even because so far the market is distorted by trade barriers. In addition, the conditions for large-scale production aiming at exporting are still precarious. On the other hand, in a hypothetical scenario of fair trade environment, the sustainable production of biofuels in Africa would be a unique opportunity for fostering economic and social development.

Keywords Biofuels trade • Sustainability • Africa • Trade barriers • Standards • Export • Local market

13.1 Introduction

Biofuels have been a focus of attention in recent years. On the one hand biofuels present a possibility to mitigate greenhouse gas (GHG) emissions, to diversify the energy matrix and to reduce oil dependency without a drastic change in the transport infrastructure. In addition, biofuels are seen as an opportunity to improve living conditions in rural areas and to allow a better income distribution. On the other hand, biofuels production has been blamed for food prices inflation and many doubts have been raised about the sustainability of its production and use.

So far, large-scale biofuels production and consumption are concentrated to a few countries/regions (US and Brazil, in case of ethanol; EU, US and Argentina, in case of biodiesel). Biofuels are still far from being commodities, firstly because of the production concentration in itself and secondly, because the main consumer markets – US and EU – have trade regimes that offset the comparative advantages of some producer countries. Furthermore, the acceptance of biofuels in these main consumer markets has been conditioned to the accomplishment of sustainability criteria. These requirements are unique in the energy market and have been more rigorous for biofuels than for many other commodities.

The aim of this chapter is to analyze the mid-term perspectives of biofuels international trade, taking into account the main driving forces for biofuels in industrialized countries and the required accomplishment of sustainability criteria. The focus of the analysis is African countries that have a reasonable potential for biofuels production, but may face difficulties in order to fulfil the required sustainability criteria imposed by the main consumer markets.

13.2 The Growth of Biofuels Production

Ethanol and biodiesel are the most important biofuels for the time being. The use of ethanol as transport fuel goes back to the origin of the automobile industry as, for instance, Henry Ford's Model T, built in 1908, used ethanol (Rosillo-Calle and

Table 13.1 Biofuels production from 2000 to 2009 in billion litres (BL)

	Fuel ethanol			Biodiesel	
Country/Region	2000[a]	2005[b]	2009[b]	2005[b]	2009[b]
USA	6.2	15.0	41.0	0.25	2.1
Brazil	10.7	15.0	26.0	–	1.6
EU		0.9	3.6	3.6	8.9
Others	0.7	2.1	5.4	0.05	4.0
Total	17.6	33.0	76.0	3.9	16.6

Sources: [a]Walter et al. (2008), [b]REN21 (2010)

Walter 2006). Currently world production and consumption is dominated by the US and Brazil, which are responsible for about 90% of the world production, with about 67 BL (billion litres) out of 76 BL produced in 2009 (see Table 13.1). On the other hand, also biodiesel has been used as substitute for mineral diesel since the beginning of the twentieth century, but in smaller quantities. From 2005 onwards biodiesel production increased significantly, spearheaded by the EU, which is currently responsible for more than 50% of the world production and about 65% of the world consumption. Table 13.1 shows bioethanol and biodiesel production in recent years.

Table 13.2 shows information on the main producers of fuel ethanol and biodiesel in 2009, and their annual average growth rates of production from 2005 to 2009. As can be seen, biodiesel production is less concentrated than ethanol. Given the very low levels of production in 2005, the growth rates of biodiesel has been faster. The average annual increase in biodiesel production indicates a rapid growth in the US, Brazil and Argentina, where the large share is based on soy, as well as in France and Spain (with the main raw materials being rapeseed and sunflower) (REN21 2010).

In relative terms, there has been rapid production growth of ethanol in Canada, France, Germany (mostly based on wheat) and in Thailand. In absolute terms, however, the bulk of the production growth has taken place in the US. As for Brazil, the country with the longest tradition in fuel ethanol production, the sector was negatively affected by the high sugarcane prices on the international market and the unfavourable weather conditions in 2009. Moreover, in this year the global economic crisis led to a considerable slow-down both in ethanol and biodiesel production in all the major producer countries (REN21 2010).

An important driver in the biofuels boom has been public policies; about 30 countries have already introduced or are interested in introducing programmes for biofuels. Their mandates vary from E2 to E25 (2–25% of ethanol blended with gasoline, volume basis) and from B2 to B8 (2–8% of biodiesel blended with mineral diesel, volume basis) (REN21 2010). Ethanol is also available as E85 blends in the US and Sweden, whilst pure hydrated ethanol is available at all Brazilian fuelling stations.

The production of bioethanol in 2009 (76 BL) was estimated at an equivalent of 4.8% of the gasoline consumption, considering a light distillates consumption of about 1,600 BL in the same year (mainly motor gasoline) (BP 2010). In 2005,

Table 13.2 Main producers of fuel ethanol and biodiesel and growth of the production from 2005 to 2009 in most important producer countries

	Bioethanol		Biodiesel	
Country	Share of production (%) (BP 2010)	Annual growth (% per year)	Share of production (%) (REN21 2006)	Annual growth (% per year)
US	52.9	28.8	12.7	70.2
Brazil	33.9	12.9	9.6	~600[a]
China	2.7	15.3	n.a.	n.a.
Canada	2.2	61.4	0.6	~0
France	1.4	63.9	15.7	44.3
Germany	1.0	46.5	15.7	8.2
Spain	0.6	11.2	3.6	56.5
Thailand	0.5	60.6	3.6	n.a.
Argentina			8.4	~200[b]
Others	4.8	37.5		52.3
Total		22.7		43.6

Sources: BP (2010), REN21 (2006, 2010)
n.a. information not available
[a]based on the production of 736 m^3 in 2005 (ANP 2010)
[b]Based on the production of 20 ML in 2005 (USDA 2009)

this figure was estimated at approximately 2.1%. In the case of biodiesel production, in 2009 its share was equivalent to 1% (16.6 BL) of the middle distillates production (1,800 BL) (BP 2010) (mainly diesel oil), whilst this figure was only 0.2% in 2005.

13.3 Biofuels Trade

The level of biomass energy trade is small compared to typical agriculture and forestry commodities (e.g. 19% for wheat and 22% for cellulose pulp in 2006) (Heinimö and Junginger 2009). In 2005, ethanol trade represented about 10% of world production and biodiesel trade was negligible (Walter et al. 2008). Based on estimations from non-consolidated figures (data from FO Licht 2010, except for Brazil where data from MAPA was used) the total volume of ethanol exported (7.8 BL; all grades ethanol, not just fuel ethanol) was close to 10% of the worldwide production of fuel ethanol (76 BL). The exports of only fuel ethanol in 2009 were preliminarily estimated as 7.8% of the world production (Szwarc 2010).

Despite a reduction of about 30% of the volume exported from 2008 to 2009, Brazil has kept the leading position as ethanol exporter (42%), followed by the Netherlands (15%), France (14%), the US (6.9%), Belgium (4.2%), Spain (3.4%) and Germany (3.4%) (MAPA 2010, FO Licht 2010). From these countries, only Brazil (3.3 BL in 2009) and France (690 ML in 2009) were large net exporters, whilst Germany (1 BL) was a large net importer (FO Licht 2010).

In case of biodiesel, the estimate shows that the total volume exported (5.3 BL) in 2009 represented about 32% of the world production (16.6 BL), with Europe as

Table 13.3 Trade figures of biofuels (ethanol and biodiesel) in 2009 – net result is the difference between imports and exports

	Imports	Exports	Consumption	Net result/consumption (%)
Ethanol (ML)				
US	1,101.9	542.5	42,315.1	1.3
EU	1,430.0	145.0	6,788.0	18.9
Brazil		3,323.5[a]	24,423.1[b]	−13.6
Biodiesel (kt)				
US	250.8	750.4	2,344.3	−21.3
EU	1,871.0	66.0	10,655.0	16.9

Sources: FO Licht (2010), except [a]MAPA (2010) and [b]BEN (2010)

main final destination[1]. Argentina was the main exporter (26% of the total volume exported), followed by the Netherlands (15%), the US (14%), Germany (13%), Belgium (11%), Indonesia (4%), Malaysia (4%) and Italy (2%). From these countries, Argentina, Indonesia and Malaysia have only exported, whilst the US (500 ML as balance) and Belgium (210 ML) were net exporters in 2009. The Netherlands, Germany and Italy were net importers.

Data of biofuels trade need to be used with caution due to many reasons, such as: (1) the different uses of ethanol, (2) the fact that biodiesel can be produced on site from imported vegetable oils (this, indeed, often happens), (3) the lack of proper codes for biofuels in trade data basis (Zarilli 2006), (4) biofuels have also been exported as blends, to overcome trade duties, and (5) some countries that have high export figures have, in fact, very small production (e.g. The Netherlands).

Table 13.3 shows 2009 imports, exports and the share of the resulting balance of biofuels trade compared with the domestic consumption of ethanol and biodiesel, in the US, EU and Brazil. It can be seen that a reasonable share of the biofuels consumption in the EU was covered by imports, whilst the opposite occurred in the US.

In 2010, the US became a net exporter of fuel ethanol and a reasonable share was traded with Europe. There is a set of reasons for this, including the growth of production in the US, the decreased demand for gasoline in the US (and, as consequence, the ethanol demand for blends), the cap of 10% ethanol that can be blended to gasoline, and the lower competitiveness of the Brazilian industry. An important driver is also the tax credit (Volumetric Ethanol Excise Tax Credit – VEETC) that companies receive for blending up to 90% ethanol to petrol even if fuel is shipped overseas. In addition, these blends are eligible for lower import duties than those applied in Europe for denatured and undenatured alcohol.

The negative impact was also perceived in Brazil, where ethanol exports fell dramatically in the last 2 years (from 4.8 BL in 2008 to about 1.8 BL in 2010). Besides the lower competitiveness due to the appreciation of the Brazilian currency against

[1] The estimate is based on FO Licht (2010) data for 20 countries (including main EU Member States and US), and on USDA (2009) for Argentina and Gardiner (2009) for Indonesia.

the US dollar and the raise of the international sugar prices, ethanol production was affected by adverse weather conditions in 2009. Yet, the main reason can be traced back to the rapid growth of domestic production in the US and the advantages given for exporting to Europe. The drop in ethanol exports from Brazil to the US, as well as the exports through Caribbean and Central American countries can be seen in Fig. 13.1 (from about 1.5 BL in 2008 to 250 ML in 2010 and from 1.2 BL to 150 ML, respectively). The large exports to Caribbean and Central American countries can be explained by the existence of specific trade agreements. CBI (Caribbean Basin Initiative) is a trade agreement which allows up to 7% of the US ethanol demand to be imported duty free even if the ethanol was originally produced outside the CBI countries (Jamaica, Trinidad, and Tobago have taken advantage of this agreement). The same occurs within the context of the USA Central America Free Trade Agreement (CAFTA), where Costa Rica and El Salvador have benefited from the duty free entry of a fixed volume of ethanol (Oosterveer and Mol 2010).

Similarly, the exports from Brazil to Europe were also reduced in the same period, from about 1.4 BL in 2008 to 380 ML in 2010. Traditionally, the main European markets for Brazilian ethanol were the Netherlands and Sweden, but straight flows to Sweden started to fall in 2005 and in case of the Netherlands a deep reduction occurred in 2009 and 2010. On the contrary, the imports to the UK have increased in recent years, but they are still relatively small (about 160 ML per year). Figure 13.2 shows the evolution of ethanol exports from Brazil to the Netherlands, Sweden and other EU Member States.

In Fig. 13.3 the evolution of ethanol exports from Brazil to CBI and CAFTA countries (mainly Jamaica, El Salvador, Costa Rica, Trinidad and Tobago and Virgin Islands) from 2004 to 2009 are presented, as well as the ethanol exports from these countries to the US in the same period. As it can be seen, a strong correlation exists between the two series.

Figure 13.4 shows the evolution of ethanol production in US from 2003 to 2009. The information is combined with imports as share of the total domestic production in the same period. It reveals that except for 2006, the percentage of imported fuel ethanol was 6%, or less. This may be explained by the US ethanol policies aiming at supporting local production that is more expensive than in some other countries (e.g. Brazil). According to data by the Energy Information Administration – Department of Energy (EIA-DOE 2010), since 2004 ethanol imports from Brazil and through the CBI and CAFTA agreements were never smaller than 90% of the total imports, but always targeting unsatisfied demand, rather than as a complement to domestic production.

To conclude, although growing, biofuels trade is still in a premature stage and the market has been distorted by trade regimes mainly imposed by the US and EU, besides unfair trade practices. Regarding ethanol, the US market – by far the largest – remains difficult to access for foreign ethanol producers, with an import tariff combined with the VEETC tax credit giving a rough 16 US$ cents/litre cost advantage to domestic US producers (REN21 2010). The EU also imposes a duty of 192 Euro/m^3 on undenatured alcohol (102 Euro/m^3 in the case of denatured alcohol), but as production costs are higher in Europe and there is a high level of unsatisfied demand, there has been more room for imports.

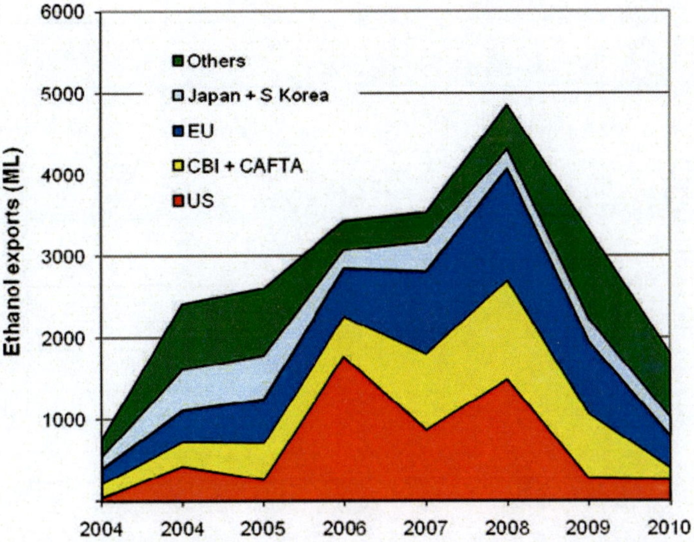

Fig. 13.1 Evolution of ethanol exports by Brazil from 2004 to 2010 (estimates for 2010) (Source: MAPA (2010))

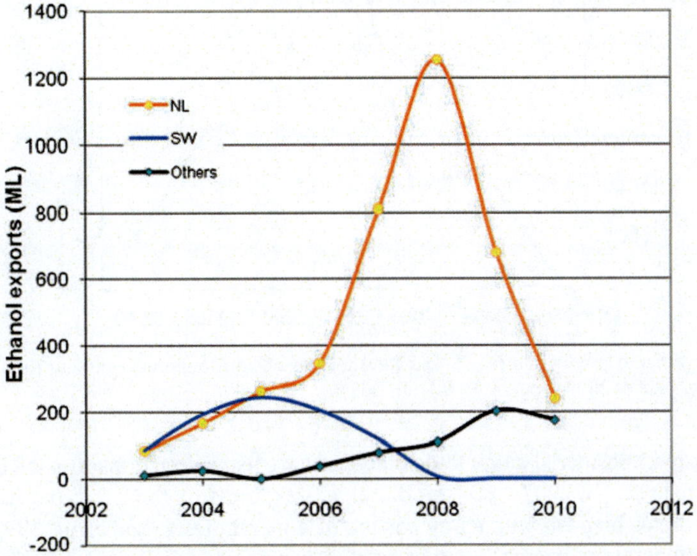

Fig. 13.2 Ethanol exports from Brazil to EU Member States (*SW* Sweden, *NL* The Netherlands) from 2003 to 2010 (estimates for 2010) (Source: MAPA 2010)

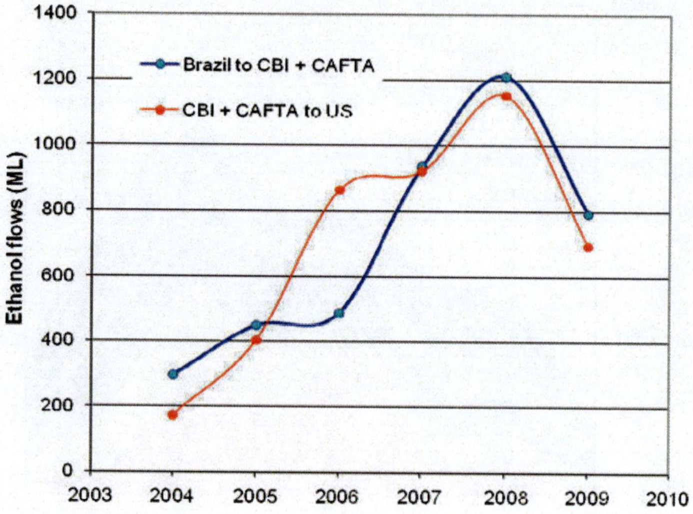

Fig. 13.3 Ethanol exports from Brazil to CBI and CAFTA countries and from these countries to US – 2004–2009 (Source: MAPA 2010 for Brazil and EIA-DOE 2010 for US)

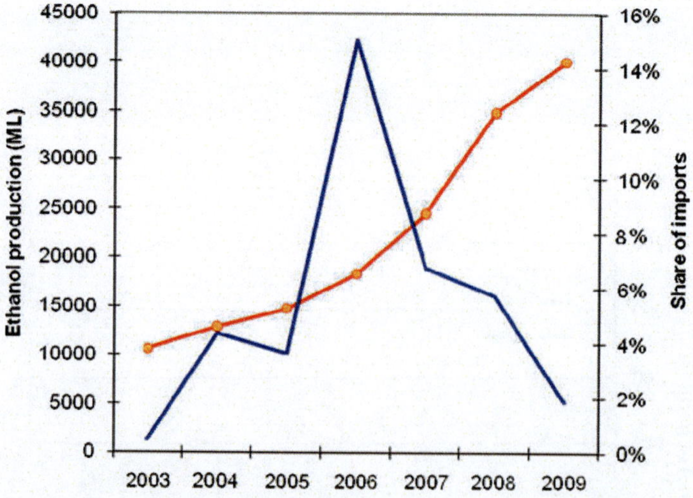

Fig. 13.4 Ethanol production in US and the share of ethanol imports regarding the domestic production – 2003–2009 (Source: EIA-DOE 2010)

In Europe, considered unfair trade practices have had a deep impact on the biofuels industry recently. In 2007–2008, biodiesel was exported by the US to the European Member States blended with a tiny portion of mineral diesel (known as B99). As in the US a federal tax exemption is granted to companies offering blends, European producers argued that the US was using unfair trade practices. As a result, in 2009, the European Commission imposed anti-dumping and countervailing measures.

However, circumvention practices were soon observed, as the US biodiesel was shipped via other countries (e.g. Canada) where the production and trade of blends were not covered by the EU duties. More recently, European producers have also blamed Argentina of taking advantage of differentiated export taxes that incentivize exports of biodiesel rather than crude soy oil (E-EnergyMarket 2010).

13.4 Overview of Africa

In order to give an overview of Africa and its current stage of development, in this section three issues are briefly addressed: firstly some differences between Africa and the world, secondly Africa's high dependence on imported oil, and thirdly the low access to energy services in African countries.

The African population amounts to 15% of the world, but its share regarding global development indicators, such as the gross domestic product (GDP), total primary energy supply (TPES), total electricity consumption and CO_2 emissions are much lower. Further, when comparing compound indicators such as energy intensity (e.g. energy per unit of GDP) and specific CO_2 emissions (e.g. CO_2 emissions per GDP) the low efficiency of African economy is highlighted. The differences between Africa and the world are even larger when the comparison is done excluding South Africa. Table 13.4 shows a summary of these indicators.

In 2009, Africa contributed 12% of the world oil production, but this production was highly concentrated (75%) to only four countries (Nigeria, Algeria, Angola and Libya) (BP 2010). Excluding them, in 2008 African countries imported 55% of their crude oil supply and 48% of their diesel oil supply (IEA 2010). Moreover, the dependence on imported fossil fuels is growing. Some countries are totally dependent of

Table 13.4 Africas main indicators (2008) and its comparison in the world; the second comparison was done excluding South Africa

Indicator	Data	Unit	Africa/world (%)	(Africa-SA)/world (%)
Population	984.3	Million	14.7	14.0
Gross domestic product (GDP)	876.2	Billion 2000 USD	2.2	1.7
GDP/capita	890	USD (2000)/capita	14.7	12.2
Total primary energy supply (TPES)	655.4	Million toe	5.3	4.2
TPES/GDP	0.75	toe/1,000 USD	250	251
Total electricity consumption	562.1	TWh	3.0	1.8
Electricity/capita	571	kWh/capita	20.5	12.7
CO_2 emissions	889.9	$MtCO_2$/year	3.0	1.9
CO_2/capita	0.9	tCO_2/capita	20.5	13.5
CO_2/GDP	1.02	$kgCO_2$/US$ (2000)	139.7	413.0

Source: IEA (2010)

imported oil, such as Kenya and Zambia, and others are totally dependent on imported diesel, such as Mozambique, Tanzania and Zimbabwe (IEA 2010). In all Sub-Saharan countries the prices of gasoline and diesel to consumers are amongst the highest in the world (GTZ 2009).

Most Africans live in rural areas and a large share of the population lives below the poverty line of 1 US$/day. In general, people living in rural areas have constrained access to energy services and both agricultural productivity and life conditions are deeply impacted by the lack of clean/potable water and limited energy supply (both electricity and liquid/gaseous fuels).

13.5 Biofuels Potential in Africa and the Challenge of Export

It has been suggested that African countries have the potential to become important producers and exporters of biomass energy, and that the bulk of the production can occur in abandoned and marginal land (Hoogwijk et al. 2009). In some countries the conditions for bioenergy production are favourable due to the suitable climate and the availability of arable land and water resources. For instance, in the context of the COMPETE project (Competence Platform for Bioenergy in Arid and Semi-Arid Ecosystems in Africa[2]), about 180 Mha land were evaluated as potentially available and suitable for bioenergy production. This land availability was identified in semi-arid and arid regions in eight countries (South Africa, Kenya, Botswana, Mali, Tanzania, Zambia, Burkina Faso and Senegal) (COMPETE 2009).

As mentioned, some African countries are highly dependent on imported fossil fuels and the access to energy services is precarious, mainly in rural areas. Here an important question is if biofuels would be able to help African countries to overcome these difficulties. And what should be the priority: small-scale production to meet local demands and improve life conditions or large-scale production for export?

Obviously there is no single answer to these questions, and priorities and strategies differ from country to country. Those who support the production aiming at the international market usually claim it is necessary to generate jobs and income and a common argument is that small-scale agricultural activities are not able to sufficiently foster the economy. But what are the opportunities for African countries in this new international market?

It was previously shown that international biofuels trade is still in a preliminary stage and that markets have been distorted by trade barriers and unfair trade practices. The main consumer markets are in the US and EU, where the main aims are to improve security of energy supply and to preserve the interests of the agricultural sector. The US market is much larger than the European, but has been less open to imports. Few countries have exported biofuels to the US (none of them from Africa) and trade priority focus is on countries with trade agreements. In Europe, the share

[2] For more information, see www.compete-bioafrica.net

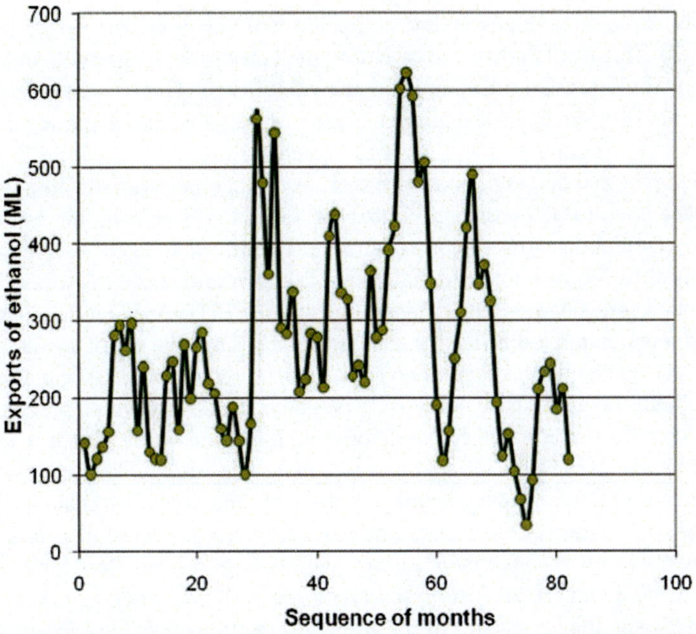

Fig. 13.5 Exports of fuel ethanol by Brazil from January 2004 to November 2010 (Source: MDIC 2010)

of imports is larger and the list of supplier countries is more diversified, but again the participation of African countries has been very small. For example, in 2009 five countries exported fuel ethanol to Europe, totaling less than 70 ML, (i.e. about 1.5% of the total volume imported by EU countries) (FO Licht 2010). By contrast, in the same year total exports of ethanol to African states were close to 280 ML, with Nigeria being the main destination and Brazil the main exporter (FO Licht 2010).

Many African countries benefit from preferential trade agreements (in particular the Generalized System of Preferences, Everything but Arms, and the Cotonou Agreement) giving them a better access to EU and US markets (Mitchell 2011). Potentially this could improve their position on the world market, but the problem of high production costs and domestic subsidies to farmers in the importing countries remains. For example, as it was previously shown, Brazil has faced difficulties competing on the international market in spite of its position as the country with the lowest ethanol production costs. Accordingly, new, less competitive producers are likely to face even more difficulties.

Another important issue is the market volatility illustrated here by the volumes of ethanol exported by Brazil from January 2004 to November 2010 (see Fig. 13.5). Differently from a new producer and exporter country, in case of Brazil the income of ethanol exports is less important as the industry in Brazil produces sugar and ethanol, both for the domestic and the international market, and some mills are also selling surplus electricity. In addition, the impact of this volatility is lower as the amount of ethanol exported is relatively low compared with the domestic consumption (see Table 13.3). This would not be the case of a small producer country with focus on the international market. On the other hand, the volatility of prices has been lower.

The volatility on the international market shows the risk a producer country would face if its domestic market is not large enough for compensating strong fluctuations of external demand, or in case there is no income diversification.

Another issue to be considered is that the market will demand the fulfilment of technical standards, and this can be a challenge. The first point to be observed is that so far there is not a common international standard, and in some cases this aspect has been pointed out as trade barrier. For instance, recently the EU defined biodiesel standards, including a maximum iodine level which in practice limits the use of soy oil and (to a lesser extent) palm oil (Oosterveer and Mol 2010). A second point to be highlighted is that developing countries have not actively participated in this discussion, and probably will be forced to accept further definitions. A third point is the doubt whether the imposed standards can be achieved by new producers at reasonable cost.

More importantly, proper planning is the first and foremost challenge for the large-scale production of biofuels, as the biomass sector has far more actors than the conventional and well-established energy sectors (e.g. oil and electricity). In addition, the development of biofuels system requires a reliable set of suppliers of goods and services and the necessities of the whole production chain have to be identified upstream. These aspects show the importance of coordinated actions before and during the implementation of a large-scale biofuels industry. Here the example of Brazil should be seen with reservation, as the sugarcane industry was already well established when large-scale production of fuel ethanol started in the second half of the 1970s. Nevertheless, even with these favourable conditions the role of the Brazilian Government at that time, coordinating and assuming some responsibilities, was crucial.

Related to this, the poor infrastructure in most African countries is a potential barrier for investments aiming at large-scale production of biofuels. Adequate roads and ports are essential, and the location of the producing mills in relation to the embarking points is an issue that needs careful consideration. In Brazil, the owners of the new sugarcane mills located in the central part of the country have been investing in pipelines to reduce logistics costs and GHG emissions. Yet, such investment is capital intensive and has been challenging for the producers.

Finally, producing countries with ambition to export will have to pay increasing attention to the discussion on sustainability criteria and certification. As they are expected to become more and more important in the future, it is essential that developing countries are participating actively in the process. Currently, there are a number of initiatives, both public and private. However, given the size of the markets, their political power and the stage of development, the mandatory initiatives in EU (Renewable Energy Directive – RED) and the US (Renewable Fuel Standard 2 at the federal level and the Low Carbon Fuel Standard in California) are likely to set the benchmark. In any case, independent certification will be required in order to prove the compliance with minimum criteria.

In all existing initiatives, life-cycle GHG emissions play a vital role, and minimum reduction targets in comparison to fossil fuels have been defined. There are still controversial aspects, such as the inclusion of emissions related to indirect impacts of land use change (ILUC). Despite the lack of solid scientific knowledge regarding this issue, it has already been decided in the US that ILUC impacts should be taken into account in the GHG balances of biofuels, and the same issue is under discussion in the EU.

In general, discussed and implemented sustainability criteria go further in Europe than in the US. For example, the potential impacts on biodiversity, water resources and food supply, as well as the respect of human, labour and land rights have been addressed in the EU Renewable Energy Directive.

Such mandatory initiatives have sometimes been blamed to work as trade barriers, aiming at constraining the production of biofuels in developing countries and, as consequence, preserving the interests of local producers in industrialized countries. Indeed, many developing nations therefore view attempts to introduce sustainability criteria as a form of "green imperialism" (Smeets and Faaij 2009). However, these initiatives can also be driven by honest concerns that large-scale production of biofuels only makes sense if their negative impacts are minimized.

Regarding sustainability requirements the following aspects need to be addressed: (1) exporting countries seeking access to the European and Northern American markets sooner or later will have to pay attention to these requirements, (2) for developing countries, compliance with sustainability criteria will require even more planning, coordinated actions, international cooperation, data gathering, capacity building and investments, (3) small-scale production for export is likely to face much more difficulties, because the costs (e.g. of certification) will be proportionally higher and also because it will be more difficult to meet some of the standards producing on a smaller scale (e.g. low GHG emissions), (4) developing countries should be actively engaged in the definition processes of sustainability criteria and standards, in order to preserve their interests, and (5) despite all good purposes, such initiatives pose the threat to establish new trade barriers.

13.6 What Options are there for Africa?

Considering the aforementioned bottlenecks and that the bulk of the population lives in rural areas, it seems that the first priority for Africa should be the production of biofuels on a small-scale focusing on domestic demand. Such a strategy could improve the living conditions of farmers and reduce some of the problems connected with industrial biofuels production. For example, Robertson and Pinstrup-Andersen (2010) point to the land tenure controversy where investors from wealthy countries have been accused of "neo-colonialism" as they are acquiring vast tracts of land in poorer nations. Whilst countries like South Africa, Mozambique and Angola have approved policies for biofuels investment, in many other African economies national biofuels strategies are still lacking (Lerner et al. 2010). Such strategies however are imperative to monitor and judge the biofuels sector. A slower approach aiming at capacity building for small production units first may be easier to implement. Outgrower schemes could be another possibility to combine small-scale farming with large-scale processing units. A number of pilot projects are currently undertaken with different feedstock (e.g. jatropha, sugarcane, palm oil) and some of them seem to be promising whilst others are struggling with the economic feasibility. For jatropha in Tanzania, for instance, it is expected that this kind of business model will remain the most important in the future, but many initiatives have faced difficulties in terms of unstable supply in combination with low yields (Caniëls and Romijn 2010). In general, it is too early to draw any definite conclusions.

In the early stages of biofuels production a logical alternative is to take advantage of existing knowledge and of the infrastructure already available. For example, in countries with tradition in sugarcane production (South Africa, Sudan, Tanzania, Zambia and Mozambique are amongst the major producer countries in Africa) the infrastructure and the sugarcane industry is more organized. Ethanol could be produced from molasses, as is the case of Malawi, where 30 ML of ethanol per year is produced and blended with gasoline, reducing fuel imports (Jumbe et al. 2009).

In addition to this, a very important issue is that ethanol from sugarcane is still the best option amongst first generation biofuels in terms of efficiency. Considering the high manufacturing costs of ethanol from corn and wheat, production in some African countries could be feasible and in accordance with international sustainability. Mozambique, for example, has decided to use its capacity and existing know-how to expand its sugarcane area for large-scale ethanol production, including exports. Two bigger projects are currently under implementation with a predicted capacity of 430 million litres per year (Janssen et al. 2009). Nevertheless, in view of the high water requirements, potential impacts on water resources should be carefully evaluated, especially where intensive irrigation is required.

To conclude, it seems more adequate for the majority of African economies to focus first on local markets rather than on exports, starting with blending at the regional level. This is in line with the conclusions of the COMPETE project, which highlight that priority should be given to small-scale projects aiming at rural electrification, water pumping and assuring the availability of transport fuels in agriculture. Nevertheless, in a second step, with improved conditions, large-scale production for blending at national level and also for export should be considered in order to capture the continent's potential in terms of land, climate and labour resources. If not, African countries run the risk of becoming single producers (and exporters) of untreated lignocellulosic biomass, at low cost, for the production of second-generation biofuels in industrialized countries. To do this, capacity needs to be developed and the production should be sustainable according to international standards.

13.7 Conclusions

Biofuels production and trade are growing rapidly, but the contribution of Africa has been limited and this situation is expected to remain for some years. Despite efforts of some African countries, the short-term opportunities for biofuels trade are constrained due to their small production, lack of infrastructure and not least existing distortions on the international market. In this sense, industrialized countries have not been pro-active in encouraging trade. In addition, recent initiatives aiming at minimum sustainability requirements of biofuels production are still working as barriers for the production in Africa rather than as useful guidelines due to (1) the lack of appropriate data and limited human skills, (2) less competitive small-scale production, and (3) exclusion of African countries from the decision process.

The potential of biofuels production in Africa is promising and production can occur with low negative impacts on the environment and on food supply, and with positive socio-economic impacts. Yet, it is a long process to exploit these opportunities.

To conclude, the route that most African countries should follow is to first foster small-scale production. The expected benefits are the improvement of living standards, reinforcement of organizational skills in small communities and the dissemination of knowledge from successful cases.

On the other hand, all African countries with potential for large-scale biofuels production should be prepared for a new context. In view of the biofuels targets in the northern hemisphere international trade will be crucial. Sustainable production of biofuels, respecting the limits imposed by the nature and taking the people's welfare as first priority, will be a unique opportunity for fostering economic development in Africa. This is the challenge to be faced.

References

ANP – The Brazilian National Agency for Oil, Natural Gas and Biofuels (2010). Information available at www.anp.gov.br

BEN – Balanço Energetico Nacional (2010). Ministerio das Minas e Energia. Brasilia

BP (2010) Statistical review of world energy 2010. Information available at www.bp.com

Caniëls MCJ, Romijn H (2010) The Jatropha biofuels sector in Tanzania 2005–9: evolution towards sustainability? Eindhoven Centre for Innovation Studies (ECIS). http://cms.tm.tue.nl/Ecis/Files/papers/wp2010/wp1004.pdf

COMPETE (2009) Competence platform for bioenergy in arid and semi-arid ecosystems in Africa, reports of the project. Available at www.compete-bioafrica.net

EIA-DOE (2010) U.S. Energy Information Administration/Department of Energy. Information available at www.eia.doe.gov

E-EnergyMarket (2010). Information available at www.e-energymarket.com

FO Licht (2010) World ethanol & biofuels report, vol. 8(18)

Gardiner N (2009) Chemical forecaster 3Q09. Drewry Publishing; Drewry Shipping Consultants Ltd, London

GTZ – Deutsche Gesellschaft für Technische Zusammenarbeit (2009) International fuel prices 2009. Available at www.gtz.de/en/themen/29957.htm

Heinimö J, Junginger M (2009) Production and trading of biomass for energy – an overview of the global status. Biomass Bioenerg 33(9):1310–1320

Hoogwijk M, Faaij A, de Vries B, Turkenburg W (2009) Exploration of regional and global cost – supply curves of biomass energy from short-rotation crops at abandoned cropland and rest land under four IPCC SRES land-use scenarios. Biomass Bioenerg 33:26–43

IEA – International Energy Agency (2010). Information available at www.iea.org

Janssen R, Rutz D, Helm P, Woods J, Chavez RD (2009) Bioenergy for sustainable development in Africa – environmental and social aspects. Paper available at www.unido.org

Jumbe CBL, Msiska FBM, Madjera M (2009) Biofuels development in Sub-Saharan Africa: are the policies conducive? Energy Policy 37:4980–4986

Lerner A, Matupa O, Mothlathledi F, Stiles G, Brown R (2010) SADC biofuels state of play study. An assessment of the Biofuel Sector Development in the Southern African Development Community. SADC Biofuel Taskforce. http://www.probec.org/fileuploads/fl11092010040316_SADC_BIOFUELS_STATE_OF_PLAY_STUDY.pdf

MAPA (2010) Ministério da Agricultura e Pecuária. Information available at www.agricultura.gov.br

MDIC (2010) Ministério do Desenvolvimento, Indústria e Comércio Exterior. Information available at www.mdic.gov.br (exports of commodities)

Mitchell D (2011) Biofuels in Africa. Opportunities, prospects, and challenges. The International Bank for Reconstruction and Development/The World Bank, Washington

Oosterveer P, Mol APJ (2010) Biofuels, trade and sustainability: a review of perspectives for developing countries. Biofuels, Bioproducts Biorefining 4:66–76

REN21 – Renewable Energy Policy Network for the 21st Century (2006) Renewable 2006 – global status report, Washington, DC

REN21 – Renewable Energy Policy Network for the 21st Century (2010) Renewable 2010 – global status report

Robertson B, Pinstrup-Andersen P (2010) Global land acquisition: neo-colonialism or development opportunity? Food Secur 2:271–283

Rosillo-Calle F, Walter A (2006) Global market for bioethanol: historical trends and future prospects. Energ Sustain Dev X(1):18–30

Smeets E, Faaij A (2009) Synthesis report on international trade opportunities in Africa. Third periodic activity report; Annex 5-3-1. Available at www.compete-bioafrica.net

Szwarc A (2010) Bioethanol: production, use and trade. Presentation at the 33rd IAEE Conference, Rio de Janeiro, June 2010

USDA – United States Department of Agriculture (2009) Gain Report – global agricultural information network. Argentina Biofuels Annual, Report number AR9018

Walter A, Rosillo-Calle F, Dolzan P, Piacente E, Cunha KB (2008) Perspectives on fuel ethanol consumption and trade. Biomass Bioenerg 32:730–748

Zarilli S (2006) The emerging biofuels market: regulatory, trade and development implications. In: UNCTAD—United Nations conference on trade and development, Geneva

Part III
Biomass Policies

Chapter 14
Keynote Introduction: Overview on Bioenergy Policies in Africa

Rainer Janssen and Dominik Rutz

Abstract Primary policy drivers for bioenergy promotion in Africa include security of energy supply, a reduction of the foreign exchange burden of oil importing countries, as well as environmental benefits such as the restoration of degraded land, reduced land abandonment, and the mitigation of greenhouse gas emissions. Furthermore, the development of modern bioenergy systems offers opportunities to diversify agricultural production and to stimulate socio-economic development. On the other hand, concerns exist that bioenergy expansion in African countries may have severe negative socio-economic and environmental impacts. In order to minimise risks and maximise benefits, in recent years several African countries have launched initiatives to establish sound policy frameworks for bioenergy in order to ensure environmentally, economically and socially sustainable production, promotion and use of bioenergy. Nevertheless, fully functional legal and regulatory frameworks have not yet been established in African countries. The most advanced regulatory frameworks for bioenergy exist in South Africa and Mozambique with the Biofuels Industrial Strategy of the Republic of South Africa (enacted in 2007) and the National Biofuels Policy and Strategy (NBPS) published in May 2009 by the Government of Mozambique. Specific activities in the field of bioenergy sustainability certification as an essential component of the regulation of the bioenergy sector are performed in Mali, Mozambique, Tanzania, and on regional level by the Southern Africa Development Community (SADC).

Keywords Africa • Bioenergy policies • Strategies • Regulatory frameworks • Institutional structures • Market regulation • Biofuel blending • Sustainability certification • Agro-ecological zoning

R. Janssen (✉) • D. Rutz
WIP Renewable Energies, Sylvensteinstr. 2, 81369, Munich, Germany
e-mail: rainer.janssen@wip-munich.de; dominik.rutz@wip-munich.de

R. Janssen and D. Rutz (eds.), *Bioenergy for Sustainable Development in Africa*,
DOI 10.1007/978-94-007-2181-4_14,

14.1 Introduction

Primary policy drivers for bioenergy promotion in Africa include security of energy supply, a reduction of the foreign exchange burden of oil importing countries, as well as environmental benefits such as the restoration of degraded land, reduced land abandonment, and the mitigation of greenhouse gas (GHG) emissions. Furthermore, the development of modern bioenergy systems offers opportunities for investment and infrastructure improvements in agriculture with the promise to diversify agricultural production and to create additional employment and thus to stimulate socio-economic development (Janssen et al. 2009).

On the other hand, concerns exist that bioenergy expansion in African countries may have severe negative impacts on biodiversity and the use of natural resources through increasing competition over land and water resources. Rising prices of agricultural commodities may negatively affect food security of the poor in developing countries and the implementation of large-scale bioenergy projects may cause negative social impacts such as conflicts over land ownership and displacement of rural communities.

In order to minimise the risks of negative impacts and to maximise the benefits in the immediate and long-term, sound legal and regulatory frameworks for bioenergy are needed to ensure environmentally, economically and socially sustainable production, promotion and use of bioenergy. Such frameworks need to be closely linked with measures aiming at environmental protection as well as economic and social development, such as the establishment of safety nets to protect the world's poorest and most vulnerable people to ensure their access to adequate food (FAO 2008).

This chapter provides an overview of existing initiatives for the formulation of regulatory frameworks for bioenergy on the African continent. After a brief introduction to key elements of appropriate regulatory frameworks, current activities on national, regional, and African Union level are presented.

14.2 Key Elements of Bioenergy Regulatory Frameworks

Regulatory tools for the promotion of bioenergy usually consist of a combination of policies and legislation including measures to encourage private investment in bioenergy industries and financial assistance to public or private investors from national, bilateral or multilateral sources (FAO 2007).

The key elements of national regulatory frameworks for bioenergy include legislation establishing institutional structures, regulating the biofuels market, creating incentives, regulating trade, introducing sustainability certification schemes, and fostering research and development.

With respect to the establishment of ***institutional structures***, national bioenergy legislation typically designates a state agency to be responsible for promoting the necessary investments and steering national bioenergy programmes. This may include technical committees responsible for setting of standards. With respect to newly created entities in African countries it is important that these agencies have sufficient technical capacities. Furthermore, efficient coordination mechanisms with other relevant state agencies need to be put in place to ensure that bioenergy policies and legislation are effectively implemented and regulation is consistent with international commitments and other government policies. Finally, broad cooperation with the civil society is necessary to facilitate widest participation and transparency in decision making.

Bioenergy policies and legislation also often contain provisions on ***market regulation*** such as biofuel blending requirements (i.e. mandated percentage of biofuels to be mixed with conventional fuels), and fixed prices for bioenergy services (including biofuels). The regulation of blending requirements, however, should take into consideration local and national market conditions in African countries.

Incentives are essential components of regulatory measures to encourage the production, use and trade of bioenergy. Such incentives may include exemptions from value added taxes, corporate taxes, and excise taxes. Furthermore, government financial institutions are required to provide financial services and benefits to local companies engaged in the bioenergy sector, and incentive schemes should focus on the provision of micro-credit facilities or low-interest loans and loan guarantees to farmers for the cultivation of crops or to build their own processing facilities. Incentives in African countries, however, should be carefully selected to stimulate investment without creating negative socio-economic (e.g. reduced labour standards, displacement of rural communities) and environmental (e.g. deforestation, soil degradation, water depletion) impacts.

Trade regulations may include import tariffs as well as mandates for bioenergy producers to buy feedstock from local farmers. Even though such regulations may distort trade, they can serve to safeguard the creation of local and national bioenergy markets and contribute to local value creation and rural development.

Sustainability certification is an important tool to ensure bioenergy development with maximised benefits and minimised negative environmental and socio-economic impacts. Since the 1990s a variety of sustainability standards and certification schemes have been developed (mainly in the EU) for the production, processing and trade of biomass and agricultural products. More recently, sustainability schemes are introduced which specifically address the production and use of (liquid) biofuels (Froger et al. 2010; Ecofys 2009; BTG 2008).

Within the European Union the development of the biofuels sector as well as exports of biofuels into Europe are mainly governed by the Renewable Energy Directive (RED 2009) which came into force in June 2009. The RED sets a target for all Member States to achieve a minimum of 10% renewable energy consumption in transport by 2020. Biofuels that count towards the target will have to meet sustainability requirements, defined in Article 17 of the RED.

In order to proof compliance with sustainability requirements, regulatory frameworks in African countries may use existing sustainability schemes such as the International Sustainability and Carbon Certification System (ISCC) originating from Germany, the Dutch sustainability scheme (NTA 8080), the RSB (Roundtable on Sustainable Biofuels) sustainability scheme, and the RSPO (Roundtable on

Sustainable Palm Oil) sustainability scheme for sustainable palm oil production. Alternatively, African countries may develop national sustainability schemes with specific reference to national framework conditions and development priorities.

Bioenergy regulatory frameworks often also contain provisions for ***research and development*** such as the obligation for an assignment of increased national funds. Enhanced efforts in bioenergy research and development are regarded as essential for the development of strong and sustainable bioenergy sectors in African countries.

14.3 National Bioenergy Regulatory Frameworks in African Countries

Currently, several African governments are in the process of developing bioenergy regulatory frameworks with the aim to promote a truly sustainable development of the bioenergy sector. The following sections provide an overview of on-going initiatives in selected African countries.

14.3.1 Bioenergy Policies in Benin

In 2007, a feasibility study was carried out on behalf of the Ministry of Mines, Energy and Water of the Democratic Republic of Benin to elaborate a strategy for the production and use of biofuels (Cocchi et al. 2009). This draft Biofuel Promotion Strategy states the general vision to develop biofuel production and supply chains as drivers for economic growth and poverty reduction with positive effects on food crops and the environment. The main objectives of biofuels development in Benin are to contribute to the revival of the agricultural sector, to improve the trade balance, to increase farmers' income, and to reduce pressure on forestry resources.

In June 2008, important steps towards the development of a national regulatory framework for biofuels were taken through the adoption of two decrees, namely Decree 360/2008 "Nomination of a National Commission for the Promotion of Biofuels" and Decree 361/2008 "General conditions for the installation of biofuel companies in Benin".

Decree 360 defines the general conditions for biofuel production in Benin aiming at maximum benefits and minimum negative impacts in the country. The export of un-processed agricultural feedstock is prohibited and the government reserves the right to identify suitable feedstock crops. Priority will be given to small-scale feedstock production by local farmers, preferably in a contract farming model. Furthermore, production, processing and trade of feedstock and biofuels in Benin are subject to authorisation by the government upon approval by the National Commission for the Promotion of Biofuels.

In the coming years, the National Commission will be responsible for the elaboration of a National Biofuel Action Plan guiding towards the establishment of a regulatory framework for the biofuels sector in Benin.

14.3.2 Bioenergy Policies in Ghana

In recent years Ghana has gained considerable experience in bioenergy technologies, namely improved cookstoves, improved charcoal production technologies, briquetting, co-generation using sawmill and palm residues, biogas using municipal/farm waste and animal dung, gasification, and biodiesel based on jatropha, oil palm, soya bean oil, and coconut oil (Ahiataku-Togobo 2009).

The main objectives for bioenergy development in Ghana include to increase access to modern energy services, to promote the use of improved cookstoves and charcoal production technologies, to support sustained regeneration of woody biomass resources through legislation and fiscal incentives, to support development of indigenous alternative transportation fuel industry based on bioenergy resources (biofuels) to replace petroleum-based fuels, and to enact legislation that will create demand for biofuels including appropriate pricing of biofuels.

The short to medium term policy actions in Ghana focus on the development of a Renewable Energy (RE) Bill to create a RE-friendly regulatory environment, to promote innovative market delivery models, to establish favourable pricing policies for RE, and to create awareness on the benefits of renewable energies including bioenergy and biofuels. Thereby, the target for this policy development is to stimulate private sector participation and to increase energy access.

In order to avoid negative impacts of bioenergy development on the local population, the Government of Ghana is taking measures to ensure that large-scale production of biofuels creates commercial benefits for the population and does not affect food production. This includes flexibility of replanting farmland for food production if food security is threatened and the allocation of degraded and arid dry lands for jatropha production. Additionally, feedstock other than jatropha is promoted such as oil palm, groundnut, cassava or other energy crops with additional economic value and existing cultivation experience in Africa.

In September 2009, the Ghanaian Ministry of Energy published a Draft Renewable Energy Bill (Government of Ghana 2009) aimed to provide an institutional and regulatory framework for the promotion, development of renewable energy for the generation and supply of electricity. This legislation will include a feed-in tariff scheme with rates (which still need to be defined) guaranteed for 20 years, purchase obligations for electricity utilities, fiscal incentives, as well as the establishment of a Renewable Energy Fund. The Renewable Energy Bill aiming at increasing renewable energies to 10% in the national energy mix was discussed by the Ghanaian cabinet in September 2010 and it is expected to be passed into law in 2011.

14.3.3 Bioenergy Policies in Kenya

Recent and on-going initiatives to develop a Kenyan biofuel policy are guided by the vision to increase access to energy through sustainable biofuel production and to reduce the import of fossil fuels by 25% by 2030. The main mission of the policy is to explore agro-energetic resources to stimulate energy diversification and to contribute to social and economic development, especially in rural areas.

The first steps of the policy development process in Kenya focused on the elaboration of a biodiesel strategy, followed by a bioethanol strategy with the final aim of establishing a National Biofuels Policy.

In 2006 the Kenyan Ministry of Energy (MoE) set up the National Biofuels Committee (NBC) to coordinate activities of stakeholders in the biofuels sector. In May 2008, a draft Biodiesel Strategy was published by NBC for comments by stakeholders, and in August 2008 the Kenya Biodiesel Association (KBDA) was registered as representation of all actors in the biodiesel value chain. By June 2009, a draft Bioethanol Strategy was finalized highlighting Kenya's general capacity to produce sufficient ethanol fuel from molasses to blend 10% ethanol in conventional gasoline.

In April 2009, the National Biofuel Policy sub-committee was formed within NBC to develop a combined biofuel policy including biodiesel, bioethanol, biogas, and solid biomass. This policy needs to be aligned with the vision and mission of the National Energy Policy (Sessional Paper No. 4 of 2004) and the Energy Act 2006 (Muok et al. 2008).

During 2010, extensive stakeholder consultations were held for the finalization of the Kenyan biofuel policy. On-going land mapping and zoning initiatives need to be refined to identify suitable areas for different feedstocks, and to allocate areas for food production and high conservation areas as unavailable for biofuels production (Davison 2010).

14.3.4 Bioenergy Policies in Mali

During recent years the Government of Mali as well as the civil society in Mali has shown considerable interest in the development of a strong and sustainable bioenergy sector. This development is embedded in several policy documents of the Government of Mali such as the Poverty Reduction Strategy with the three main objectives for the period 2007–2011 to develop infrastructures and the productive sector, to consolidate structural reforms, and to strengthen the social sector with respect to education, health, and water access.

The achievement of these objectives addresses several priority areas of which three are closely interlinked with the development of a sustainable bioenergy sector, namely food security and rural development, development of small and medium size enterprises, and the protection and sustainable management of natural resources.

In 2006, the National Strategy on Renewable Energy was published by the Ministry of Energy and Water (MEE) stating the targets of 10% reduction in fossil fuel imports by 2014, 15% by 2019, and 20% by 2024. This strategy includes the following main objectives: (1) improve access to energy especially from renewable sources, (2) rationalise the use of existing energy sources, (3) increase efficiency of the use of existing natural resources to produce energy, (4) promote the sustainable use of biomass resources through the conservation and protection of forests, and (5) strengthen government capacity and streamline administrative procedures within the energy sector.

Biofuels are foreseen to play a major role to achieve the objectives of the National Strategy on Renewable Energy. The National Agency for the Development of Biofuels (ANADEB), legally established on 5th June 2009, was set-up as the implementing agency of the National Strategy on Biofuels in Mali. The main responsibilities of ANADEB include the establishment of a centralised and harmonised framework for biofuel promotion, the increase of technical and regulatory capacities, and the creation of a dialogue between main public and private actors. Furthermore, ANADEB is concerned with the enacting of production licensing requirements and technical quality standards for biofuels as well as the promotion of trade between international partners in biofuels.

Thereby, the National Strategy on Biofuels states the importance of ensuring the environmental, economic and social sustainability of the development of the biofuels sector in Mali, and ANADEB is currently involved in the elaboration of national sustainability criteria and a biofuel certification scheme suitable for the specific framework conditions in Mali.

In the framework of the project *Mainstreaming Sustainability in the Biofuel Sector in Mali*, co-funded by the Global Sustainable Biomass Fund administered by NL Agency, The Netherlands, an intensive stakeholder consultation has been launched with the establishment of cross-sector multi-stakeholder working groups. On 22–23 July 2010 a stakeholder workshop was organised in Bamako in order to elaborate initial recommendations on the development of a sustainability scheme for Mali. It was recommended to establish a committee within ANADEB responsible for the coordination of stakeholder contributions for the formulation of the Malian sustainability scheme. Future work includes the identification and prioritisation of potential negative impacts and a concise list of sustainability concerns which need to be addressed to guarantee the sustainability of the biofuels sector in Mali (Janssen and Rutz 2010). The national sustainability scheme in Mali is foreseen to be implemented by 2012 under the guidance of ANADEB.

14.3.5 Bioenergy Policies in Mozambique

The Government of Mozambique is very actively encouraging the introduction of bioenergy (biofuels) in order to save foreign currency, to reduce environmental problems of the increasing transport sector, to reduce dependence on unpredictable and volatile world market oil prices and to contribute to rural development through generating employment and increasing income opportunities (Mataveia 2009).

The main aim of bioenergy development in Mozambique is to foster large-scale production of biofuels for national consumption and exports. This shall be supported by the gradual introduction of biofuels blending with petrol and diesel starting with low blends of 5–10%. Furthermore, the Government of Mozambique supports biofuel based rural electrification projects and places high priority on increasing access to energy for the (rural) poor by promoting modern energy services such as jatropha oil, gelfuel and modern wood-stoves for lighting and cooking.

The conditions for the development of the bioenergy sector in Mozambique are favourable due to its suitable climate for the cultivation of sugarcane and other energy crops. Mozambique has 7 million ha of available arable land, abundant labour and water resources for the production of bioenergy without threatening food production and food security.

On 21 May 2009 the Government of Mozambique published a National Biofuels Policy and Strategy (NBPS) in its official journal (Government of Mozambique 2009). This policy states the clear vision to establish the country's biofuels sector to contribute to energy security and socio-economically sustainable development. The institutional framework will include the creation of a National Programme for Biofuel Development to give financial support to activities and projects. Furthermore, a National Commission for Biofuels (CNB) will be set up to supervise the implementation of the biofuel strategy.

In order to set up a national market for biofuels in Mozambique a Biofuel Commercialisation Programme (PCB) will be established to purchase ethanol and biodiesel for blending with fossil fuels. With respect to biofuels export Mozambique will act as exporter of processed biofuels (i.e. biodiesel instead of vegetable oil) to enable local producers to add value to their production. It is expected that biofuel expansion (450,000 ha, compulsory blending of E10, B5) will generate substantial macroeconomic benefits including 150,000 new jobs.

Prior to the elaboration of the biofuels policy the Government of Mozambique has embarked in a detailed resource assessment and research on promising feedstock options. It was concluded that biofuel production in Mozambique shall be based on sugarcane and sweet sorghum for ethanol, and jatropha and coconut for biodiesel.

Furthermore, the Government of Mozambique performed an agro-ecological zoning initiative to specifically identify land available for food and for bioenergy production. The government will place strict limitations on land approval and it will identify selected agro-ecological areas for biofuel production which will be the only areas permitted for production. Thereby, guiding principles will be to avoid the use of basic food crops and monocultures, and to favour biofuel development that enhances biodiversity. The land zoning from 2008 identifying about 7 million ha of land available for commercial agricultural activities was performed at a scale of 1:1,000,000 (IIAM and DNTF 2008), and a more accurate zoning is currently carried out on a scale of 1:250,000.

With the approach of agro-ecological zoning a potential conflict between food and fuel production can be avoided. It is now in the responsibility of the government to implement this land allocation method even though potential biofuel producers will be likely to object and challenge this limitation to their business development.

In order to ensure the environmental, economic and social sustainability of biofuels production, the policy states that all actors participating in the Biofuel Commercialisation Programme need to be certified to prove that their operations are in line with national norms and criteria. Priority is given to the development of a common approach to sustainability criteria in line with global initiatives such as the Round Table for Sustainable Biofuels (RSB) and the sustainability criteria included in the recently adopted European Renewable Energy Directive. The Government of Mozambique has set up a Working Group within the national Biofuels Task Force on "Sustainability Criteria and Development Models" to elaborate national sustainability principles. This initiative is supported by the bilateral SADC (Southern African Development Community) Programme for Biomass Energy Conservation (ProBEC) of the German Technical Cooperation (GTZ). First drafts of sustainability principles have already been elaborated (Lerner 2009).

In conclusion, Mozambique is amongst the leading countries regarding bioenergy development in Africa. A concise overview of recent biofuel developments in Mozambique is presented by Schut et al. (2010) highlighting the need for adequate policy tools and instruments to ensure that biofuels will create maximum opportunities and benefits for the population in Mozambique.

14.3.6 Bioenergy Policies in South Africa

South Africa was the first country in Africa to initiate a regulatory framework for renewable energies with the launch of a White Paper in 2003 targeting 4% of renewable energy by 2013 (Government of South Africa 2003). In order to reach this target the White Paper recognises the need to create an enabling environment through the introduction of fiscal and financial support mechanisms, the development of physical infrastructure for grid access, and the creation of an appropriate legal and regulatory framework to encourage the entry of multiple Independent Power Producers into the current electricity sector and stimulate RE market creation.

The approach to RE taken in South Africa foresees the facilitation of "early win" investments in commercially proven technologies that can demonstrate the benefits of renewable energy at low level of national subsidies.

During 2009, the National Energy Regulator of South Africa (NERSA) announced the South Africa Renewable Energy Feed-in Tariff programme (REFIT) to further stimulate the renewable energy sector in South Africa. NERSA also published regulatory guidelines, a draft power purchase agreement (PPA) and rules on selection criteria for renewable energy projects under the REFIT programme.

In September 2010, the government procurement process for renewable energy based electricity production was officially launched. Private project developers are invited to respond to a Request for Information (RFI) by submission of feasibility studies for wind, solar, biomass, biogas, landfill gas and small hydropower projects.

Complementing abovementioned legal and regulatory framework for the electricity sector with the aim to contribute to the RE target of the White Paper, the government enacted the Biofuels Industrial Strategy of the Republic of South Africa in 2007 (Government of South Africa 2007). With respect to the 2006 draft strategy document the penetration level of biofuels by 2013 was revised down from 4.5% to 2% mainly due to food security concerns. Crops proposed for biofuel production include sugarcane and sugarbeet for ethanol and sunflower, canola and soy beans for biodiesel, whereas maize and jatropha have been excluded due to negative impacts on food security and potential invasiveness.

A detailed analysis of the implications and results of the Biofuels Industrial Strategy is presented in Chap. 16 of this book.

14.3.7 Bioenergy Policies in Tanzania

The vision set forth by the Government of Tanzania for the development of the bioenergy sector is to contribute to the replacement of fossil transport fuels and to stimulate socio-economic development through rural electrification projects. Benefits expected from the implementation of bioenergy projects are to improve energy security, reduce oil imports and foreign exchange burdens, as well as to provide alternative markets for farmers creating new jobs and income generation opportunities (Kiwele 2009).

The conditions for the development of the bioenergy sector in Tanzania are favourable due to its suitable climate, as well as available arable land and water resources. In 2005, the Government of Tanzania established a Biofuels Taskforce under the lead of the Ministry of Energy and Minerals with the participation of a variety of government ministries and institutions. The National Biofuels Taskforce has the mandate to develop the National Biofuels Policy and to elaborate interim Biofuel Guidelines that will be used until the Biofuel Policy is fully developed. The institutional difference between the Biofuel Guidelines and the Biofuel Policy is that the guidelines are approved by the cabinet and that the policy is to be approved by the parliament (FAO 2010).

The Biofuel Guidelines were approved in December 2009 and address the key issues of institutional framework, application procedure for investors, land acquisition and use, contract farming, and sustainability of biofuel production. The guidelines focus on ensuring socio-economic sustainability of bioenergy development, the avoidance of food-fuel conflicts, and sufficient value creation for the local rural population. Different land acquisition and tenure systems are introduced for bioenergy projects including shorter leasing periods of 5–25 years, the possibility to use land as equity, and mandatory villager shares in projects. Furthermore, investors will be required to use part of the land allocated for the production of food crops. Finally, contract farming, outgrower schemes and community engagement will be promoted, and Environmental and Social Impact Assessments (ESIA) will be required for the implementation of bioenergy projects.

As in the case of Mozambique, these measures are suitable to avoid conflict between food and fuel production. However, the Government of Tanzania will have to strictly implement these measures, if necessary against the pressure of biofuel investors.

The guidelines furthermore identified necessary improvements of the institutional framework such as the establishment of a Biofuels One Stop Centre under the Tanzanian Investment Center (TIC), a Biofuels Steering Committee chaired by the Ministry of Energy and Minerals and assisted by a Biofuels Technical Advisory Group (BTAG).

Currently, the Government of Tanzania is engaged in the implementation of a Biofuels Action Plan including a thorough review of existing policies and legal and regulatory frameworks with the aim to develop the National Biofuel Policy in the coming years. This action plan includes agro-ecological zoning of land available for bioenergy production, capacity building programmes for government departments and other stakeholders, as well as awareness creation initiatives for the population.

More details on biofuel policies in Tanzania are presented in Chap. 15 of this book.

14.3.8 Bioenergy Policies in Zambia

The vision of the Government of Zambia is to ensure environmentally sustainable exploitation of biomass resources in order to realise supply security and stable prices of transport fuels, to increase investment in the agricultural sector and to contribute to socio-economic development. Thereby, specific policy goals are to improve the management of woodlands for sustainable firewood production, to improve the efficiency of charcoal production, and to promote alternatives to firewood (Kalumiana 2009).

Activities under preparation by the Government of Zambia include the elaboration of appropriate financial and fiscal instruments for stimulating production and use of biomass, the implementation of public awareness campaigns, and the development of policies and a regulatory framework for biomass.

Also in Zambia the conditions for the development of the bioenergy sector are favourable. The current dependency of the country on food imports is mainly caused by the lack of infrastructure and investment in the agricultural sector. Therefore, bioenergy is seen as an excellent opportunity to significantly enhance the production potential of feedstock for both food and biomass production.

The Ministry of Energy and Water Development is elaborating a long-term Energy Strategy (2009–2030) that includes biofuels as priority sub-sector. Current policy measures include detailed assessments of available resources and market demand for bioenergy, the elaboration of an efficient legal and institutional framework, and the formulation of incentives to stimulate investments in the bioenergy sector.

Furthermore, in order to establish a national market for biodiesel, a blending mandate of biodiesel in diesel is currently under negotiation within the Government of Zambia.

So far, no specific initiatives have been implemented on how to guarantee socio-economic sustainability of bioenergy projects. In this field close cooperation with the Environmental Council is foreseen and Environmental and Social Impact Assessments (ESIA) will be required for bioenergy projects. Further activities to ensure sustainability (such as measures under implementation in Mozambique and Tanzania) are thus urgently needed to provide guidance to the development of the bioenergy sector.

14.3.9 Bioenergy Policies in Other African Countries

Due to the limitations of the present book chapter, this section presents very brief information about bioenergy policies in other African countries (Lerner et al. 2010; Government of Senegal 2007; Sambo 2008).

In ***Angola*** the parliament has approved the proposed Biofuel Policy and Biofuel Strategy in 2010. The former Government of ***Madagascar*** had elaborated a bill concerning the transformation, transportation, storage and use of biofuels, which has not yet been introduced in the parliament. ***Malawi*** is using the comprehensive Energy Act from 2004 to govern biofuels development. ***Namibia*** has published a National Bio-Oil Road Map strategy document. Bioenergy policies in ***Nigeria*** include the Renewable Energy Component of the National Energy Policy (2003), a draft Renewable Energy Masterplan (2007), as well as a pilot fuel ethanol programme by the Nigerian National Petroleum Corporation based on cassava and sugarcane as feedstock. In July 2007, the Ministry of Agriculture of the Government of ***Senegal*** launched the Special Biofuel Programme aiming at large-scale jatropha cultivation to contribute to energy security and sustainable socio-economic development. ***Swaziland*** has drafted both a strategy and action plan but the documents are yet to be adopted. Finally, biofuels development in ***Zimbabwe*** has occurred through actions by individual ministries or parastatals as the National Energy Policy makes no specific reference to biofuels.

14.4 Regional and African Union Level Policy Initiatives

The importance of integrating national initiatives in African countries to develop a regulatory framework for bioenergy on a broader regional and pan-African level is underlined in Chap. 18 of this book. Such integration is necessary to streamline African interests and to ensure maximum benefits and minimum negative environmental and socio-economic impacts of the development of global bioenergy markets in the African context.

14.4.1 Southern Africa Development Community (SADC) Policy Initiatives

The SADC biofuel strategy completed in 2005 stresses the need for exploring the potential of implementing biofuel initiatives in SADC countries and calls on all members to develop biofuels national strategies taking into account opportunities and constrains on the sector (SADC 2005).

A SADC Biofuels Task Force was established in 2008 with the mandate to develop a competitive and sustainable biofuels industry within southern Africa. Biofuels will be used as a platform for the sustainable development of the SADC region, for enhancing energy security, renewable energy development, contributing to tackling climate change and poverty alleviation. The Task Force also aims to assist in resolving possible conflicts between production of crops for fuel versus for food.

SADC actively supports the development of national biofuel strategies through its technical advisors and it has commissioned a policy development support tool (crop decision making tool) which can be used by Member States for the development of biofuel policies and implementation strategies.

In 2010, the SADC Biofuel State of Play Study was published with support of the German development agency GTZ under the Programme for Basic Energy Conservation (ProBEC). The study provides a profound assessment of the biofuel sector development in southern African countries as well as the approved SADC Sustainability Framework for Biofuels (Lerner et al. 2010). The following recommendations for SADC Member States are stated:

- Develop a nationally appropriate sustainability framework to ensure that benefits of biofuel production and use are realised and national agricultural productivity is increased.
- Develop enabling national and regional policies to ensure political commitment and a long-term integration of biofuels into the overall development plans and policies.
- Continue and expand support for research, development and dissemination to develop SADC appropriate technologies, feedstock varieties and production models.
- Develop win-win benefit sharing structures and partnership models to ensure that local communities benefit more than only engaging as wage labour.
- Promote investment in biofuels and crop diversification that includes biofuel crops around existing farming activities without compromising food needs.

14.4.2 West African Policy Initiatives

Commissioned by UEMOA (Economic and Monetary Union of West Africa) and the Rural Hub for West and Central Africa the “Sustainable Bioenergy Report in

UEMOA Member Countries" published in October 2008 concluded that donor and host country investments in bioenergy can reduce the exposure of West African countries to high food and oil prices and open up new economic opportunities in clean energy development (UEMOA 2008). A summary of findings of this report on bioenergy opportunities in the UEMOA nations Benin, Burkina-Faso, Cote d'Ivoire, Guinea Bissau, Mali, Niger, Senegal and Togo is presented in Chap. 17.

In the field of financing, the African Biofuels Renewable Energy Fund (ABREF) was set up to help overcome barriers and to facilitate flows of investment into African biofuel and renewable energy projects to promote sustainable development and contribute to the reduction of greenhouse gas emissions. It is first phase of implementation, ABREF target countries will be the ECOWAS (Economic Community of West African States) countries Benin, Burkina Faso, Cabo Verde, Côte d'Ivoire, Gambia, Ghana, Guinea, Guinea Bissau, Liberia, Mali, Niger, Nigeria, Senegal, Sierra Leone and Togo.

Finally, the Government of Senegal has taken the leadership in the Association of Non-Oil Producing Countries ("Pays Africains Non-Producteurs de Pétrole" (PANPP), or the so-called Green OPEC), established in 2006 uniting 15 African countries to exchange experiences, technologies and knowledge for the development of biofuel industries. One of the main objectives of PANPP constitutes the creation of institutional, organisational and financial instruments to promote the large-scale introduction of biofuel production in the entire continent.

14.4.3 African Union Level Initiatives

In July 2007 the African Union Commission (AUC), the Government of Brazil and the United Nations Industrial Development Organisation (UNIDO) jointly organised The First High-level Biofuels Seminar in Africa in Addis Ababa, Ethiopia, under the theme 'Sustainable Bio-fuels Development in Africa'.

This seminar was organised in the framework of the 2004–2007 Strategic Plan of the AUC in the field of new and renewable energies. The 8th Assembly of the African Union Commission meeting in Addis Ababa in January 2007, endorsing the measures adopted by the African Ministers in charge of Hydrocarbons (oil and gas) at their 1st Conference held in Cairo on 14 December 2006, requested the AUC to elaborate policies and strategies for the development of clean, new and renewable energies, particularly biofuels, as an alternative solution to hydrocarbons, in response to the rise in oil prices which has adverse effects on the economies of African countries.

The seminar key objectives were to (1) brief policymakers, the private sector, regional institutions and other stakeholders on the potential, risks and trade-offs of developing biofuels in Africa; (2) facilitate sharing of experiences in developing biofuels amongst countries in Africa and between Africa and Brazil, other countries and regions; (3) explore the potential and challenges to the dissemination of priority

The Addis Ababa Declaration on Sustainable Biofuels Development in Africa (Recommendations)

1. Develop enabling policy and regulatory frameworks for biofuels development as a matter of priority, taking into account the following aspects
   a. Link to overall development policies;
   b. Promote equity, including gender equity;
   c. Ensure participation of all stakeholders;
   d. Promote local consumption; and
   e. Enhance energy security
2. Commit to include biofuels in broad energy related frameworks;
3. Take lead on those aspects on biofuels research and development along the biofuels value chain that have significant implications for Africa;
4. Harmonise national biofuels policies, strategies and standards through regional economic communities to ensure economies of scale and access to international markets;
5. Increase the capacity of key players along the biofuels value chain;
6. Participate in the global sustainability discussions and develop immediate guiding principles on biofuels development to enable Africa to compete internationally;
7. Call upon public financing institutions to support biofuels;
8. Call upon development partners to assist countries in keeping abreast with developments in the biofuels sector through North-South and South-South cooperation;
9. Minimise the risks associated with captive markets for small-scale producers;
10. Formalise the organization of similar high-level seminars at continental and regional levels;
11. Establish a forum to promote access to information and knowledge on biofuels related sector (e.g. best practices in policy development, technology-transfer, investment promotion, trade, capacity building etc); and
12. Commit themselves to implement the attached plan of action and call upon the African Union Commission to present the Addis Ababa declaration to the upcoming ministerial conferences on sectors relevant to biofuels.

Fig. 14.1 The Addis Ababa Declaration on Sustainable Biofuels Development in Africa (Source: African Union 2007)

biofuels technologies; and (4) consult key stakeholders towards developing a programme of action for sustainable biofuels development.

Based on the plenary and thematic sessions and Ministerial Roundtable discussions of the First High-level Biofuels Seminar in Africa, the seminar participants agreed upon the following recommendations laid down in The Addis Ababa Declaration on Sustainable Biofuels Development in Africa (see Fig. 14.1).

The Addis Ababa Declaration underlines the importance attributed on AUC level to the sustainable development of a biofuels sector in Africa. However, information on recent AUC activities in the field of biofuels is difficult to access.

14.5 Conclusions and Way Forward

In recent years several African countries have launched initiatives to establish sound policy frameworks for bioenergy in order to ensure environmentally, economically and socially sustainable production, promotion and use of bioenergy. Nevertheless, fully functional legal and regulatory frameworks have not yet been established in African countries posing the danger of negative environmental and socio-economic impacts of un-regulated development of the bioenergy sector.

The most advanced regulatory frameworks for bioenergy exist in South Africa and Mozambique with the Biofuels Industrial Strategy of the Republic of South Africa (enacted in 2007) and the National Biofuels Policy and Strategy (NBPS) published in May 2009 by the Government of Mozambique.

Specific activities in the field of bioenergy sustainability certification as an essential component of the regulation of the bioenergy sector are performed in Mali, Mozambique, Tanzania, and on regional level by the Southern Africa Development Community (SADC). In Mali, a sustainability scheme suitable for the specific national framework conditions in elaborated under the guidance of the National Agency for the Development of Biofuels (ANADEB) founded in June 2009. The Government of Mozambique has set up a Working Group within the national Biofuels Task Force on "Sustainability Criteria and Development Models" to elaborate national sustainability principles. In Tanzania, sustainability aspects are addressed in the Biofuels Guidelines (approved in December 2009) until the national Biofuels Policy is fully developed. On SADC level a Sustainability Framework for Biofuels was approved in 2010 with the mandate for SADC Member States to translate this framework into nationally appropriate sustainability schemes.

On the way forward towards the development of a sustainable bioenergy sector in Africa it is recommended that before an African country develops new legislation relating to bioenergy, it should have a well-considered and clear policy on the subject. Next, the existing legal framework should be analysed, the gaps and weaknesses identified and the challenges, threats and opportunities examined.

Thereby, it is of crucial importance to carefully integrate policies for land use, agriculture and energy and align them with policies for rural development, transport and finance. Furthermore, bioenergy development in African countries will only find its proper environmental context and agricultural scale if convergence with biodiversity, GHG emissions, and water use policies is achieved.

References

African Union (2007) The addis ababa declaration on sustainable biofuels development in Africa. The first high-level biofuels seminar in Africa 'sustainable bio-fuels development in Africa', 30 July–1 Aug 2007. Available at: http://www.unido.org/fileadmin/user_media/Services/Energy_and_Climate_Change/Renewable_Energy/Conference_Ethiopia/Declaration.pdf

Ahiataku-Togobo W (2009) Bioenergy policy implementation in Ghana. Presentation at COMPETE conference, Brussels, Belgium, 24–25 Nov 2009. Available at: http://www.compete-bioafrica.net/

BTG (2008) Sustainability criteria & certification systems for biomass production. Report prepared for EC DG TREN, Final report, Feb 2008

Cocchi M, Grassi G, Onibon A, Grassi A (2009) State of the art of the development of a National biofuel promotion strategy in the Democratic Republic of Benin. In: Proceedings of the 17th European biomass conference and exhibition, Hamburg, pp 2484–2486

Davison SC (2010) Biofuels policy development in Kenya. Presentation at World Biofuels Market, Amsterdam, the Netherlands, 17 Mar 2010

Ecofys (2009) Development of feedstock sustainability standards. Report commissioned by the Renewable FA, Nov 2009

FAO (2007) Recent trends in the law and policy of bioenergy production, promotion and use, FAO Legislative Study 95. Food and Agriculture Organisation of the United Nations, Rome. ISBN 978-92-5-105870-1

FAO (2008) The state of food and agriculture 2008, biofuels: prospects, risks and opportunities. Food and Agriculture Organisation of the United Nations, Rome. ISBN 978-92-5-105980-7

FAO (2010) Bioenergy and food security – the BEFS analysis for Tanzania. Environment and natural resources working paper 35, Climate, Energy and Tenure Division (NRC) publications, Food and Agriculture Organisation of the United Nations, Rome, 2010, ISSN 2071–0992

Froger E, Paz A, Vissers P (2010) Selection of a sustainability standard for pilot assessments of Jatropha producers in Mozambique. Report of project 'Towards sustainability certification of Jatropha bio-fuels in Mozambique' co-financed by NL Agency, Utrecht, May 2010

Government of Ghana (2009) Draft renewable energy bill. Final draft report of consultancy services for renewable energy policy and regulatory framework and renewable energy law for Ghana – Electricity Generation, Sept 2009

Government of Mozambique (2009) Resolucao n.º 22/2009, Aprova a Politica e Estsategia de Biocombustiveis, Boletim da Republica, Publicacao Oficial da Republica de Mozambique, 3.º Suplemento, 21 May 2009

Government of Senegal (2007) Biofuels in Senegal – Jatropha program 2007–2012. Ministry of Agriculture, July 2007

Government of South Africa (2003) White paper on renewable energy. Department of Minerals and Energy, Nov 2003

Government of South Africa (2007) Biofuels industrial strategy of the republic of South Africa. Department of Minerals and Energy, Dec 2007

IIAM and DNTF (2008) Zoneamento Agrario—Resultados do exercıcio de validação de resultados de terra disponıvel a nıvel local (fase 2). IIAM and DNTF, Maputo, p 16

Janssen R, Rutz D (2010) Development of sustainability criteria for biofuels – international initiatives and expertise. Report of project 'mainstreaming sustainability in the biofuel sector in Mali' co-financed by NL Agency, May 2010

Janssen R, Rutz D, Helm P, Woods J, Diaz-Chavez R (2009) Bioenergy for sustainable development in Africa – environmental and social aspects. In: Proceedings of the 17th European biomass conference and exhibition, 29 June–3 July 2009, Hamburg, Germany, pp 2422–2430

Kalumiana OS (2009) Bioenergy policy implementation in Zambia. Presentation at COMPETE policy conference, Lusaka, Zambia, 26–28 May 2009. Available at: http://www.compete-bioafrica.net/

Kiwele P (2009) Bioenergy policy implementation in Tanzania. Presentation at COMPETE policy conference, Lusaka, Zambia, 26–28 May 2009. Available at: http://www.compete-bioafrica.net/

Lerner A (2009) Sustainability criteria for biofuels – an alternative for Southern Africa? COMPETE Newsletter, (4):7, Apr 2009

Lerner A, Matupa O, Mothlathledi F, Stiles G, Brown R (2010) SADC biofuel state of play study – an assessment of the biofuel sector development in the Southern African Development Community

Mataveia MA (2009) Biofuel development in Mozambique. COMPETE Newsletter, (4) 10, Apr 2009

Muok BO, Kirui S, Theuri D, Wakhungu JW (2008) Policies and regulations affecting biofuels development in Kenya. PISCES Policy Brief No. 1, Dec 2008. Available at www.pisces.or.ke

RED (2009) Directive 2009/28/EC of the European Parliament and of the Council of 23 April 2009 on the promotion of the use of energy from renewable sources and amending and subsequently repealing Directives 2001/77/EC and 003/30/EC. Official J Eur Union: L 140/16–L 140/62

SADC (2005) Feasibility study for the production and use of biofuels in the SADC Region, Gaborone, Botswana, Aug 2005

Sambo AS (2008) Renewable energy options for the environment and sustainable development in Nigeria. Presentation at National Workshop on Energy Investment, Abuja, Nigeria, 4–5 June 2008

Schut M, Slingerland M, Locke A (2010) Biofuel developments in Mozambique update and analysis of policy, potential and reality. Energ Policy 38(2010):5151–5165

UEMOA (2008) Sustainable bioenergy report in UEMOA Member Countries. Oct 2008. Available at: www.unfoundation.org

Chapter 15
Biofuel Policies in Tanzania

Sarah Mohamed, Gavin Fraser, and Estomih N. Sawe

Abstract Tanzania's energy consumption is dominated by biomass, most of which is consumed by households. Apart from biomass, the country has extensive hydropower, coal and natural gas resources. The overwhelming dependence on biomass for basic cooking and heating needs by the majority in Tanzania, is a major cause of poverty and several other deleterious effects. As a consequence of the harmful effects of fossil fuels, biofuels are seen as a source of sustainable growth to help alleviate poverty in Tanzania. In this chapter, present policies and other framework conditions influencing the development of the biofuel sector in Tanzania are investigated. This includes the Tanzanian Energy Policy, Land Act, National Forest Policy, National Environment Policy, Agricultural Sector Development Policy, and the Tanzanian Transport Policy. The necessity for policies directly related to biofuel production is discussed. The Tanzanian government has established a Biofuels Task Force (BTF) in order to produce guidelines for the design of a set of appropriate policy initiatives. Due to the lack of formal biofuel policies, several external agencies have made recommendations on policies that the Tanzanian government should adopt. It is concluded that the policies existing in Tanzania are indirect and thus biofuel producers lack a reliable framework. Due to increasing biofuels, especially amongst small-scale farmers, the government must seek to provide these smallholders with incentives to grow their productions along with large-scale producers.

Keywords Tanzania • Biofuel policy • Sustainability • Economic growth • Small-scale farmers

S. Mohamed • G. Fraser (✉)
Rhodes University, PO Box 94, Grahamstown 6140, South Africa
e-mail: sarahm31@hotmail.com;g.fraser@ru.ac.za

E.N. Sawe
TaTEDO, Mpakani A, Plot No. KJM/MPA/98, Near Institute of Social Works,
Kijitonyama, PO Box 32794, Dar es Salaam, Tanzania
e-mail: edirector@tatedo.org

R. Janssen and D. Rutz (eds.), *Bioenergy for Sustainable Development in Africa*,
DOI 10.1007/978-94-007-2181-4_15,

15.1 Introduction

Tanzania's energy consumption is dominated by biomass (Fig. 15.1), most of which is consumed by households. The remainder is composed of petroleum products (8%) and electricity and other sources (2%).

In addition to biomass, the country has considerable hydropower, coal and natural gas resources. Hydropower potential is nearly 5,000 MW, of which currently 560 MW is exploited in large plants and 4 MW in small plants. Coal reserves are estimated at 1,200 million tons, whilst limited amounts are mined and used for electricity generation in the south of the country. Natural gas reserves are estimated at 45 billion m^3 exploited for electricity production and industrial use, e.g. in the cement industry (Norges Naturvernforbund 2009). Only petroleum products need to be imported.

Electricity production for the national grid totalled 4,185 GWh in 2007, mostly from hydropower (60%) and natural gas (36%) (International Energy Agency 2007). Electricity is also produced from coal (3%) and diesel (1%). Small amounts of electricity are imported from Uganda and Zambia. In addition, several agro-industries operate biomass-fuelled co-generation plants to produce electricity and heat for their own processes with a total installed electrical capacity of about 35 MW.

Grid connection rates are very low, i.e. only 12% of the total population and 2.5% of the rural population has access to electricity (Rural Energy Agency 2009).

The overwhelming dependence on biomass for cooking and heating needs by the majority in Tanzania is a major cause of poverty and several other deleterious effects. These include deforestation and soil degradation from charcoal production and the use of fuel-wood. Charcoal production alone causes loss of forest cover at a rate of more than 100,000 ha per year with an increasing tendency (Sawe 2009b).

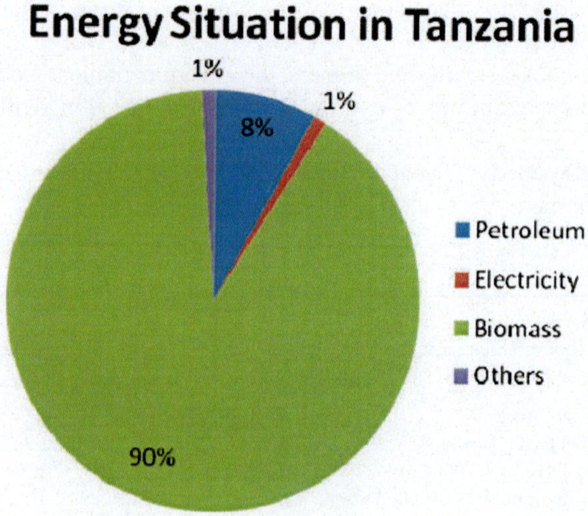

Fig. 15.1 Energy sources in Tanzania (Source: Sawe 2009b)

In addition, desertification is occurring at an increasing rate and marginal agriculture is being affected due to limited water resources during droughts (Developing Renewables 2006).

15.2 Reasons for Biofuel Policy

Bioenergy developments are high on the agenda of many countries today in an effort to improve energy access, energy security and global greenhouse gas emissions. Furthermore, bioenergy offers enormous potential to boost agricultural growth and reduce poverty in developing countries like Tanzania. Biofuel developments in Tanzania could provide an important opportunity to improve agriculture by mobilizing investors and local capital to enhance energy access and improve productivity.

Pressure on Tanzania's electricity supply is increasing. This is due to growing economic activities and improving ability and willingness of the population to buy electricity appliances both in rural and urban areas. In addition, the demand from neighbouring countries Kenya, Malawi, Uganda and Zambia onto the Tanzanian national grid is growing. The growth of electricity-intensive sectors such as the mining sector is intensifying this pressure. Eighty percent of the Tanzanian population is rural, largely isolated from access to the national grid, and hence deprived of opportunities to get access to modern energy services. Only 2.5% of Tanzania's rural population has access to grid electricity (Sawe 2009b). The energy needs in rural and urban areas are met by fuel-wood, resulting in deforestation. Wood from rural areas is used for charcoal production. The development of a sound biofuel policy would not only provide a source of energy for rural areas and reduce deforestation, but with appropriate policy incentives, it would allow its establishment on a large-scale and help Tanzania meet its increasing energy demands (African Rural Energy Enterprise Development 2010).

A more general reason for the need to formulate a biofuel policy is for the country's overall development. The Millennium Development Goals emphasize ensuring a sustainable environment and reducing poverty. Energy is an essential service whose availability and affordability determines levels of development. Tanzania faces the challenge of accelerating its economic growth. This has been the stimulus for formulating and reviewing policies and strategies aimed at reducing poverty in the context of the Development Vision 2025 that envisages sustainable socio-economic development by the year 2025. The success of all these ambitions depends on reliability, affordability and sustainability of modern energy services (Kamanga 2008).

According to Dufey (2006), one of the most critical conditions for the implementation of biofuel production is a public policy that makes production competitive in the earliest stages of industry development. Dufey claims that policy tools such as tax benefits, subsidies and loan guarantees are essential in order to encourage the production and consumption of biofuels and the development of the market. Support is needed to account for its high cost compared to conventional fuels to exploit the many positive externalities associated with biofuel production (Dufey 2006). Currently, high costs of production discourage producers from entering biofuel production. If a mechanism exists in the form of a public policy to encourage individuals to engage in the production of biofuels, the market for biofuels would attract more producers, thereby increasing competitiveness.

Although public policy is deemed vital to encourage individuals to engage in the production of biofuels, it may not be beneficial to all. Biofuel production has developed in industrialized countries due to protectionist policies defending the domestic biofuels industry from international competition. Protectionist policies have proved to be costly for developing countries that could potentially become efficient producers or exporters of biofuels. Low-income consumers pay high prices for food staples as their prices rise in world markets. This rise is largely associated with distortional policies (World Development Report 2008). Even though public policy for the inducement of domestic production of biofuels may be a pre-requisite, it may not be beneficial or efficient on a global level.

The World Development Report (2008) further cautions that it may not be wise for governments of developing countries to devote a large amount of resources to the creation of incentives, displacing alternative activities with higher returns (World Development Report 2008). Thus, developing countries engaging in the production of biofuels with large support from their governments may be in danger of diverting resources away from sources of comparative advantage and not utilizing their resources in the most efficient ways.

Although Dufey (2006) recommends the use of public policy to encourage the production and consumption of biofuels due to their potential benefits, it is acknowledged that governments need to assess whether the benefits outweigh the costs.

On the other hand, a report by the United Nations (UN) notes that governments have an incentive to support small-scale biofuel production because this may reduce social welfare payments. The increased employment resulting from the development of the biofuels industry will mean that fewer people are unemployed seeking unemployment benefits. Thereby, governmental expenses are reduced and tax revenues from employed workers increase. In addition, through multiplier effects, money earned and spent will be an overall economic stimulus (United Nations Energy 2007).

15.3 Necessity of Biofuel Policy

Currently there is no formal biofuel policy in Tanzania. Investors are operating in a policy vacuum. As a result, the Tanzanian Biofuels Task Force (BTF) has been established to produce guidelines for the government in order to design appropriate policy initiatives (Sumbi 2009). The BTF falls under the Ministry of Energy (Janssen 2006). Other aims of the BTF include ensuring close co-operation between the different government bodies involved in the development of policies for biofuels and providing an information channel between the stakeholders, namely, the government, industry, farmers' associations and non-government organizations (Janssen et al. 2005). To date, the BTF has performed a SWOT analysis, prioritized strategic action, produced draft guidelines for biofuels development and devised a comprehensive action plan for biofuel development (Tanzanian Ministry of Agriculture, Food Security and Co-operatives 2009).

15.3.1 Factors Inhibiting Biofuel Production

Tanzania's Energy Policy encourages switching from traditional fossil fuels to alternative environmentally friendly fuels. However, it does not emphasize biofuels because at the time the policy document was drawn up biofuels were not discussed as alternative source of energy (Haikam 2009). This shows that Tanzanian legislation has not yet caught up with the current needs of the population and this hinders the further development of biofuel production.

Moreover, due to the weak legal institutional framework in Tanzania, contracts between smallholder farmers and large-scale biofuel producers result in the non-enforcement of standards, and often both parties do not live up to their commitments. This dissuades small-scale farmers from entering into agreements with large-scale producers for fear of being exploited, whilst large-scale producers do not believe that small-scale farmers will be able to meet their obligations (Marwa 2008).

Furthermore, the National Forest Policy seeks to protect the ownership and management of forest land by communities. Investment in biofuel production can have direct impacts on forests, creating conflict between the National Forest Policy, which protects the land of those not wishing to engage in biofuel production, and other policies that indirectly encourage the production on biofuels (Haikam 2009).

At present, formal policies are not inhibiting the production of biofuels, but the lack of a direct biofuel policy is an inhibiting factor. Existing policies do not impede biofuel production. However, the absence of direct policies or guidelines causes lower interest of foreign and domestic investors towards investments in expanding biofuel production.

15.3.2 Policies Promoting Biofuel Production

A concern that may seem to hinder the development of biofuel production is the Land Act. This Act promotes the optimum use of land resources, community ownership and it protects communities facing land acquisition pressures for large-scale commercial biofuel production (Haikam 2009). However, rural communities can use the land they own to develop small-scale biofuel production initiatives without the threat of having their land being taken away. Protection of ownership ensures rural communities with long-term ownership and hence, the willingness to invest their limited resources into the production of biofuel on a small-scale.

The National Environment Policy endorses economic development but not at all costs with regard to potential environmental impacts. Consequently, this policy explicitly suggests the development of alternative energy sources with reduced environmental impacts to support growth (Haikam 2009). As Tanzanian policy advocates the development of alternative sources of energy, individuals will not be discouraged to invest in the production of biofuels.

The Agricultural and Livestock Policy acknowledges the fact that agriculture includes the production of both, edible and non-edible oils, with non-edible oils regarded as potential feedstock for biofuels (Haikam 2009). This acknowledgement gives owners of land unsuitable for the farming of food crops, an incentive to continue the cultivation of crops that yield non-edible oils that can be bought by large-scale biofuel producers or used by small-scale biofuel initiatives in their own communities.

The Agricultural Sector Development Strategy of 2001 aims to create an environment that is conducive to the improvement of productivity and profitability of the agricultural sector (Tanzanian Ministry of Agriculture and Food Security 2001). The strategy states that the role of the government is to create an environment for medium and large-scale investors to make use of the abundant land resource of the country. In 2006, almost 50% of Tanzania's land area was available for agriculture (Janssen 2006). As this strategy is not directed at food crops only, it provides an environment in which biofuel production can grow.

In addition, Tanzania's Transport Policy states that the environmental impacts of transport infrastructure and services must be taken into account (Haikam 2009). This shows that the Tanzanian government recognizes the negative effects of fossil fuels on the environment and is willing to consider alternative sources of energy. This provides an incentive to the transport sector to use biofuels, stimulating demand and encouraging individuals to engage in production.

Furthermore, based on the recognition that Tanzania is a country well endowed with natural resources, which can be used to grow its agricultural industry, the Tanzanian Investment Centre (TIC) has created strong incentives to attract investment from foreign and domestic investors to work with Tanzanian farmers (Charles 2009). Incentives include (Tanzanian Ministry of Agriculture, Food Security and Co-operatives 2009):

- Zero-rated duty in capital goods, farm inputs including fertilizer, pesticides and herbicides
- Favourable investment allowances and deductions on agricultural machinery and tools
- Deferment of VAT payment on project capital goods
- Import duty exemption on raw materials for inputs for exports
- Zero-rated VAT on agricultural exports and for domestically produced agricultural inputs
- Indefinite carry-over of business losses against future profit for income tax
- Reasonable corporate and withholding tax rates on dividends

These investment incentives are attractive for large-scale farming. However, small-scale farmers find it difficult to acquire finances to obtain the necessary equipment to begin ventures. Therefore, it is recommended that the incentives provided by the TIC should have a larger focus on facilitating the acquisition of finances by small-scale farmers and farmers' co-operatives.

15.4 Conclusions and Recommendations

Due to the lack of a formal biofuel policy, several external and internal agencies such as GTZ, FAO, University of Dar es Salaam, and TaTEDO have made recommendations about the kind of policy that the Tanzanian government should adopt. These recommendations include (carbon-based) fuel-taxes, vehicle taxes and subsidies to stimulate demand, incentives for investment in biofuel production facilities to stimulate supply, initial protection of local manufacturers against cheaper imports in order to develop a strong national biofuels industry and the introduction of technology standards that require compatibility with specific mixtures of biofuels (Janssen et al. 2005). There is a need to support and empower small-scale farmers and local investors to produce biofuels for local and export markets.

Although existing policies indirectly encourage individuals to engage in the production of biofuels, there is no formal policy framework for their development of the biofuels sector. As biofuels are becoming an important source of energy, Tanzanian legislation needs to catch up and provide the appropriate incentives to encourage their production and use.

Furthermore, in the development of policies, the government must keep in mind small-scale biofuel producers. When developing policies, governments often focus on providing incentives to large-scale producers and large multinational companies, whereas small-scale farmers require special incentives. These special incentives could include funds, grants and assistance in the establishment of farmers' associations in which small-scale farmers come together and pool their efforts to be able to compete with large-scale multinationals. As financing is important for the growth of biofuel production local financing institutions should be encouraged to support small-scale farmers. Financing is especially vital for the acquisition of equipment by rural communities to process oil crops into fuels to meet local energy needs (Sawe 2009a).

References

African Rural Energy Enterprise Development (2010) The energy sector. United Nations Environment Programme. http://www.areed.org/country/tanzania/energy.pdf. Accessed 5 Jan 2010

Charles K (2009) Agricultural Mechanization in Tanzania. Precision Air, Oct–Dec, pp 30–31

Developing Renewables (2006) country energy information Tanzania. Renewable energy in developing countries: current situation, market potential and recommendations for a win-win-win for EU industry, the environment and local socio-economic development. http://www.energyrecipes.org/reports/genericData/Africa/061129%20RECIPES%20country%20info%20Tanzania.pdf. Accessed 6 Jan 2010

Dufey A (2006) Biofuels production, trade and sustainable development: emerging issues. Report for the International Institute for Environment and Development, London

Haikam 2009. What other policies say about biofuel in Tanzania. Citizen Journalism in Africa. http://www.citizenjournalismafrica.org/node/1349. Accessed 16 Sept 2009

International Energy Agency (2007) Electricity/heat data for Tanzania. http://www.iea.org/stats/electricitydata.asp?COUNTRY_CODE=TZ. Accessed 3 May 2010

Janssen R (2006) Opportunities for biofuels in Tanzania. WIP – renewable energies, Germany. http://www.gfse.at/fileadmin/dam/gfse/fse%206/PLENARY_IV/5.__WIP_GFSE-Presentation__Rainer_Janssen.pdf. Accessed 16 Sept 2009

Janssen R, Sawe EN, Woods J, Pförtner R (2005) Liquid biofuels for transportation in Tanzania: potential and implications for sustainable agriculture and energy in the 21st century, WIP – Renewable Energies, Themba Technology, TaTEDO, Integration Umwelt und Energie GmbH, commissioned by GTZ. http://www.gtz.de/de/dokumente/en-biofuels-for-transportation-in-tanzania-2005.pdf. Accessed 16 Sept 2009

Kamanga KC (2008) The Agrofuel Industry in Tanzania: a critical enquiry into challenges and opportunities. Research report. Land Rights Research and Resources Institute (LARRRI) and Joint Oxfam Livelihood Initiative for Tanzania (JOLIT). Dar es Salaam, Tanzania

Marwa NW (2008) The economic potentials of liquid biofuels production: challenges and opportunities for smallholder farmers in Tanzania. http://www.irst.ac.rw/IMG/pdf/MARWA_PRESENT-ATION.pdf. Accessed 8 Jan 2010

Norges Naturvernforbund (2009) Sustainable energy solutions in East Africa: status, experiences and policy recommendations from NGOs in Tanzania, Kenya and Uganda. Report 2/2009, Norges Naturvernforbund, Oslo

Rural Energy Agency (2009) Rural energy agency: role and mandate. Presentation for the NORAD small hydropower development working session, 28 Oct 2009

Sawe EN (2009a) Liquid biofuels: pertinent issues, smallholder farmers and biofuels development potential in Tanzania. Tanzania Traditional Energy Development and Environment Organisation. http://www.tatedo.org/news/biofuels.htm. Accessed 16 Sept 2009

Sawe EN (2009b) Sustainable charcoal production for poverty reduction in Tanzania (some issues and options). In: Bioenergy for sustainable development in Africa Conference, Brussels

Sumbi P (2009) WWF's perspectives on biofuels and the current situation in Tanzania. World Wide Fund for Nature. http://cmsdata.iucn.org/downloads/08_wwf_tanzania_regional_biofuel_workshop_nairobi_20_april_2009_small.pdf. Accessed 16 Sept 2009

Tanzanian Ministry of Agriculture and Food Security (2001) Agricultural sector development strategy. Tanzanian policy document. Ministry of Agriculture and Food Security, Dodoma

Tanzanian Ministry of Agriculture and Food Security and Co-operatives (2009) Tanzania Government Perspective on Biofuels. Ecole Polytechnique Federale du Lausanne. http://cgse.epfl.ch/webdav/site/cgse/shared/Biofuels/Regional%20Outreaches%20&%20Meetings/2009/East%20Africa/Govt%20Tanzania.pdf. Accessed 16 Sept 2009

United Nations Energy (2007) Sustainable bioenergy: a framework for decision makers. United Nations. http://esa.un.org/un-energy/pdf/susdev.Biofuels.FAO.pdf. Accessed 1 Apr 2009

World Development Report (2008) Biofuels: the promise and the risks. World Bank. http://siteresources.worldbank.org/INTWDR2008/Resources/2795087-1192112387976/WDR08_05_Focus_B.pdf. Accessed 8 Mar 2009

Chapter 16
Biofuel Policy in South Africa: A Critical Analysis

Thapelo Letete and Harro von Blottnitz

Abstract In 2007 the South African government released the country's National Biofuels Industrial Strategy targeting a biofuels market penetration of 2% of liquid road transport fuels by 2013. Contrary to the international situation, the main driver for the development of a biofuels industry in South Africa is neither the economic threat of increasing oil prices nor mitigation of greenhouse gas emissions, but the need to create a link between the country's first and second economies. Specifically, the government hopes to stimulate economic development and to alleviate poverty through the promotion of farming in areas previously neglected by the apartheid system. Before the release of this strategy, commercial sugar producers and maize farmers represented the majority of the parties looking to drive the South African biofuels industry. But, 2 years after its release none of the ventures by these stakeholders have been able to take off, mainly due to the Strategy's restrictions on the type and source of feedstock as well as on the type of farmers whose participation in the industry would be subsidised. This chapter presents a critical scientific-based analysis of the implications and results of South Africa's National Biofuels Industrial Strategy. Firstly, an update is presented on the state of the biofuels industry in the country, highlighting the current production statistics and the major investment activities, and how these were affected by the release of the Strategy. Then the ambiguities in the Strategy are outlined and critically analysed with reference to

T. Letete (✉)
Energy Research Centre, University of Cape Town, P. Bag X3, Rondebosch 7701, South Africa
e-mail: t.letete@uct.ac.za

H. von Blottnitz
Chemical Engineering Department, University of Cape Town, P. Bag X3, Rondebosch 7701, South Africa
e-mail: harro.vonblottnitz@uct.ac.za

R. Janssen and D. Rutz (eds.), *Bioenergy for Sustainable Development in Africa*,
DOI 10.1007/978-94-007-2181-4_16,

the current state of the biofuels industry in the country. The chapter concludes with lessons leant from the South African experience by those African countries which are yet to develop their respective biofuel policies.

Keywords Industrial biofuels strategy • Underutilised land • Former homelands • Bioethanol • Biodiesel • Emerging black farmers

16.1 Introduction

In 2007 the South African government released the National Biofuels Industrial Strategy proposing a biofuels market penetration target of 2% of national liquid fuel supply by 2013. Before the release of the Strategy, commercial sugar producers and maize farmers represented the majority of the parties looking to drive the South African biofuels industry. However, 2 years after the release of the Strategy none of the ventures that were planned by these stakeholders have been able to take off.

This chapter presents a critical analysis of the key issues of South Africa's Biofuels Industrial Strategy and a discussion on how these could have contributed to the relatively slow development of the biofuels industry in the country. First a summary of the Strategy is given, followed by a critical analysis of specific issues of the strategy in connection with the current status of biofuel production in the country.

16.2 Summary of the Strategy

The National Industrial Biofuels Strategy initially aims at a short-term focus (5 year pilot) to achieve a 2% penetration of biofuels in the national liquid fuel supply, or 400 million litres per year to be based on local agricultural and manufacturing production (DME 2007). This target represents about 30% of the national renewable energy target for 2013, set in the White Paper on Renewable Energy (DME 2003).

Contrary to the international situation, the main driver for the development of a biofuels industry in South Africa is neither the economic threat of erratically increasing oil prices nor a mitigation agenda for anthropogenic climate change, but the need to create a link between the country's 1st and 2nd economies.[1] Specifically, the government hopes to stimulate economic development and to alleviate poverty through the promotion of farming in areas that were previously neglected by the

[1] South Africa's 1st and 2nd economies: the 1st is integrated with the global economy through modern industrialisation and produces the bulk of the country's wealth, whilst the 2nd is isolated from the first and global economies and is characterised by poverty, underdevelopment and marginalisation resulting from years of apartheid rule.

apartheid system, and in areas of the country that did not have market access for their produce, most of which are in the former homeland areas[2] (DME 2007).

The Strategy targets new and additional land, and estimates that about 14% of arable land in South Africa, mainly in the former homelands, is currently under-utilised and that only about 1.4% of arable land will be required to achieve the 2% target. The government will support the development of the under-utilised land to a level that will compete commercially, and only agricultural products grown in the previous homelands by historically disadvantaged farmers will qualify for support (DME 2007).

The Strategy proposes that support be done through existing agricultural support programmes as well as through support of investments made by the project in agricultural development. These are support programmes of the Department of Agriculture, including programmes for small-scale and emerging farmers. According to the Strategy, these programmes can be targeted to support farmers in crop selection, hedging, agricultural methods, logistics, infrastructure, research and development, and in negotiating contracts with biofuels manufacturers. The government will also ensure the training and capacity building of previously disadvantaged communities and emerging entrepreneurs to maximise transformation and the benefits in the biofuels industry (DME 2007).

The Strategy further proposes a producer support mechanism to be used to balance the difference in fuel tax support to bioethanol and biodiesel by setting a fixed margin price. It is proposed that bioethanol receives a 100% petrol tax exemption, which amounts to about R 1.21 per litre,[3] whereas biodiesel should receive a 50% diesel fuel levy exemption, effectively amounting to R 0.53 per litre (based on 2007 prices) (DME 2007).

The Strategy recommends that the 2% target be achieved through B2 or 2% blending level for biodiesel and an E8 or 8% blending level for bioethanol. Mandatory blending is not recommended for this incubation phase (DME 2007).

For food security reasons, the Strategy proposes that for the production of bioethanol, maize should be excluded in the initial phases of the strategy implementation. It is envisaged that the use of maize will only be considered "*once certainty on the ability of the currently under-utilised land to produce has been ascertained and the necessary measures are in place to guard against extreme food inflation*" (DME 2007).

16.3 Analysis of Key Points

The 2009 report of the South African Biofuels Chair admits that progress, especially investment, in the development of the country's biofuels industry has been very modest (Chair of Energy Research 2009). According to the report, the

[2] Homeland areas (homelands) were semi-autonomous areas that were set aside for black South Africans under the apartheid regime, and only became integrated fully with the rest of South Africa in 1994.

[3] Average 2010 Rand to 1US$ exchange rate = 7.33 (X-rates 2010).

only real activities to date have been the approval of 3.2 billion Rands (437 M US$) by South Africa's Industrial Development Corporation (IDC) and Energy Development Corporation (EDC) for two bioethanol plants that should collectively produce about 190 ML per year of bioethanol from sugarcane and sugarbeet, and the planned erection of a 1.1 Mt per year soybean crushing facility that will produce and distribute about 228 ML biodiesel by Rainbow Nation Renewable Fuels Ltd.

Indeed no commercial biofuel plants have been established in the country. Only biodiesel is currently being produced for the transport market, and this is from the more than 200 small-scale initiatives that use recycled vegetable oil, most of which were established long before the Strategy was released in 2007.

Looking back at the historical biofuel developments in the country, it is clear that the approval of funding for the two planned bioethanol plants and the soybean crushing facility is, in no way, assurance that commercial biofuel production will actually happen. There have been plans for commercial biofuel plants of this magnitude in South Africa before, but none of them actually took off. The question that must be asked is: Why is the establishment of the agriculture-based biofuel industry envisaged by the Biofuels Industrial Strategy of South Africa not taking off? The Central Energy Fund (CEF) Energy Development Corporation project manager, Sibusiso Ngubane, who was closely involved with the investigations into finding the right locations for the proposed biofuel plants, points to the exclusion of maize as feedstock for bioethanol production, whilst South Africa's Biofuels Chair seems to believe that it is because of the absence of obligatory nationwide blending (Chair of Energy Research 2009; IPS News 2008).

In the following sections key points of the Strategy are discussed that have the potential to render the country's biofuels industry inactive, or at least slow down its development.

16.3.1 Blending

There are ambiguities in the Strategy on how a B2 blending level for biodiesel and an E8 blending level for bioethanol will achieve the targeted 2% liquid fuel penetration or 400 million litres.

In 2007, the national gasoline consumption in South Africa was 11,558 million litres, whilst diesel consumption was 9,757 million litres (SAPIA 2009). In volumetric terms, 2% of the total liquid fuel consumption in 2007 was therefore 426 million litres, which is well within the 400 million litre target. A B2 blending level requires about 195 million litres of biodiesel; whilst an E8 blending level requires 925 million litres of bioethanol. A combination of B2 and E8 blends amount to 1,120 million litres of biofuels or a national liquid fuel penetration of 5.3% on volume basis or 3.4% on energy basis, both of which are much higher than the proposed 2%.

The absence of mandatory national blending levels implies that only environmentally conscious individuals and institutions will be willing to use biofuels in their vehicles. This means that the actual market for biofuels in the country is therefore very limited and fragile.

As regards blending of ethanol into gasoline, this appears to be heavily intertwined with a slowly evolving policy discussion on the issue of cleaner fuels. The government is caught between conflicting interests of the motor manufacturers (who want to deploy engine technologies that rely on the advanced Euro standards), and of the fuel's industry (who would have to invest heavily into modernising their refineries). Ethanol as oxygenate is important to future investment choices of the refineries. However, until these are formally requested by government with clear regulatory obligations to the fuels industry, any local investment into fuel bioethanol would remain in the risk of not securing a local market. This issue exposes a key shortfall of a supply-side driven policy intervention.

All of these issues regarding blending levels proposed by the Strategy can potentially contribute to the inactivity of the biofuels industry in the country, with respect to investment. Without clear understanding of major policy points and without a secure and sizeable market, investors will tend to be overly cautious, if not totally reluctant, to engage in any business venture.

16.3.2 Participation: Farmers and Crops

Long before the release of the Strategy, commercial maize farmers were amongst the first groups of stakeholders to lobby for bioethanol production in the country. The most predominant early venture in this respect was that of Ethanol Africa, a corporation between maize farmers, technologists and specialists in the clean technology market. The main focus of Ethanol Africa was to unlock the value contained in maize through the conversion of maize to ethanol. In 2005 the company released its 8 year plan to build eight grain processing bioethanol plants from 2005 to 2012 around the central and north-eastern part of the country, starting with Bothaville in the heart of South Africa's maize triangle (South Africa Info 2009). By early 2007 the company had secured funding in the tune of 110 million US$ for the facility in Bothaville and only awaiting the release of the final Biofuel Strategy which, to their disappointment, had not only excluded the use of their targeted feedstock, but had excluded the whole farming area of Bothaville as well as the majority of their farmer shareholders who did not come from previously disadvantaged communities (Business Report 2007). Needless to say, the venture could not take off.

The Strategy's requirement of only land that is in the former homelands implies that there cannot be any agricultural participation in this industry in two of the country's nine provinces; the Northern Cape and the Western Cape. In the Free State province, in which Bothaville is located, only the former homeland of Qwaqwa and part of the former Bophuthatswana homeland areas can participate according to the strategy.

During a study tour in the former homeland of Qwaqwa in the Free State, Letete (2009) discovered furthermore that not only do most of the black emerging farmers in these previously disadvantaged areas know nothing about biofuels, but they are also usually very sceptical of such new ventures and generally not willing to engage in crops that they are not familiar with.

All in all, the combination of the Strategy's requirements of participation in the agricultural sector by only previously disadvantaged farmers and the exclusion of maize as a feedstock has undoubtedly slowed down the establishment of an agriculture-based biofuels industry in South Africa.

16.3.3 Arable Land to be used

The Strategy also has ambiguities regarding the type of land that is targeted for biofuels production. The Strategy talks about "*new and additional*" and "*currently under-utilised*" land which is to be found in the former homelands, but there is no attempt anywhere in the Strategy to specify which land is being referred to.

In an attempt to understand the nature of land referred to in the Strategy, Letete investigated the agricultural area of Qwaqwa and discovered that there are three types of land in this area that could be classified as currently "*under-utilised*" (Letete 2009):

- *Land owned by emerging black farmers*: Since the late 1990s, the South African government has been awarding agricultural land to emerging black farmers, through various schemes, as a means of land reform. Because of lack of financial, management and, in some cases, technical skills, most of these farmers have been struggling to operate the farms effectively, sometimes even leading to total abandonment of the farms.
- *Communal land*: This is generally composed of a number of large pieces of land in the rural areas that are used by the whole community for agricultural purposes. All farming carried out in this land is purely subsistence in nature.
- *State land*: In the former homelands there are usually areas of state-owned land that are of agricultural quality which were meant to be demarcated for agricultural purposes, but were never demarcated under the apartheid regime. This type of land is usually left unused, illegally used by the community for grazing purposes or used for cultural activities.

The use of the first two types of land above for biofuel production is bound to conflict with the food industry. Whilst the little produce that the emerging farmers are able to achieve at the moment is currently being sent to regional silos that feed into the national food industry, the land used for communal subsistence farming is vital for survival in these communities and in many cases the community cannot afford to use it for anything else. The formal demarcation of state-owned land, on the other hand, is usually a lengthy process wherein decisions rest with the highest national department authorities.

It is therefore unclear to many stakeholders which land is referred to by the Strategy as new, additional and currently underutilised, and this can potentially slow down the involvement of many interested parties in the biofuels industry.

16.3.4 Government Support

The extent of government support for biofuel production is also an issue in the Strategy.

Three recent studies, by the Bureau for Food and Agricultural Policy (BFAP) (2007), Nolte (2007), and Letete (2009), show that the economic viability of biofuels from locally grown energy crops may not be favourable. Whilst the studies show that the economic incentives proposed by the Strategy for bioethanol production are enough to make sugarcane ethanol economically viable, they clearly show that the proposed biodiesel production incentives are definitely insufficient to make agriculture-based biodiesel ventures viable. Working with 2008 commodity and energy prices, Letete (2009) has shown that the proposed biofuel tax support leads to losses for the biofuel producer of R 3.70, R 3.13 and R 0.36 per litre of soybean biodiesel, sunflower biodiesel and canola biodiesel, respectively. This is mostly due to the high prices of virgin vegetable oil in the country, which can be as high as three times the current retail price of diesel, like in the case of soybean. This is one of the major reasons why the production of biodiesel in the country has to date been primarily from recycled vegetable oil.

Furthermore, agricultural land owned by previously disadvantaged farmers in South Africa is mostly land acquired through the national Land Reform programme, and since the inception of the programme in the 1990s there have been support programmes in place to assist beneficiaries – emerging black farmers – to operate the farms sustainably, but these programmes have to date been very unsuccessful. From the late 1980s to the early 1990s, 114 black emerging farmers in the eastern Free State were awarded some 96,000 ha of land under the guidance of the Agricultural Development Agency of Qwaqwa (Agriqwa) (Business Trust and DPLG 2007). The Agriqwa scheme also included test farms which acted as support centres for these black farmers in terms of technical advice, training, planning and funding. In 1995, however, this scheme was terminated by the government, leaving the beneficiaries without any support. In 2008 more than half of these farmers were burdened with debt to the extent that some could no longer operate the farms (Letete 2009).

The Comprehensive Agricultural Support Programme (CASP) and the Micro Agricultural Financial Industrial Scheme of South Africa (MAFISA) are the two existing support programmes that have been established to support agricultural land reform. CASP was set-up as an initiative of the Department of Agriculture in 2004 to provide post-settlement support to beneficiaries of the land reform and other producers who have acquired land through private means, whilst MAFISA is a pilot initiative aimed at providing funding for on-lending to target markets in the quest to address the financial service needs of entrepreneurs in the second economy and to strengthen the developmental agricultural micro-finance system for their benefit. At a parliamentary Monitoring Group briefing by the Department of Agriculture in 2008 regarding these two programmes, it emerged that both programmes were dismally failing to adequately support emerging farmers (Parliamentary Monitoring Group 2008).

If government agricultural support for emerging farmers has never been successful since the inception of the land reform programme, there is no reason for farmers to believe that the support programmes will be effective for the establishment of the biofuel industry.

In summary, the proposed financial support for biofuel producers is insufficient, especially for biodiesel, and it is uncertain how real the agricultural support will actually be.

16.4 Conclusion

This experience of South Africa offers many lessons to other African countries that are yet to develop or are in the process of developing appropriate biofuel policies. These lessons can be summarised as follows:

- There should be a thorough analysis of the required support from government, both in terms of value and quality, to ensure that support is well-suited for the establishment of the biofuel industry in the country. This analysis should also include a transparent assessment of government capacity to actually deliver on such promises.
- There should be a detailed consultation process with targeted stakeholders in developing biofuel policies, and this should be accompanied by proper information dissemination and awareness campaigns, especially for farmers who are often isolated from cities and information technology, and have poor education backgrounds.
- Key terms of the policy should be clearly defined and main points clearly laid out, especially for lay farmers and investors for whom clarity is absolutely necessary if they are to engage in such ventures.
- The implications of voluntary blending versus mandatory blending should be clearly analysed, especially with respect to quality control issues and market creation for the biofuels.
- It is finally also vital to the success of the biofuel industry that any outstanding issues and conflicts between key players and stakeholders in the biofuel industry are addressed before the policy is put in place.

References

BFAP (2007) Modelling the impacts of macroeconomic variables on the South African biofuel industry. Agricultural Economics Association of South Africa, Pretoria

Business Report (2007) Ethanol Africa gets $110 m injection for first plant. 2007 [cited 2009 10/12/2009]. Available from: http://www.busrep.co.za/index.php?fSectionId=&fArticleId=3733546

Business Trust and DPLG (2007) Nodal economic profiling project – Maluti-a-phofung, Free State

Chair of Energy Research (2009) Biofuels, South Africa biofuels: IEA Taskgroup 39 progress report. In: van Zyl E, Prior B (eds) University of Stellenbosch, Stellenbosch, South Africa

DME (2003) White paper on renewable energy, Pretoria, South Africa
DME (2007) Biofuels industrial strategy of the Republic of South Africa. Pretoria, South Africa
IPS News (2009) Maize farmers lobby to supply biofuel industry. ENERGY-SOUTH AFRICA. 2008 [cited 2009 07/11/2009]. Available from: http://www.ipsnews.net/africa/nota.asp?idnews=40945
Letete TCM (2009) Multiobjective modelling of biofuel supply systems. Department of Chemical Engineering, University of Cape Town, Cape Town, p 168
Nolte M (2007) Commercial biodiesel production in South Africa: a preliminary economic feasibility study. Department of Chemical Engineering, University of Stellenbosch, Stellenbosch
Parliamentary Monitoring Group (2008) Comprehensive agriculture support & Mafisa Funding Programmes: Agriculture Department briefings. [cited 2010 12/02/2010]. Available from: http://www.pmg.org.za/report/20080220-comprehensive-agriculture-support-mafisa-funding-programmes-agriculture-department-b
SAPIA (2009) Product demand: sales of major petroleum products in South Africa 2007. [cited 2009 12/06/2009]. Available from: http://www.sapia.co.za/stats/demand.htm
South Africa. Info (2006) Ethanol Africa forges ahead. [cited 2009]. Available from: http://www.southafrica.info/about/sustainable/ethanol-100506.htm
X-rates (2010) South African Rands to 1 USD – Historical, 2010. http://www.x-rates.com/d/ZAR/USD/hist2010.html

Chapter 17
Promoting Biofuels in West Africa: An Engine for Development

Mamadou Dianka

Abstract This chapter shows how biofuels can contribute to improve energy supply in West African countries and to alleviate poverty by creating jobs for young people who are presently very committed to emigrate to Europe. In the frame of the fight against deforestation, reduction of energy dependency, poverty and greenhouse gas emissions, the UEMOA Commission intends to stimulate the development of the biofuels sector for household energy, transportation and electricity generation with special focus on unlocking the potential of ethanol, gelfuel and biodiesel. The first phase of the UEMOA project involved the execution of a regional study on biofuels production potential and markets. This study enabled to determine and quantify market opportunities for the set-up of a supply system and the assessment of the technological, economic and financial feasibility of ethanol/gelfuel production units in the UEMOA region. In the Member States with little potential to produce ethanol (from sugarcane, cashew tree and cassava), the possibility of producing biodiesel from jatropha was evaluated. This chapter demonstrates that the region has a large potential to become a leading producer of many types of biofuels given the availability of raw materials, climate conditions, land availability and production costs. The economical and political stability of the Union favour the development of a common biofuels market that can become a substantial exporter of biofuels to the EU. The role of bioenergy in West African countries can be considerable. The application of modern bioenergy technologies can contribute to efficient energy use, reduction of dependency on imported fossil fuels, increased access to electricity, job creation and development of business opportunities between Africa and Europe.

Keywords West Africa • Exodus • Rural development • Youth • UEMOA • Ethanol

M. Dianka (✉)
Union Monétaire et Economique Ouest Africaine (UEMOA), BP 543,
Ouagadougou, Burkina Faso
e-mail: mdianka@uemoa.int

R. Janssen and D. Rutz (eds.), *Bioenergy for Sustainable Development in Africa*,
DOI 10.1007/978-94-007-2181-4_17,

17.1 Introduction

Since several years, West African countries, particularly the UEMOA countries, face severe challenges such as land degradation, desertification, lack of access to energy and climate change. The current energy and agricultural sector has highlighted the daunting challenges of the UEMOA region. The strong dependence of the region on oil imports absorbs most export income, whilst rising commodity prices aggravate the already fragile food security. Access to renewable energy in UEMOA countries is limited, the rural electrification rate remains below 5%. Despite the rapid urbanisation of UEMOA countries, people living in urban areas use little modern fuels. Firewood accounts for more than 70% of final energy consumption in the region, which puts pressure on the environment with a negative impact on the health of the population. The ability of small farmers to improve their living conditions is further compromised by high world prices for energy and grain over the past 2 years.

With a population of more than 90 million inhabitants and an area of three million square kilometres, more than 200,000 young people annually migrate to the capitals in search of employment.

In order to face these challenges, various measures will be taken at different levels to combat rural poverty and improve food and energy security. Biofuels could thereby not only increase agricultural production and industrial productivity, but also diversify and develop the rural economy through the creation of new products and new markets.

Many opportunities exist for West African countries to meet their needs for social, economic and ecological services by integrating bioenergy and sustainable energy into their energy sectors. The UEMOA countries have the potential to grow both food and biofuel crops, as well as develop non-food sources of biomass for energy production.

17.2 The Availability of Suitable Raw Material

From a regional point of view, the agricultural production potential for the ethanol sector mainly comprises (1) the humid areas of Ivory Cost, Guinee Bissau, Benin and Togo where rain fed sugarcane, cassava and cashew tree are cultivated and (2) the geographical zones around the Niger, Senegal and Gambia rivers with intensive irrigation of sugarcane and rain fed oils seeds such as cotton, jatropha and castor. Co-operation with the following sub-regional organisations charged with the development of Sahel zones is important for the implementation of biofuel projects:

- OMVS (which includes Senegal, Mali, Mauritania and Guinea Conakry): the establishment of more than 600,000 ha for biofuels production is possible.
- OMVG (which includes Senegal, Gambia, Guinea Conakry and Guinea-Bissau): 50,000 ha could be exploited.
- The Niger River Office in Mali: a minimum of 800,000 ha is available.

The preliminary options for each country are the following:

Benin – the most suitable raw material for the production of ethanol gelfuel is cassava. With an average production of 2.8 million tons of cassava per year, Benin could produce 20,000 m^3 of ethanol by using just 5% of its annual harvests (no competition with food supply needs).

Burkina Faso – new cultivation of sugarcane seems to be the most accessible raw material for the production of ethanol at present. If the 5,000 ha owned by SOSUCO are used for this purpose, an ethanol production potential of 20,000 m^3 per year can be estimated. The energy required for the conversion of sugarcane juice to ethanol could be provided by bagasse. Another potential source is sweet sorghum if the plantation envisaged in the Sourou Valley becomes a reality. Furthermore, SN CITEC (Dagris Group) plans to build a factory in the short-term with a production capacity of 10,000 tons per year based on cotton seeds.

Ivory Coast – the country has a large potential (19,000 m^3/year) to produce ethanol as a result of extensive availability of cheap molasses, enabling the profitable production of ethanol and gelfuel. Production costs are estimated at 121, 165 and 122 FCFA/l for ethanol, gelfuel and biofuel, respectively (based on the assumed conversion rate of 1 Euro equal to 655 FCFA).

Guinea Bissau – the cashew tree apple currently seems to be the most suitable raw material for the production of ethanol. The annual production is estimated at 400–600 thousand tons, of which only 30% are used for the production of juice, wine and spirits. If the remaining 70% could be used to produce ethanol, the ethanol production potential would be approximately 8,400–12,600 m^3/year.

Mali – the real production potential depends mainly on the new sugar mill in Markala. The envisaged output of 170,000 tons of sugar per year will result in an availability of 61,000 tons of molasses per year, which can be converted into 18,000 m^3 of ethanol.

Niger – the ethanol production potential is very low in Niger due to the absence of sugarcane production and low precipitation. However, there is particular interest to produce biodiesel from jatropha oil. Initial calculations based on cost estimates indicate that biodiesel could compete with (fossil) diesel.

Senegal – the ethanol production potential in Senegal is considerable. The Senegalese Sugar Company (CSS) produces roughly 35,000 tons of molasses with a high sugar content per year, which they plan to convert into 2,500 m^3 of industrial ethanol (96%) and 10,000 tons (12,500 m^3) of anhydrous ethanol for use as biofuel. Furthermore, several biofuel project promoters were identified including 30,000 ha of sugarcane and 20,000 ha of jatropha.

Togo – in spite of the presence of a small sugar industry, the immediate potential for the production of ethanol is low unless new sugarcane plantations are developed to replace industrial pineapple plantations. However, like in Niger, the private sector has particular interest to produce jatropha oil as a feedstock for biodiesel. Initial calculations based on cost estimates of the various production factors indicate that biodiesel could compete (5% lower prices) with fossil diesel, if competitive production of jatropha seeds can be ensured.

17.3 Feasibility of Ethanol Production

Regarding the feasibility of ethanol as household fuel, Table 17.1 shows the calculated price levels per MJ.

The figures in Table 17.1 indicate that under the current market conditions household fuels based on ethanol cannot compete with butane. The price levels of wood and charcoal, on an energy base, are definitely lower than those of butane, (not included in the overview). Gelfuel production costs are generally 20–30% higher than those of ethanol.

However, in Senegal and in Ivory Coast ethanol could compete with butane gas if subsidies on butane are eliminated or if equivalent subsidies are introduced for ethanol. In addition, in all studied countries, the production of anhydrous ethanol (see Table 17.2) for use as motor fuel is advantageous. In fact, according to the

Table 17.1 Prices per MJ of ethanol and household fuels

Country	Raw material	Ethanol (FCFA/MJ)	Real price of butane (FCFA/MJ)	Real price of butane (%)[a]	Non-subsidised price of butane (FCFA/MJ)	Non-subsidised price of butane (%)[a]
Benin	Cassava	17.8	8.7	103	13.0	37
Burkina Faso	Sugarcane	17.7	6.3	183	12.7	39
Ivory Coast	Molasses	9.1	5.5	67	13.0	−30
Guinea-Bissau	Cashew tree apples	25.2	11.7	117	11.7	117
Mali	Molasses	14.4	7.0	106	12.0	20
Senegal	Molasses	11.7	6.0	94	12.6	−7

[a]Price gap. The percentage shows how much more expensive (or cheaper) ethanol is

Table 17.2 Price per MJ of anhydrous ethanol, biodiesel and fossil fuels

Country	Raw material	Product	Price (FCFA/MJ)	Gasoline price (FCFA/MJ)	(%)[a]
Benin	Cassava	Anhydrous ethanol	15.0	13.6	11
Burkina Faso	Sugarcane	Anhydrous ethanol	15.1	18.2	−17
Ivory Coast	Molasses	Anhydrous ethanol	9.3	18.6	−50
Guinea-Bissau	Cashew tree apples	Anhydrous ethanol	22.6	22.2	2
Mali	Molasses	Anhydrous ethanol	12.2	18.6	−34
Niger	Jatropha	Biodiesel		15.2[b]	−11
Senegal	Molasses	Anhydrous ethanol	10.2	19.5	−48
Togo	Jatropha	Biodiesel		14.1[b]	−5

[a]Price gap. The percentage shows how much more expensive (or cheaper) ethanol (or biodiesel) is
[b]Diesel

results of the financial analyses, the value of ethanol as household fuel is 36–70% below the value of ethanol as transport biofuel.

With the exception of Benin and Guinea-Bissau the local production of anhydrous ethanol can compete with gasoline. Feasibility in Benin suffers from illegal imports of hydrocarbons from Nigeria whilst production costs in Guinea-Bissau are high due to high raw material costs and low capacity utilisation as a result of the limited seasonal availability of cashew apples. In these countries, modest support measures (for example tax exemptions) could render the production of anhydrous ethanol viable.

On the other hand, the production of anhydrous ethanol as fuel substitute for imported hydrocarbons is favourable due to cost savings and should be stimulated in Ivory Coast (–50%), Senegal (–48%), Mali (–34%) and, Burkina Faso (–17%). These countries could save on the import of hydrocarbons by developing local resources.

Regarding the production of biodiesel in Niger and Togo, preliminary calculations indicate that this fuel can compete with (fossil) diesel. Biodiesel production costs are 5–11% less than those of diesel. However, these costs are highly sensitive to the price of jatropha seeds.

In conclusion, ethanol-based household fuels will only be able to compete with butane gas if policies with market introduction incentives are put in place in each country. The use of ethanol as transport fuel will be much more profitable. Table 17.3 summarises opportunities that are considered a "programme d'urgence" for short-term implementation in several UEMOA countries.

Table 17.3 Potential projects in the UEMOA countries

Country	Type project/potential	Scale of units	Investment	Remarks
Benin	20,000 m^3/year ethanol based on cassava	1,000 – 10,000 m^3/ year	FCFA 337 Mio (small cut) FCFA 2.6 Mld (large cut)	Biofuel is 11% more expensive than gasoline
Burkina Faso	20,000 m^3/year ethanol based on sugarcane	20,000 m^3/year	FCFA 5.3 Mld	Biofuel is 17% less expensive than gasoline
Ivory Coast	19,000 m^3/year ethanol based on molasses	10,000, 5,000 et 4,000 m^3/ year	FCFA 2.1 Mld FCFA 1.2 Mld FCFA 966 Mio	Biofuel is 50% less expensive than gasoline
Guinea-Bissau	~10,000 m^3/year ethanol based on cashew tree apples	1,000 m^3/year	FCFA 652 Mio	Biofuel is 2% more expensive than gasoline
Mali	18,000 m^3/year ethanol based on molasses	18,000 m^3/year	FCFA 4.8 Mld	Biofuel is 34% less expensive than gasoline
Niger	Biodiesel based on jatropha	10,000 m^3/year	FCFA 500 Mio (factory) FCFA 3.5 Mld (plantation)	Biodiesel is 11% less expensive than diesel

(continued)

Table 17.3 (continued)

Country	Type project/potential	Scale of units	Investment	Remarks
Senegal	15 000 m³/year ethanol based on molasses	15,000 m³/year	FCFA 3.2 Mld	Biofuel is 48% less expensive than gasoline
Togo	Biodiesel based on jatropha	10,000 m³/year	FCFA 500 Mio (factory) FCFA 3.5 Mld (plantation)	Biodiesel is 5% less expensive than diesel
Total General	Ethanol (93 000 m³/ year) et biodiesel (20 000 m³/year)	Ethanol: 1,000 – 20,000 m³/ year; Biodiesel: 10,000 m³/a	Ethanol: FCFA 23.2 Mld (EUR 35.4 mio) Biodiesel: FCFA 8.0 Mld (EUR 12.2 Mio)	Substitution of 57,100 m³ of gasoline and 19,000 m³ of gasoil. Forex savings FCFA 20.6 Mld (EUR 31.4 mio)

Mld Milliard, *bl* Billion, *Mio* million

17.4 Recommendations

The principal recommendation is to continue the development of a biofuel sector in the UEMOA region. The potential to produce anhydrous ethanol and/or biodiesel from local feedstock is promising in all Member States. However, a strong will to reform the hydrocarbons sector and to introduce incentives with the aim to enable investments is necessary. Specific recommendations for the steps to be followed are:

- Continuation of the dialogue on identified projects between biofuel stakeholders and decision makers. Attention of public authorities and financial institutions should be drawn to the need to support the sector with financial and institutional incentives.
- Adoption of Community directives to develop the market. The adoption of such directives would encourage Member States to take appropriate legal and fiscal measures to promote the production and local consumption of biofuels.
- Implementation of a support programme to promote awareness, to assist the private sector to carry out technical and financial feasibility studies and to prepare investment dossiers, to facilitate technology transfers and to promote RTD of local specialised institutes.
- Implementation of agro-energy policies aimed at the long-term development of the large potential to produce ethanol and biodiesel. This policy will also have to settle the land question in order to facilitate private investment. The European Union and other donors could provide support to tackle energy, agriculture and job creation issues.
- Establishment of bioenergy sector development funds, intended to stimulate a favourable investment climate, and to provide direct finance to the private sector. UEMOA could associate with sub-regional financial institutions.

17.5 Conclusion

It is obvious that West Africa, with a relatively young population and large potential of agricultural raw materials and land, has important assets to ensure food and energy security through biofuels. In addition to these advantages, the sector is a generator of jobs that can help to reduce the rural exodus and immigration to Europe. In this context, strong policies are needed acknowledging the role of biofuels as an engine for developing social inclusion for rural youth. Thereby, biofuels could contribute to stopping immigration and addressing the challenges of energy independence, food security and climate change in West Africa.

References

Dianka M (1995) Les problèmes des énergies domestiques en zone sahélienne et les différentes stratégies. IEPF/LEF, Dakar, Senegal

Dianka M (2001a) Challenges in meeting the biomass energy needs in West Africa. PNUD/IEPF, Dakar, Senegal

Dianka M (2001b) L'approvisionnement en énergie traditionnelle et impact dans les villes africaines. Colloque francophone de Beyrouth, Liban

Dianka M (2002) Initiative de politique ouest africaine dans le secteur des Energies traditionnelles pour un environnement durable – Esquisse d'un plan d'actions régional biomasse-énergie. RPTES Banque Mondiale, Dakar, Senegal

Dianka M (2003) Sustainable and participatory energy management project in Senegal: moving towards alleviating environmental problems and poverty, Dakar, Senegal

Dianka M (2007) Défis et opportunités des bioénergies en Afrique subsaharienne

Dianka M (2008) Défis et opportunités des énergies renouvelables en Afrique subsaharienne Conférence de l'ONUDI Dakar Sénégal

Dianka M, Ouedraogo (2006) Vision et stratégie régionale de valorisation énergétique de la biomasse

EPM BTG (2007) Etude sur le développement de la filière Ethanol/Gelfuel comme Energie de cuisson dans l'espace « UEMOA »

UEMOA (2006) LIVRE BLANC UEMOA-CEDEAO: Pour une politique régionale axée sur l"accés aux services énergétiques

UN Foundation Hub Rural October (2008) Sustainable bioenergy development in UEMOA Member Countries, Ouagadougou, Burkina Faso

Chapter 18
Strategies for a Sustainable Pan-African Biofuels Policy

Charles B.L. Jumbe and Michael Madjera

Abstract Biomass is common in people's daily lives and would not require any special policy if it would not have received a new dimension in the recent past: the production of CO_2 neutral energy and in particular biofuels. Triggered by the highly industrialized countries especially in Europe, this has developed fast and like any development, has both positive and negative effects. In particular, the production of biofuels from biomass has become part of the international process of combating CO_2 emissions especially in highly industrialized countries, but also in view of the volatility of fossil fuel supplies and the dwindling fuel reserves. Today, biofuels remain a key part of the global energy solution and Sub-Saharan Africa (SSA) represents a fascinating marketplace of untapped opportunities amid unique challenges. Due to the cost, land, and climate restrictions of producing biofuels in Europe and America and the current challenges Asian production is facing, many investors are turning to Africa for growing energy crops for biofuels production. The interest in biofuels production in SSA countries is gaining momentum such that the production of energy crops will become an integral part of the energy sector development for the future. Nevertheless, such a process will require long-term thinking and far-sightedness since bioenergy will change its position within the energy mix. The future energy mix will be most probably dominated by solar energy and biomass will mainly form the basis for sustainable chemical production processes. The growing international and domestic demand for biofuels will necessitate that SSA countries develop a Pan-African biofuels policy to address possible diverging or even

C.B.L. Jumbe (✉)
Centre for Agricultural Research & Development, Bunda College of Agriculture, University of Malawi, P.O. Box 219, Lilongwe, Malawi
e-mail: charlesjumbe@yahoo.com

M. Madjera
Evangelical Church in Middle Germany, P.O. Box 1424, 39004 Magdeburg, Germany
e-mail: michael.madjera@onlinehome.de

R. Janssen and D. Rutz (eds.), *Bioenergy for Sustainable Development in Africa*,
DOI 10.1007/978-94-007-2181-4_18,

conflicting international and SSA interests and to adopt the same policy or similar stipulations of fundamental principles for biofuels in the SSA region. This will also help to counter the possibility that SSA countries out-compete against each other or that SSA producers benefit little from private investments in the sector. This chapter discusses critical issues surrounding biofuels development and proposes the way forwards for the development of a Pan-African biofuels policy to ensure that biofuels are produced in a sustainable manner and that the negative effects of biofuels development on environment, economy and society do not outweigh the expected beneficial outcomes of energy security and poverty alleviation in Africa.

Keywords Biofuels • Energy crops • Biomass • Poverty alleviation • Energy security • Biofuels policies • Food production • Biodiversity • SADC • Sub-Saharan Africa

18.1 Introduction

Biomass is the main source of primary energy in Sub-Saharan Africa. In some countries, wood-fuels (firewood and charcoal) account for more than 90% of the total energy consumption. Apart from meeting the energy needs of the rural population, tens of thousands of people derive their livelihood from direct and indirect involvement in biomass production and marketing. Biomass also contributes millions of dollars to local economies in the form of revenues, taxes, levies and incomes. However, compared to other sectors, the biomass sector is considered as "traditional" such that it is seldom given high priority in energy policies and poverty alleviation strategies. Whilst most countries have developed policies and strategies in the power and petroleum sectors, there are few comparable strategies for the traditional use of biomass in the energy sector. This has implications for the biomass sector development.

Although Sub-Saharan Africa (SSA) with its comparatively low energy demand has a limited interest in the production of biomass for energy or fuel production for own consumption, it has, compared to the limited availability of arable land in highly industrialized countries and in particular in Europe, ample arable land with potential for biomass production. This, combined with lower production costs in SSA, has resulted in a growing international interest for investments into the production of energy crops in SSA countries. Considering that the development of this sector unfolds both positive and negative effects, there is need for developing a comprehensive policy to regulate the sector including appropriate mechanisms to harness the positive outcomes whilst mitigating the negative effects. Specifically, such a policy is needed to address the negative side effects such as an imbalanced development of the national economy, an imbalanced use of land to the disadvantage of food crop production, and the destruction of nature. This requires a clear definition of the targets which in turn requires a thorough assessment and analysis of the relevant factors.

As such, whilst countries may implement a country-specific policy to regulate the sector, there is need for a trans-regional or a Pan-African policy that will lay out the framework for the production and trade in biofuels. This is required considering that SSA countries strive to improve the living conditions of the poorest sectors, and their economies. For most SSA countries, the production of energy crops and biomass should result in the improvement of the welfare of the people and should also have a stimulating effect on the development of national economies. However, any such positive achievements must avoid negative effects to the detriment of some sectors of the population.

Last but not least, the development of a trans-regional or Pan-African policy would be important to avoid that SSA countries are played off against each other.

18.2 The Interest of SSA in the Production of Energy Crops

The recent political interests in promoting biofuels and the resulting increase in the demand for arable land for growing energy crops has fuelled the debate of the consequences of the bioenergy sector development in Africa. In fact, developments in the biofuels sector offer both promises and challenges for developing countries in SSA. In general, the production of biofuels in relatively small quantities is of little environmental or economic benefit. However, from a policy perspective it is of particular interest for national governments to ascertain whether biofuels development beyond subsistence use will yield beneficial outcomes of enhancing rural incomes and poverty alleviation or may rather lead to alienating land for biofuels development at the expense of food crops or hastening environmental degradation, water scarcity and loss in biodiversity due to large-scale production of energy crops for biofuels. Great care must therefore be exercised in land use and farming systems to ensure that the production of energy crops for biofuels do not degrade land, water or natural eco-systems or undermine food production.

18.2.1 Potential of Biofuels in Economic Development

The potential benefits of biofuels will not materialise if there are no markets for biofuel products. In particular, domestic markets are most important for countries intending to indulge in commercial biofuels development since such markets can be internally controlled. In contrast, international markets are disputed areas, predominantly controlled by the highly industrialized and emerging countries. These countries have an influence on the markets by subsidizing large-scale biofuels producers and lifting the international standards for biofuels products beyond the reach of most developing countries due to the current low state of technologies for biofuels production. As a result, developing countries are often not competitive despite having comparatively low production costs. In addition, most industrial countries still highly protect their own agriculture that significantly reduces Africa's competitiveness potential. An example is the breakdown of the Ministerial Meeting in Cancun in 2003 where Mali, Burkina Faso and other African countries left in protest because their cotton production was not competitive due to massive national subsidy policies (Gilton et al. 2004). In February 2001 the US support to the cotton sector was 2.3 billion US$. The EU's support (to Greece and Spain) totalled 700 million US$ and China provided 1.2 billion US$. Subsidies encouraged surplus cotton production, which was then sold on the world market at subsidized prices (Aksoy and Beghin 2005). This has depressed world cotton prices, damaging those developing countries which rely on cotton for a substantial component of their foreign exchange earnings. The expansion of cotton cultivation in many developing countries has played an important role in reducing poverty, where the scope for replacing cotton by other crops is limited. These gains have been threatened by the fall in world prices for cotton.

The process of energy crop production could face a similar fate. Although the global economy offers great export market opportunities for SSA biofuels, penetration of biofuels products into international markets presents a major challenge. Besides huge subsidies provided to biofuels producers, a number of countries in the developed world have put in place tariff barriers. For example, the USA levies a 54% tariff on ethanol imported from Brazil (Lanely 2006).

This calls for countries in SSA to develop a Pan-African biofuels policy to harness the market potential within the SSA region. With the total population estimated at 836 million people, SSA has an enormous consumer potential that ought to be developed.

Such a development will require the assessment of existing national demands and capacities to satisfy demand from own resources. Any export production has to be adapted with regard to the national economy in a way that it is in balance with necessary imports. In many instances, goods are imported by African countries although they could be produced locally and from local materials such as oil seeds.

Common markets in SSA have been already established, such as Economic Community for West African States (ECOWAS), Common Market for Eastern and Southern Africa (COMESA), Southern Africa Development Community (SADC) or Communauté Économique et Monétaire de l'Afrique Centrale i.e., Economic and Monetary Community of Central Africa (CEMAC). But the challenge remains to "develop" the potential consumers into "consumers with a potential". In 1999, out of a population of 180 million in the SADC region, only one of ten people had a job in the formal sector and only about 20% of the economically active population had jobs. Official unemployment rates in 2008 were 7.6 per cent in sub-Saharan Africa and 10.1 per cent in North Africa. While the rate of unemployment is relatively low in sub-Saharan Africa, the proportion of workers in vulnerable employment is about 77 per cent of the labour force (Economic Commission for Africa/African Union 2010).

The real value of energy crops for SSA are the manifold with different products that can be produced for regional markets and own consumption. In 2008, the overall demand of fuel for SSA excluding South Africa was estimated at 5.7 million tons and accounted for only 0.2% of the world's demand. In 2009. the global oil consumption declined by 1.2 million barrels per day (b/d), or 1.7%, the largest decline since 1982 (BP 2009, 2010).

Despite low wages in SSA the production costs for biodiesel are substantially higher than for fossil diesel, whereas ethanol could be competitive in some cases.

For example in Botswana biodiesel would be competitive with a reduction of levies by 75% that currently apply to fossil diesel and 50% for ethanol (Kgathi and Mfundisi 2009). This also applies to South Africa (Nolte 2007) and Europe. However, high production costs of biodiesel are not to be seen as a major constraint, but requires a careful assessment in order to develop a balanced policy.

There are opportunities that SSA countries can benefit from the growing international interest in biofuels considering that most African countries have ample land and good climate for growing different energy crops. However, the promotion of biofuels should be weighed against the negative effects of uncontrolled expansion of the biofuels industry. As such, SSA countries may be the producer of the necessary raw materials and other biofuels products. But this has to be done in well calculated proportions, i.e. in relation to its own demand and, in case of excess production for export, in a balanced use of arable land without a disrupting food production.

The main emphasis by SSA countries should be on a balanced portfolio of goods from energy crops involving as many industrial sectors as possible. A number of products can help to improve the living conditions in rural areas: edible oil as food supplement, clean burning fuel for kitchen stoves reducing health hazards (especially respiratory diseases amongst women and children), fodder for life stock, fuel for agricultural machinery and generators, fertilizer, building materials etc. Various goods can be industrially produced involving at least six different industrial sectors: food, chemical including fuel, agricultural, building material, pharmaceutical and cosmetics. Biofuels are used in rural areas of Mali and Tanzania to generate electricity (Jumbe et al. 2009). Stimulating the development of these industries would result in a sustainable development of the national economies. This would have a positive impact on the overall social development by generally improving rural living standards, a higher rate of employment, an improvement of health standards, a reduction of crime, drug abuse and other social problems.

A diversity of goods from energy crops is also envisaged in view of a change to be expected within the next 30–40 years concerning the position of bioenergy within the energy mix. According to the OECD/EIA (2010), the use of biofuels for transport is projected to rise rapidly from 1 million barrels per day in 2010 to 4.4 million barrels per day in 2035 owing to the rising oil prices and government support. The energy mix will be most probably dominated by solar energy and biomass will mainly form the basis for sustainable chemical production processes (Pfennig 2007). A widespread production portfolio would reduce the risk of destabilized economic developments.

18.2.2 Areas of Concern and Possible Negative Effects

Due to the different interests and priorities of the various players, it is inevitable that there will sometimes be a clash of interests. But also short sighted approaches could have negative side effects in the long run, which should be avoided. As such, the pre-calculation of possible clashes of interest and/or negative side effects are the challenges for a sound policy and its mechanisms.

Besides a positive impact on community development and the direct improvement of living conditions in rural areas, the production of energy crops and the processing of goods will provide good business opportunities. Although this is generally to be seen as a positive aspect it may also be accompanied by negative side effects. Since the driving force behind business is profit, it has to be expected that for the sake of profit, other "values" from biofuels development may be ignored or even sacrificed. From a commercial perspective, it is the common principle that the lower the production costs the higher the profit margins. As such, technologically advanced firms have a strategic and comparative advantage as they have the leverage to undercut their competitors by cutting prices in order to gain a greater market share thereby driving high-cost firms out of business.

Although some countries in SSA such as South Africa, Mali and Tanzania have been involved in biofuels production, the new entrants should be cautious of the potential risks of fast-tracking biofuels development in their countries. It will therefore be important for each country to analyze the prevailing circumstances thoroughly because of the complexity of the biofuels industry.

18.2.2.1 Neglect of Food Crop Production

From a purely economic reasoning, should the production of energy crops become an interesting business proposition to generate good income for producers, it might be more attractive for producers to grow energy crops at the expense of food crops. With the income from the sale of energy crops, the food needed can be purchased. However, it has to be considered that increasing production of biofuels from food crops may result in a substantial increase in the food price worldwide. The neglect of food crop production will result in dependency on the market for food by food deficit countries. This will then expose consumers to the vagaries of food insecurity. Those who are likely to suffer would be the net food importers particularly the resource poor households including women and the low income earners.

18.2.2.2 Balanced Use of Land

To secure the necessary production of food crops requires a balanced use of available arable land. This applies to the area to be used as well as to the quality of soils. Bearing in mind that investors are business orientated, it is obvious that they will try to get access to land that is of the best quality, to operate at the easiest conditions (topographic structure, infrastructure) and to have a secure water supply (rain fall, irrigation). Optimal conditions will result in minimizing the production costs with optimal harvesting results and, therefore, get maximum profit. If the most suitable land for crop production is used for energy crops, food production on second or third degree land will require higher input costs (land preparation, fertilizer) with lower productivity. As a consequence less food will be produced at a higher price making food in general more expensive.

Any decision concerning a controlled and balanced use of arable land requires a comprehensive assessment about and cataloguing of the available arable land, its position, soil quality, access, topographic structure, and water availability.

It will be less complicated to exercise control over land use on government owned land than over privately owned land. It will be important to find appropriate mechanisms that enable the necessary controls without intruding on the rights of citizens or communities. Besides controlling mechanisms it would be a good starting point to make private land owners understand the necessity of a balanced use of land which would require appropriate campaigns based on well considered policies.

18.2.2.3 Social Component

Labour costs are a major part of the production costs. Most entrepreneurs will try to offer conditions of employment which are advantageous to them, but not necessarily to the employees. These may include low wages and salaries, long working hours, inexpensive working conditions, minimum social security such as sick leave, unemployment insurance, pension fund, medical aid etc. It has to be kept in mind that at the beginning of the industrial revolution, responsible entrepreneurs accepted their social responsibility to look after the welfare of their employees and even built villages for them. Such private social initiatives have in the meantime this did not happen in all industrialised countries. Social protective mechanisms have been promoted during the course of the development of the social market economy in order to prevent exploitation and social injustice. It is the responsibility of national governments in SSA to develop regulatory frameworks (e.g., labour laws) to guard against exploitation of workers by private investors through contract farming arrangements and joint ventures. There are three major stabilizing factors to prevent exploitation and social injustice:

- Labour and social legal regulations
- Trade unions
- High employment rates

In most African economies, supportive and protective mechanisms for the work force are generally underdeveloped and on a low level. This results almost automatically from an economy which is predominantly characterized by subsistence-oriented operations rather than by extensive business operations. Subsistence production is characterized by small production quantities due to limited resources (land, capital and labour). The owner of the business and often family members are the core of the work force. It is therefore difficult to have well developed and efficient labour and social protective mechanisms such as laws or trade unions.

In addition, the employment situation has no stabilizing effect. In cases of a high employment rate, job seekers are scarce and employers will have to make lucrative offers if they want to find suitable candidates. A high unemployment rate such as in SSA with an oversupply of job seekers tends to make labour a cheap commodity.

Ample experience with regard to social justice or at least acceptable conditions of employment during the last century has shown that employees need some kind of protection. Studies have shown that governmental protection often had a paralyzing effect on economic development and growth.

With regard to the SSA countries, there is need to carefully assess whether the existing labour and social protective mechanisms are sufficient to protect the labourers. This is particularly important in view of possibly large and in particular international investments to produce millions tons of feedstock for biofuels production.

On the other hand, it has to be accepted that an investor from Europe for example could not pay wages applicable in Europe due to local labour laws. Wages and salaries are an integral part of the national economy and reflect its economic power. Because of the difference in the price of goods to be imported by wealthy countries and the comparatively low production costs in a developing country, the likelihood of exploitation of workers presents a challenge if no guiding regulations are established.

Although international agreements for the import of bioenergy products are under preparation which, inter alia, will also make reference to the International Labour Organization (ILO) regulations. Nonetheless, the reasons for concern remains. It will not be too difficult to bypass such regulations if they are not firmly anchored in the individual countries by ways of corresponding national regulations. It will be important that the individual SSA countries pass the necessary laws and regulations thus making them judicable at national levels to protect local citizens against exploitation.

18.2.2.4 Nature Protection

In particular large-scale energy crop production may be in conflict with nature protection. This may happen when land, being valuable from a biological point of view as biotope or for biodiversity, could also be used as fertile arable land.

If not protected by respective regulations, biologically valuable areas may be cultivated and thereby destroyed. A very well known example is the clearing of jungles in Asia for the growing of oil palms, whereby the positive effects on climate and environment due to the production of CO_2 neutral energy is by far outweighed by negative effects. Resulting damages would not only harm nature but also other areas of the national economy and in particular tourism.

18.2.2.5 Possible Conflict of Interest Amongst African Countries

Due to differences in technological advancements, there is a possibility that African countries may play against each other in order to get the best biofuels market opportunities. With regard to energy crops, it may well happen that a large investor approaches a country with tempting investment proposals. However, the expected terms and conditions might not match with the adopted policy of that country. It will therefore have to refer to restrictive conditions which would make the investment less lucrative. The investor may try to apply pressure in order to have his conditions accepted with the threat that he would go to a neighbouring country. Such situations are exploitative and need to be avoided. Unfortunately, most SSA countries do not have the capacity to resist such pressure. There are examples in Africa where large pieces of land have been purchased for the growing of energy crops by foreign investors.

It is therefore important that, countries with a common interest adopt a common policy. The SSA countries would strengthen their position if they develop and adopt a common policy including the necessary regulations. This might sound optimistic, however it would be worthwhile SSA should see the potential of energy crops as a chance to proceed on the way towards the African Union Common Agricultural policy, like the common market in Europe paved the way towards the EU. The second best would be to have a policy for a group of states, e.g. the SADC. A policy at national level which is very crucial is the third best option.

18.3 Considerations for Developing a Pan-African Biofuels Policy

The foregoing discussion is a basis for developing a comprehensive policy for SSA countries that addresses the following issues:

- What type of energy crops or biofuels feedstock is best suited for African needs and at the same time can secure an export market and incomes over time without changing the traditional farming system?
- What scale of production of biofuels can best promote and protect the interests of African farmers? Are there opportunities for African farmers to cooperate with large-scale producers to create a win-win situation?
- What type of production system is best suited for African farmers without interfering with the rights of farmers to land and water?
- Can biofuels programs be designed so that they can simultaneously generate food, fodder, various industrial products and fuel in an environmentally friendly manner? Is intercropping of energy crops with other food crops possible to minimize or avoid the reduction in biodiversity or help in mitigating climate change?

The complexity of the issues will require a political framework comparative to a legal framework that can accommodate guidelines and regulations specific to the situation in individual countries. The strategic concept proposed in this chapter needs to be orientated towards all levels, namely, (1) national SSA level, (2) regional SSA level, (3) SSA level, and (4) SSA – European level.

1. **National biofuels policy development:** It is important that each country conducts the necessary research and analysis to identify opportunities and challenges for the development of a biofuels industry. This is important considering the diversity amongst countries with respect to resource endowment, level of technological

advancement and market development. For example, due to differences in land resources, the proportionate use of arable land for food and energy crop production will depend mainly on the total area of arable land available, the soil quality, the size of the population and the annual national food requirements. Each of these factors will differ from country to country.

Since the results are essential to formulate the basis for a common policy, it will be advisable to specify the items to be investigated including relevant criteria of each factor to be addressed in a guideline to be used in every country. Thereafter, it will be possible to extrapolate the results and to find the criteria of the common policy. Main areas of concern are a balanced use of arable land for food crops and energy crops, preventing an exploitation of labour, the protection of biotopes and biodiversity and over-concentration of energy crop production at the expense of other crops.

Particular areas to be supported are the development of national markets, the involvement of various industries and the integration of energy crops and its potential to directly improve the living conditions especially in the rural areas. On the one side, this will require analyzing the potential of energy crops and biomass as well as the existing and latent demand of goods produced. In particular, the inclusion of the industry on a broad level will be of utmost importance for a sustainable economic development of each country. Likewise, national markets as the backbone of national economic developments will have to be developed. An appropriate policy will have to support the results aimed at developing suitable mechanisms. On the other side, negative effects will have to be anticipated and identified and thereafter appropriate controlling mechanisms designed.

2. **A regional biofuels policy development:** There is need to establish a task force or commission on regional levels such as SADC, COMESA, etc. with representatives from each country. Such commission may be given the task to collect the results of the research in each country and to draft a policy based on the similarities and differences among countries. To coordinate the research in each country to obtain the information to be given to the regional committee, each country could appoint a commission or it could entrust the central coordination to one ministry. With reference to the guideline mentioned above, it will be necessary to appoint the regional commission at least at the same time as the national ones although it would be even better to first establish the regional commission. It could give already valuable advice to individual countries with regard to the appointment of the national committee or coordinator.
3. **Pan-African biofuels policy development:** With regard to the common SSA policy as the ultimate goal, it will be important to appoint a SSA commission, possibly subjected to the AU. Members of such a commission should come from the regional commissions. The task of the SSA commission should be to collect the results of the work of the regional commissions and to incorporate the same in a draft of a SSA policy to be submitted to the AU. The Pan-African biofuels policy needs to ensure that international interests do not conflict with the paramount interests and targets of African countries to improve the living conditions of its citizens and to develop their markets for a sustainable economic development.

It has to prepare itself for controlling the process in order to get the results wanted. For this, it is necessary that the policy includes appropriate mechanisms to implement the policy.

4. **SSA-European biofuels policy development:** It was mentioned earlier that the energy crop production in SSA has an international or rather transcontinental dimension that concerns – or should concern – in particular Europe. It is, therefore, recommended to establish a joint SSA – European commission to develop guidelines for a win-win biofuels policy. Due to the historic developments, but also due to the geographical position, Europe and SSA have a lot of common experiences and opportunities but also challenges that should form a solid basis for a long-term partnership. From a European perspective, SSA problems are to be seen as European problems. Although SSA should concentrate on the SSA internal significance of energy crops, it also has to consider the international perspective. The EU target to introduce 10% of alternative fuels by 2020 may open up export markets for SSA although the access to arable land in eastern Europe makes such export developments by no means certain. The historical developments and the geographical position of Europe and Africa have established a relationship that would and will serve as a stabile basis for cooperation with regard to biofuels development.

   A common African-European Biofuels policy has the potential to mutually harness the potential of biofuels to economic development in both Africa and Europe. On one hand, Europe is both a major exporter and the world's largest importer of food, mainly from developing countries including Africa. This will broaden African access to European market; Again, the European farming sector uses safe, clean, environmentally friendly production and processing methods that could enhance the productivity of African farmers.

   On the other hand, Africa has ample land, cheap labour and good environment for the growing different types of energy crops for biofuels production. All these point to the need for a common African-European biofuels policy.

References

Aksoy MA, Beghin JC (2005) Global agricultural trade and developing countries. The World Bank, Washington DC

BP (2009) Statistical review of world energy, Houston

BP (2010) Statistical review of world energy. www.bp.com/statisticalreview

Economic Commission for Africa/African Union (2010) Economic Report on Africa 2010. Promoting high-level sustainable growth to reduce unemployment in Africa. ECA, Ethiopia

Gillson I, Houlton C, Buncombe K, Page S (2004) Understanding the impact of cotton subsidies on developing countries. Working Paper, MPRA Paper No. 15373. Unpublished

Jumbe CBL, Msiska F, Madjera M (2009) Biofuels development in Sub-Saharan Africa: are the policies conducive? Energy Policy 37(11):4980–4986

Kgathi DL, Mfundisi K (2009) Potential impacts of the production of liquid biofuels on ecosystems and so sustainability: The case of Botswana. Paper presented at the workshop on Energy, Ecosystems and Livelihoods, 28–29 Apr 2009, Pretoria, South Africa

Lanely K (2006) Biofuels: promises and constraints. International Food and Agriculture Trade Policy Council, IPC Discussion Paper, Dec 2006, Washington, DC
Nolte M (2007) Commercial biodiesel production in South Africa: a preliminary economic feasibility study. MSc Thesis. University of Stellenbosch, South Africa
OECD/EIA (2010) World Energy Outlook 2010. International Energy Agency, France
Pfennig A (2007) Supporting debottle-necking of global human processes by applying appropriate balances. Biotechnol J 2:1485–1496. www.biotechnology-journal.com

Part IV
Sustainability of Biomass Production and Use

Chapter 19
Keynote Introduction: Sustainability Considerations for Biofuels Production in Africa

Rocio A. Diaz-Chavez and Jeremy Woods

Abstract Different drivers have been identified for the promotion of bioenergy crops in both developed and developing countries. The interest in bioenergy projects and biofuels has been driven by a combination of factors, but primarily by initiatives on climate change to reduce GHG, to reduce dependency on oil fuels and by the potential for socio-economic development. At the same time this has led to an increasing concern with the wider implications of biofuels production, particularly if grown at large-scale. Such concerns include environmental sustainability, green house gases (GHG), land use changes and impacts on food prices. Verification systems offer a possibility to assure sustainable production, but have not fully addressed the perception of communities and the effects at different scales. This chapter focuses on sustainability considerations in Africa for bioenergy crops, the certification and verification implications and what is seen regionally as good practice.

Keywords Sustainability assessment • Biofuels Africa • Renewable energy • Sustainable development • Certification and standards

R.A. Diaz-Chavez (✉)
Centre for Environmental Policy, Imperial College London, Exhibition Road 313A Mech Eng Bld South Kensington, London SW7 2AZ, UK
e-mail: r.diaz-chavez@imperial.ac.uk.

J. Woods
Porter Institute, Imperial College London, Exhibition Road 313A Mech Eng Bld South Kensington, London SW7 2AZ, UK
e-mail: j.woods@imperial.ac.uk

R. Janssen and D. Rutz (eds.), *Bioenergy for Sustainable Development in Africa*,
DOI 10.1007/978-94-007-2181-4_19,

19.1 Introduction

Different drivers have been identified for the promotion of bioenergy crops in both developed and developing countries. The interest in bioenergy projects and biofuels has been driven by a combination of factors, but primarily by initiatives on climate change to reduce GHG, to reduce dependency on oil fuels and by the potential for socio-economic development. The growing interest in bioenergy projects and particularly biofuels has led to increasing concern with their wider implications, particularly if grown at large-scale. Such concerns include environmental sustainability, greenhouse gases (GHG), land use changes and impacts on food prices.

Particularly in the EU and the USA where targets have been imposed to achieve GHG reductions, the development of the biofuels industry has seen many arguments in favour of the use of biomass (e.g. security of energy supply, diversification of energy sources, low-carbon emission, an alternative market for agricultural products, land rehabilitation, amongst others). Furthermore, it has been considered that importing biofuels from developing countries will not only allow developed nations to diversify their energy mix and meet environmental requirements, but could also create new economic opportunities for rural areas, allowing low-income countries to become vital producers and exporters of a valuable new good.

Nevertheless, some possible negative social and environmental implications have also been raised for both developed and developing countries, especially on issues such as land competition, the questionable reduction of emissions, 'the fuel versus food' debate and the indirect effects of land use change (Diaz-Chavez and Woods 2008). To counterbalance the possible negative effects some measures are being put in place to ensure their sustainability (e.g. certification, accreditation, and traceability) that will have a major impact, either positive or negative, on the development of the biofuels industry.

For this, different standards and criteria have been developed. Main environmental and social criteria have been included in a number of proposals for standards such as the Roundtable on Sustainable Palm Oil (RSPO), the Roundtable on Sustainable Biofuels (RSB), the Round Table on Responsible Soy (RTRS) and some country initiatives such as the meta-standard from the Renewable Fuel Agency of the UK which considers the use of available standard systems (i.e. Forest Stewardship Council, Rainforest Alliance), or the Netherlands Cramer Criteria.

Despite the efforts from some organisations (e.g. the Roundtable for Sustainable Biofuels (RSB) and the Global Bioenergy Partnership (GBEP)) to include the opinion of developing countries, where a major production is expected to provide enough biofuels for the EU to comply with its targets, some aspects of the criteria developed have different relevance in developing countries (such as Brazil and a number of African countries). This is the case for instance for socio-economic criteria and some environmental issues of biofuel production.

19.2 Sustainability Assessment in the Biofuels Context

The main objective of a sustainability assessment is to evaluate the sustainability performance of the economic, environmental, social and political processes or products (in this case bioenergy projects or initiatives). For an effective sustainability assessment there must be clear delineated principles and if possible decision criteria based on well integrated understanding of the key requirements for sustainability including international agreements, policies or directives which reflect consensus on the approaches to achieving sustainability. The framework adopted included the "traditional" environmental, social and economic themes and an added one on policy and institutions. The links between the different themes of sustainability also contribute to the better understanding of the sustainability process and assessment (Diaz-Chavez 2006).

Considerable amount of work has been focused on sustainability issues, but not on a global scale. A more systematic approach with greater acceptance is necessary. The problem is how to create a global standard that allows for national and global activities, given the complexity of many of the issues involved. Any sustainability standard must include the three key components: economic, social and environmental aspects. In addition, a political and institutional new pillar has to be included since many of the issues implied in sustainability are of political nature (e.g. targets) (Diaz-Chavez 2008).

A number of issues need to be considered to ensure both a sustainable production and use of biomass oriented towards energy needs and reducing GHG emissions. Amongst these are environmental and social concerns, which bring into consideration the area of land required from energy crops for producing electricity and biofuels for transport. Additionally, other concerns include the effects that the large-scale cultivation of energy crops and use of residues may have on biodiversity, soils, hydrology and landscape (Diaz-Chavez and Woods 2008).

One of the main difficulties in assessing sustainability is the implication of "whose sustainability" is being addressed. This implies a review of the concepts and understanding of different societies and also the main goals for the achievement of sustainable development. Additionally, it must be assessed how sustainable development can be measured or reviewed and which tools shall be used (Sheate 2009). Some environmental tools are more appropriate for particular objectives or activities (e.g. Strategic Environmental Assessment (SEA) for policies, plans, and programmes) whilst others are diverse but at the same time limited to specific activities or products (e.g. Life Cycle Assessment (LCA)).

The case of biofuels has also produced a review of which available environmental management tools can be used in a quite new cross-cutting and sectoral area. One opportunity is to use at good practices to work towards improvement of available tools.

19.3 Good Practice Definition

The Food and Agriculture Organisation (FAO) defines **Good practices** as: "Any collection of specific methods that produce results that are in harmony with the values of the proponents of those practices. In agriculture, applies available knowledge to addressing environmental, economic and social sustainability for on-farm production and post-production processes resulting in safe and healthy food and non-food agricultural products" (FAO 2005).

According to the International Organisation for Standardisation (ISO 2006) the definition of a standard is: "*A normative document, developed according to* ***consensus*** *procedures*". The ISO also provides definitions for all activities related to the standardisation process (see Box 19.1).

In practice, a certificate is issued when a producer of a product (or process) has answered a set of standardised questions categorised by the principles that make up the standard as follows:

- '**Principles**' which are defined as 'general tenets of sustainable production'
- '**Criteria**' 'Conditions to be met to achieve these tenets' and which help to define the indicators to be answered.
- '**Indicators**' the individual questions that show how a farm, producer or company could prove that a particular criterion is met.

Therefore, it is the indicators that need to provide sufficient detail to ensure that the principles underpinning the standard are being adhered to. However, in complex systems a 'value judgment' may be necessary to set the detail, total number and

Box 19.1 Definitions from ISO on Standards (ISO 2006)

A '***Standard***' refers to a set of principles and criteria to be used consistently as rules, guidelines, or definitions of characteristics to ensure that materials, products, processes and services meet their purpose. The 'standard' will also define indicators and methods that are used to measure compliance with principles and criteria.

'***Certification***' refers to the issuing of written assurance (the certificate) by an independent, external body – a certification body – that has audited an organisation's management system and verified that it conforms specifically to the standard.

'***Accreditation***' refers to the formal recognition by a specialised body – an accreditation body – that a certification body is competent to carry out certification.

'***Assurance scheme***' generally refers to the overall framework relating to the development of a standard, the accreditation of certification bodies, and the certification of products and services (Woods and Diaz-Chavez 2007).

complexity of the indicators. Too much detail and the certification procedures become too unwieldy, expensive and difficult to administer. Too little detail and serious doubts will be raised about the ability of the scheme to assure that its products meet the standard (Woods and Diaz-Chavez 2007).

This balance between coverage, detail and simplicity can only be met by a transparent decision process that uses a 'representative' set of stakeholders encompassing a 'balance of interests' to define, the principles, criteria and indicators of the standards. For the case of the (COMPETE 2009) project a review of different standards and schemes was produced (Diaz-Chavez 2010a) and an alternative set of principles and criteria was developed based on the opinion of different stakeholders, considering sustainability assessment in a wider context and available cases of good practice.

19.3.1 Selection of Principles

Different activities were conducted in order to arrive to a selection of principles, amongst them: a benchmarking process of different sustainability schemes, interviews and consultation with different stakeholders including partners in the COMPETE project, high level conferences and meetings with policy makers of different African countries. During the meetings and conferences, different stakeholders ranging from producers (companies and farmers), policy makers, NGOs, academics and consultants have provided their opinion, which were compiled in different documents and reports available at the project website (www.compete-bioafrica.net).

Countries that have participated in the different conferences and meetings included: Botswana, Burkina Faso, Ethiopia, Ghana, Kenya, Malawi, Mali, Nigeria, Senegal, South Africa, Tanzania, and Zambia.

One noticeable point is that some of the principles and criteria widely discussed in the EU and the USA context, do not present the same relevance for developing countries. An example is the greenhouse gas (GHG) emissions calculation, land availability, the debate of indirect land use change (ILUC) and the food versus fuel debate. This last one is considered more in terms of competition of feedstocks and land availability which in several African countries is not meaningful due to large potential availability of land.

Social issues may also be valued differently in the North and South. For instance, child labour concerns are not seen in the same context as long as children collaborate (not under exploitation circumstances) with the family tasks in the farms and do not neglect their studies.

19.3.2 Boundaries and Usefulness

The principles were selected within the sustainability framework mentioned above and according to the definitions of principles with the intention to provide a clear and balanced guideline for *Good Practices* (Diaz-Chavez 2010b). The intention of the principles and criteria was to provide an alternative guideline that could be used by different stakeholders when considering:

- to initiate or assess a bioenergy proposal or project
- to assess the sustainability of a feasibility report for a bioenergy proposal or project
- to review policy guidelines and assist in the decision-making process of a bioenergy proposal or project
- to review and/or assess an ongoing bioenergy proposal/project

Finally, the principles are not exhaustive and may differ under different frameworks, projects, experts, countries or any other stakeholders' opinion.

19.4 Selected Principles

This section presents and briefly explains the principles selected in the framework of the COMPETE project. Some of the principles have clear links amongst them, especially regarding the compliance with policies and regulations (Principles 10 and 11). Table 19.1 presents the 12 principles and the topic they are related to (environmental, social, economic, policy and institutions).

Table 19.1 Principles for sustainability assessment for bioenergy initiatives

No.	Principle	En	S	Ec	P
1.	Good agro-ecological and forestry practices (biodiversity, soil)	✓			
2.	Not affecting water supply and quality	✓			
3.	No land use change that detrimentally affects food security	✓			
4.	Community participation (from planning)		✓		
5.	Women's participation (from planning)		✓		
6.	Skills transfer (management, business, agriculture)		✓		
7.	Community inclusion in business or economic model (contract with investor or NGO)			✓	
8.	Added value in the community (individual, money, assets, land, co-products)			✓	
9.	Improvement in services and infrastructure(energy supply, health) reinvestment of revenue within the community			✓	
10.	Compliance with National and/or guidelines for bioenergy policy in place				✓
11.	Compliance with Local programmes, regulations and/or plans in place				✓
12.	Respect land rights and avoid displacement				✓

Source: Diaz-Chavez (2010b)
Note: *En* Environmental, *S* Social, *Ec* Economic, *P* Policies

19.4.1 Environmental Principles

Principle 1: Good agro-ecological and forestry practices (biodiversity, soil)
This principle considers that the basic environmental characteristics to grow bioenergy crops will be followed according to the agro-ecological and forestry conditions of each country, region or community. They include: land use type, soil conditions (adequate for the selected bioenergy crop), soil management and protection, no negative effect on biodiversity, good agricultural practices (e.g. use of fertilisers and pesticides), good forestry practices (e.g. conservation and management).

Principle 2: Not affecting water supply and quality
This principle seeks to consider that especially in areas where water is constrained it will not be used for bioenergy crops or be limited or managed according to the good agricultural practices. It also seeks to avoid the pollution of water that negatively affects its quality (e.g. due to overuse or bad use of fertilisers and pesticides). The principle considers the use of water first for human consumption and for food crops.

Principle 3: No land use change that detrimentally affects food security
Land use for bioenergy crops should be considered within the national policies and agro-regionalisation along with other policy instruments and guidelines (if available). The change of current agricultural land for bioenergy crops should not affect food security.

19.4.2 Social Principles

Principle 4: Community participation
It considers the community participation in the bioenergy project, programme or plan since the early stages of the planning process. Community participation is not only part of a sustainability process but will also contribute to the success of the project and will allow the community to participate in the decision-making process. Additionally, it provides a feeling of "belonging" and "being recognised" by the community.

Principle 5: Women's participation
This principle looks for women participation in bioenergy initiatives from the early stages of the planning process. Including women since the beginning will allow to provide direct gender benefits and empower women in activities directly related to them.

Principle 6: Skills transfer (management, business, agriculture)
Transfer of skills is related to the added value of growing bioenergy crops. This includes different stages of the business cycle and it applies to the different production and scale schemes (e.g. out-growers, small, medium and large-scale). It also includes productive areas (agriculture), transformation (e.g. extraction of oil from seeds), management, and business skills (e.g. revenue and trading).

19.4.3 Economic Principles

Principle 7: Community inclusion in business or economic model (Contract with investor or NGO)
The inclusion of the community in the business or economic model will prevent the exploitation of its members and will provide the mechanisms to comply with other principles such as Principles 6 and 8.

Principle 8: Added value in the community (individual, money, assets, land, co-products)
The added value from the bioenergy initiative can be translated not just in terms of an increment in the income of the community and at individual level (e.g. savings or additional income), but also with additional assets (e.g. animals, food production), land (e.g. individual or communal land) and co-products (e.g. income from soap making).

Principle 9: Improvement in services and infrastructure (energy supply, health) and/or reinvestment of revenue within the community
At community level the possibilities of reinvesting the revenue to improve services and infrastructure (if previously agreed within the community) is considered as main objective. These services can be related to energy supply or better access to health services.

19.4.4 Policy and Institutions Principles

Principle 10: Compliance with national policies and/or guidelines for bioenergy policy in place
Where national policies or guidelines regarding bioenergy production exist, these should be followed by all stakeholders involved in the bioenergy initiative such as the proponent, the community, national and international consultants and developers, investors, NGOs amongst others.

Working with the national, regional and local authorities is considered to be important especially for developers and investors to look at cross-cutting sectors (e.g. environment, social, industry, agriculture sectors).

Principle 11: Compliance with local programmes, regulations and/or plans in place
When working with the national, regional and local authorities it is considered important to be aware of all programmes, plans and regulations at local level. Compliance with them will strengthen the bioenergy initiative and avoid conflicts with different stakeholders and regulators.

Principle 12: Respect land rights and avoid displacement
The debate on land rights in developing countries (mainly in Africa) led to this principle to be considered by communities, governments and investors. Displacement

needs to be avoided as far as possible or regulated by the decision of (local) authorities after thorough studies with the involvement of all affected stakeholders. Adequate compensation and further studies for relocation need to be included according to international practice (e.g. such as the guidelines from the World Bank 2008).

19.5 Assessing Bioenergy Initiatives

The proposed principles described above were included in a form to assess bioenergy initiatives or projects. The assessments were conducted following a review of basic information given from the initiative or project and field visits. The information for the assessment needs to include as minimum the following points:

- Type of initiative (e.g. private, government, community, NGO, other)
- Agreements or reviews of the initiative with local, regional and national authorities (e.g. for the compliance with regulations and policies of Principles 10 and 11)
- Type of land use for the bioenergy project (agricultural land, forest, grassland, other)
- Type of feedstock (e.g. cassava, jatropha, palm oil, sugarcane, other)
- Production scheme (community, out-growers, cooperative, private, other)
- Scale of the plantation (number of hectares of agricultural land at the farm, community or regional level[1])
- Contract or agreement type with the farmers or out-growers (e.g. fixed contract, employment, number of years, fixed price or alternative price model)
- Final use of the feedstock, co-products and sub-products
  - feedstock to be sold without any treatment,
  - oil or fuel processed on-site,
  - oil or fuel to be used on-site for electricity,
  - oil or fuel to be sold for local, regional, national use,
  - fuel for export
- Community participation in the business scheme since planning (meetings organised, community decisions, women's included in scheme)
- Overall benefits for the community (individual income including new assets such as animals or land, electrification scheme, cooperative benefits)

The assessment was conducted with the form following a qualitative score system. Although it is recognised that the assessment will be of a subjective nature, in order to reduce bias it is expected that the assessment will be conducted by a stakeholder or expert with experience on bioenergy projects and independent from the proposed project.

[1] Determining the scale of the plantation on small, medium and large will vary according to the farm, community or region where the initiative is proposed and to the type of feedstock. This is the reason why it is considered here as a relation between the number of hectares and the total agricultural area in the community or region.

The assessment was based on a qualitative scale for each principle according to the following scores:

1. The project does not consider this principle (0%)
2. The project covers this principle partially (<30%)
3. The project covers partially this principle in (30–70%)
4. The project covers partially this principle in (<70%)
5. The project fully covers the principle (100%)

The assessor validated the score by writing short statements on the reasons for giving the score. Thereby, subjectivity is reduced and decision-maker are provided with more information for improvements in the initiative. A maximum of 60 points can be scored. The project needs to be reviewed if it gets less than 35 points.

19.6 Assessment of COMPETE Examples

In order to provide aid for the use of the guidelines, a series of examples of current projects or initiatives were assessed according to the principles of these guidelines. Table 19.2 presents the main findings of the assessment (see COMPETE website).

The assessments of the initiatives allowed seeing different perceptions of the local production. Most of the examples in Africa had jatropha as a feedstock. There were some other examples reviewed with sugarcane, but because of the sensitivity of the cases it was decided not to publish them in this chapter.

Examples also vary with respect to type of production and scale. There were initiatives of NGOs as the one developed by TaTEDO in Tanzania, but also private initiatives as the one from Diligent in Tanzania. The importance of the initiatives for the communities are obvious as the main goal is the access to energy which provided additional benefits such as the use of mills or the use of electricity in business (Janssen 2009). Another important community initiative was the one developed by Winrock India in Ranidehra, India, where the major benefit was also access to electricity. This is a good example for community involvement in a project and the process to agree on the conditions for putting the project in place. This project benefited 105 households and 553 people. Additional benefits included the installation of a rice de-husking machine and streetlights (Khatun and Diaz-Chavez 2009).

As one of the main concerns in Africa is land tenure, many of the private initiatives had to deal with this. In the case of Diligent in Tanzania, Farioli and Portale (2009) reported that because of the alternative form of production with outgrowers the company managed to solve the problem and at the same time to provide an input to local livelihoods.

The case of sugarcane analysed by Watson (2009) showed the involvement of the local community as the private initiative Kilombero Sugar Company Ltd (KSCL) initiated a partnership with out-growers to stimulate community development. The partnership led to the establishment of the Kilombero Community Trust (KCT) which (1) set up a farm to demonstrate best agricultural practices to growers, (2) provides

Table 19.2 Assessment of four case studies

No.	Name of initiative	Organisation	Country	Objective	Feedstock	Type of investment	Score
1.	Ranidehra, Chattisgarh	Winrock International India (WII)	India	Electrification village	Jatropha	NGO	48
2.	Integrated Sustainable Energy Services for Poverty Reduction & Environmental Conservation	Tatedo	Leguruki, Tanzania	Energy services	Jatropha	NGO	49
3.	Out-growers Scheme for oil production	Diligent	Tanzania	Out-growers scheme for biofuel	Jatropha	Private	45
4.	Sugar producer	Kilombero Sugar Company Ltd (KSCL)	Kilombero, Tanzania	Sugarcane cultivation and sugar production. Ethanol and bioenergy plants by 2012	Sugarcane	Private/Community/ Government	49

regular capacity building workshops on management and business skills required by growers, and (3) mapped growers farms and entered information on their biophysiographic and production characteristics in a Management Information System in order to assist growers in planning their harvesting and delivery schedules to the factories.

This case study shows that private initiatives have the potential to create benefits for local communities by increasing energy access and contributing to improved livelihoods.

The next step in the proposed COMPETE methodology is to generate applicable indicators that can be applied to this framework. Sbiti and Diaz-Chavez (2010) started work on this for jatropha plantations in southern Mexico. This will contribute to the better understanding of the framework in a different region.

19.7 Conclusions

The principles presented in this chapter are meant to provide guidelines for a general assessment and are regarded as "What is important" in developing countries, particularly in Africa where most of the opinions were collected for the selection and definition of the principles.

The assessment of the bioenergy initiatives provide further information in terms of the sustainability reliance of the project or proposal to the different stakeholders including decision-makers, investors, NGO's and community leaders.

These guidelines for Good Practice do not attempt to replace any kind of regulation, verification, or certification system that is in place within national interpretations. Nevertheless, they can prove useful for decision-making and policy implementation in those countries where these are not yet established or produced and where bioenergy production is foreseen.

A further step is to produce indicators which could be used at local level. One of the challenges for this is to find adequate data for implementing and testing the indicators.

References

COMPETE (2009) Competence platform on energy crop and agroforestry systems for arid and semi-arid ecosystems – Africa. http://compete-bioafrica.net/

Diaz-Chavez RA (2006) Measuring sustainability in peri-urban areas. In: McGregor D, Simon D, Thompson D (eds) The peri-urban interface in developing areas: approaches to sustainable natural and human resource use. Earthscan, London

Diaz-Chavez RA (2007) Comparison of draft standards- contribution to ECCM, 2006. Environmental standards for biofuels. A report commissioned by the LowCVP. The Edinburgh Centre for Carbon Management, IIED, ADAS, Imperial College London, London

Diaz-Chavez RA (2008) Best sustainability report. Report on deliverables D 9.11 and D 9.15. Imperial College London, London

Diaz-Chavez RA (2010a) Understanding and implementing certification, COMPETE. http://compete-bioafrica.net/sustainability/COMPETE-WP3-D3.5-Implementing%20Certification-Final.pdf

Diaz-Chavez RA (2010b) Good practice guidelines to project implementers. COMPETE. http://compete-bioafrica.net/sustainability/Annex3-3-3-COMPETE-032448-3rdReport2009-D3.4-Good%20practices%20report-Final.pdf

Diaz-Chavez RA, Woods J (2008) Sustainability assessment of biofuels in practice. In: 5th biofuels conference. Winrock International, New Delhi

FAO (2005) FAO/Netherlands conference water for food and ecosystems. http://www.fao.org/ag/wfe2005/glossary_en.htm

Farioli F, Portale E (2009) Good practice assessment for bioenergy projects. Diligent Tanzania Jatropha oil production. COMPETE

ISO (2006) ISO in brief – International standards for a sustainable world. International Standards Organisation. Available at: http://www.iso.org/iso/en/prods-services/otherpubs/pdf/isoinbrief_2005-en.pdf. Cited Jan 2006

Janssen R (2009) Good practice assessment for bioenergy projects. Tatedo Leguruki Project. COMPETE

Khatun K, Diaz-Chavez RA (2009) Good practice assessment for bioenergy projects. India Jatropha electrification initiative of Winrock International India (WII). COMPETE

Sbiti L, Diaz-Chavez RA (2010) The sustainable feedstock challenge: a case study of *Jatropha curcas L.* in Mexico. Draft paper under review

Sheate W (2009) The evolving nature of environmental assessment and management: linking tools to help deliver sustainability. In: Sheate, W (ed). Tools, techniques and approaches for sustainability: collected writings in environmental and assessment policy and management. World Scientific (Singapore), pp 1–32

Watson H (2009) Good practice assessment for bioenergy projects. Kilombero Sugar Company Ltd (KSCL), Tanzania, COMPETE

Woods J, Diaz-Chavez RA (2007) The environmental certification of biofuels. Report for the OECD, Paris

Chapter 20
Environmental Impacts of Biofuel Production in Africa

Kingiri Senelwa, Lazare Etiégni, Odipo Osano, Kirongo Balozi, and Moses Imo

Abstract Although total bioethanol production in Africa in 2006 was less than 500 million litres, the potential is considered high. South Africa and East Africa alone are estimated to have an annual potential of 7.3 and 1.3 billion litres, respectively. In addition, there is intense interest in biodiesel with large-scale projects being developed across the continent. The land considered for biofuels is significant, estimated at 5.5 million hectares, some of which will be located in fragile ecosystems and in ecologically sensitive environments. In Mali, Ghana, Sudan, Ethiopia and Madagascar, up to 2.5 million hectares of land has already been allocated to foreign investors for biofuel production. The motivating policy goals for biofuels production and the high degree of biodiversity and diverse climatic conditions notwithstanding, increased production, different agricultural practices especially on highly degraded land due to long-term agricultural mismanagement portend key environmental issues associated with land requirements and farming systems, conversion technologies and scale of operation. Impacts include destruction of habitats and biodiversity, deforestation, and declining water quantity and quality. Whilst these impacts may not be apparent in Africa, they present important researchable questions for planning as Africa gears for increased participation in the international biofuels markets. Tools to define and assess areas suitable for sustainable biofuels production exist, and should be used by governments to include biofuels into an overall energy, climate, land-use, water and agricultural strategy. This shall benefit society, the economy and the environment as a whole.

K. Senelwa (✉) • L. Etiégni • K. Balozi • M. Imo
Department of Forestry & Wood Science, Chepkoilel University College, Moi University, P.O. Box 1125, Eldoret 30100, Kenya
e-mail: ksenelwas@yahoo.co.uk; lazetiegni@amatala.org; balozibk@hotmail.com; imomoses@yahoo.com

O. Osano
School of Environmental Studies, Moi University, P.O. Box 3900, Eldoret 30100, Kenya
e-mail: odipo@africaonline.co.ke

R. Janssen and D. Rutz (eds.), *Bioenergy for Sustainable Development in Africa*,
DOI 10.1007/978-94-007-2181-4_20,

Keywords Biofuels • Africa • Environmental impacts • Life cycle • Greenhouse gas balances

20.1 Introduction

Biofuel development policies have been driven by concerns over energy security, the need for convenient alternatives to fossil fuel, and a desire to reduce greenhouse gas emissions. From an African perspective, additional themes include the need to (1) provide opportunities to diversify agricultural production, enhance employment in rural areas, raise rural incomes and improve quality of life, (2) diversify fuel supply sources and develop long-term replacements for fossil oil whilst reducing expenditure on imported fossil energy, and (3) boost technological developments in countries with favourable conditions for biofuels production. Unfortunately, biofuels production and use can be a major source of serious environmental impacts which can threaten the overall social and economic development objectives (Lazarus et al. 1995). Thus, incorporation of environmental considerations is an important new area for energy planners.

A key challenge in pursuing this objective is to incorporate strategies to limit negative impacts of biofuels production. In assessing the impacts, the objectives should be to inform biofuels development and utilization policies whilst promoting sustainability standards and criteria for the production of environmentally benign biofuels. In the context of Africa, key questions that need to be addressed in assessing the environmental impacts of biofuels include:

- What are the positive environmental impacts of biofuels production and processing?
- Will biofuels help reduce greenhouse gas emissions and mitigate climate change?
- Do biofuels threaten habitat integrity, biodiversity, soils and water flows and cycles?
- How will enhanced biofuels production impact on agricultural land and other land use alternatives?
- Can environmentally sustainable biofuels production be achieved in Africa?

An additional theme would consider how to ensure that the production and use of biofuels is sustainable.

20.2 Biofuels in Africa: A Historical Perspective and Trends

In Africa, interest in ethanol from biomass to reduce oil imports dates back to the 1970s. In Kenya, efforts to blend ethanol with gasoline declined in the mid 1990s due to conflicting interests amongst petroleum distributors and a lack of concerted policy initiatives. Renewed interest has seen current ethanol production capacity increase to about 125,000 l/day. Zimbabwe began to produce bioethanol to supply a 5% mix in road fuel in the early 1980s whilst Malawi's experience dates back to 1986 with a current installed bioethanol capacity of 32 million litres with plans to double the capacity. Other countries with active liquid biofuels development programmes include South Africa and Ghana where the respective governments have pledged to fund biofuels projects (GRAIN 2007), Nigeria, Tanzania, Burkina Faso, Cameroon, Lesotho and Madagascar.

Although the total bioethanol production in Africa was less than 500 million litres in 2006, the potential is considered to be high. For instance, the potential in South Africa and East Africa alone is estimated at 7.3 and 1.3 billion litres, respectively. In addition, there are intense interests in biodiesel with large-scale projects being developed across Africa. The land considered for biofuels, estimated at 5.5 million hectares, is significant with more than 119,000 ha in 97 large-scale projects in East Africa by early 2008. In Mali, Ghana, Sudan, Ethiopia and Madagascar, up to 2.5 million hectares of land has been allocated to foreign investors for biofuels development.

20.3 Environmental Impacts of Biofuels Production in Africa

Environmental impacts of biofuel production depend to a large extent on the selection of areas that are used for production, the scale of operation, the crops cultivated and the farming practice. From a life cycle assessment perspective, biofuels production can have both positive and negative environmental impacts depending on the type of feedstock and the fuel production pathways and technologies (Rösch and Skarka 2009). The pathway contains five major stages: (1) feedstock production, (2) feedstock transportation, (3) fuel processing, (4) fuel blending and distribution, and (5) fuel use. Although impacts are generated at all stages of the cycle, the most significant consequences occur in the first, third, and fifth stages. Further, the nature and extent of impacts varies according to climatic and production conditions (Berndes 2002), scale of production, cultivation, and land-management practices.

20.3.1 Energy Balances of Biofuels

The contribution of biofuels to energy supply depends both on the energy content of the biofuel and on the energy used for its production which defines the fossil net energy balance (NEB). A fossil NEB of 1 means that the energy needed to produce 1 l of biofuels equals its energy content (no net energy gain or loss), whilst a balance of 2 means that 1 l of biofuel contains twice the amount of energy than required for its production. It was shown that the NEB of biodiesel produced from different feedstocks can vary from 0.46 to 6.35 (Licht 2007). Petrol and diesel have NEB values of 0.8–0.9, implying net energy losses of 19.5% and 15.7%, respectively, because energy is consumed in refining crude oil into usable fuel and transporting it to markets. If biofuels have NEB values exceeding 0.8–0.9, they contribute to reducing dependence on fossil fuels. The favourable NEB for some biofuels illustrates the potential to displace a significant proportion of fossil fuels in Africa. However, in developing biofuels, effort should be taken to optimize energy balances by learning from past experiences and best case scenarios. The selection of crops for large-scale biofuels projects in Africa should be guided by favourable energy balances and crops with low NEB should be discouraged.

20.3.2 Greenhouse Gas Balances of Biofuels

Greenhouse gas (GHG) balances are the result of a comparison of biofuels with fossil fuels with respect to all GHG emissions throughout the entire life cycle. An analysis of greenhouse gas emission impacts of commercial biodiesel production using imported soybean oil from Brazil in South Africa showed that upstream total emissions from the biomass feedstock production, transport and manufacture were less than 16% of fossil baseline emissions. Whilst upstream emissions were significant, they did not outweigh the substantial emissions reductions, even when the oil feedstock is transported for long distances by ship. Fargione et al. (2008) and Righelato and Spracklen (2007) estimated the carbon emissions avoided by various ethanol and biodiesel feedstocks grown on existing cropland (i.e. sugarcane, maize, wheat and sugarbeet for ethanol, and rapeseed and woody biomass for diesel) and found that more carbon would be sequestered over a 30-year period by converting the cropland to forest. Thus, extensive deployment of biofuels will have a positive effect on climate change associated with greenhouse gas emissions in the energy sector.

20.3.3 Biodiversity and Habitat Integrity

High-profiled biofuels development projects in Africa include for example (1) the plan to turn a third of Uganda's Mabira forest (7,100 ha) to sugarcane for the production of electricity and ethanol, (2) the plan to cut down thousands of hectares of rainforest on two islands of Lake Victoria (Kalangala and Bugala) for conversion into a palm oil plantation, (3) Tanzania's model of large-scale, monoculture production of biofuels in regions that can attract investment including Ruipa, Usangu plains and Wami Basin, and (4) Benin's humid areas of Oueme, Plateau, Atlantic, Mono, Couffo and Zou which have been earmarked for palm oil expansion.

Although these developments will intensify deforestation and enhance biodiversity loss, the data and analysis needed to assess the extent and consequences are still lacking. The lack of data notwithstanding, these examples illustrate that the frontiers for increased biofuels production are the remaining wetlands, sacred and communal forests, and fallow lands with rich biodiverse ecosystems comprised of woodlands and forests. Mabira forest for instance, one of the key water catchment sources for the Nile River and Lake Victoria, is estimated to store 3,905,000 ton of carbon dioxide.

Conversion to sugarcane plantation threatens 312, 287 and 199 species of trees, birds and butterflies, respectively, some of which are endemic to the forest.

Although some biofuel crops (e.g. jatropha and castor) have grown in Africa under cultivation and in the wild for decades, the introduction of large plantations of such crops raises serious questions about potential impacts on native ecosystems, especially since certain crops are considered invasive in some parts of the world. Invasion of native vegetation by biofuel crops is a major concern for some feedstock types. Several of the short-cycle woody plants that hold promise for second generation biofuels could also be invasive (Kartha 2006). Such crops must be studied with respect to their potential invasiveness.

Some biofuel plantations will be based on single species characterized by low levels of genetic diversity (such as sugarcane) which increases the susceptibility of these crops to new pests and diseases. Further, a preference for genetically modified crops will create risks mainly associated with the increased use of herbicides, affecting soil micro-organisms and birds. The replacement of local crops with large-scale mono-cropping for biofuels might (1) lead to the simplification of agro-ecosystems, and (2) increase the susceptibility of agro-ecosystems to diseases and pests, making such systems more dependent on pesticides. These processes make farming systems less stable, less robust, and unsustainable, and reduce the resilience of the systems to both bio-physical and socio-economic shocks including pathogen infestation and uncertain rainfall (Lambrou and Laub 2006).

Positive effects on biodiversity have been noted in marginal areas where new perennial mixed species have been introduced to restore ecosystem functioning. Experimental data from degraded and abandoned soils (Tilman et al. 2006) show that low-input high-diversity mixtures of native grassland perennials offer a range of ecosystem services, including wildlife habitat, water infiltration and carbon sequestration (IFAD/FAO/UNF 2008).

20.3.4 Land-Use Change, Land Conservation and Soil Erosion

In discussing the impacts of land-use change, land conservation and intensification resulting from biofuels production, it should be recalled that most of the land in Africa is highly degraded, e.g. due to agricultural mismanagement. Thus, it might be difficult to single out potential impacts of biofuels production, especially at large-scale. Care should be taken not to associate all existing environmental impacts to biofuels.

Planting of perennial biofuel crops such as jatropha on marginal lands has the potential to prevent soil erosion and regenerate agricultural potential, providing shade and nutrients for other crops (Becker and Francis 2003). Similarly, some crops such as rapeseed are commonly used as rotational crops to provide soil cover in between other harvests whilst other crops like jatropha and croton can help to reforest degraded areas replenishing soils and local hydrology over time. Similarly, intercropping, rotations, use of nitrogen-fixing plants, windbreaks, wildlife corridors, conservation tillage and use of organic fertilizers can all improve environmental impacts.

Although large-scale estimates have been made for future expansion of biofuels production without damage to existing agricultural systems or natural ecosystems based on the use of "unused", "fallow", "marginal" and "waste" lands in some countries such as Kenya (GOK/GTZ 2008), it is important to note that advancing the agricultural frontier into natural habitats is an international environmental concern (Biofuelwatch 2007). Biofuels may also displace other economic and land use activities, leading to displacement effects which are often not well understood or ignored (Turner et al. 2007).

20.3.5 Water Resources

Hoogeveen et al. (2009) assessed the impact of increasing demand for biofuels on global water resources in the coming decade and estimated that around 1% of all water withdrawn for irrigation was used for the production of bioethanol, mainly for irrigated sugarcane and maize. It was shown that in 2017, the amount of water to be withdrawn for biofuel production would increase by 74% if agricultural practices remain the same.

Whilst the potential for expansion of irrigated areas may appear high in Africa, the actual scope for increased biofuels production under irrigated conditions on existing or new irrigated lands in the near to medium term is limited by infrastructural requirements to guarantee water deliveries and by land-tenure systems that may not conform with commercialized production systems. Furthermore, many irrigated sugar-producing regions in southern and eastern Africa including Awash, Limpopo, Maputo and Nile river basins are already operating near the hydrological limits of their associated river basins. Thus, current low levels of irrigation water withdrawals in Africa will increase only slowly. In regions with scarce water resources the start-up or extension of biofuels production will lead to problems concerning drinking water abstraction (Berndes 2002; De Fraiture et al. 2008). Only the "blue" water[1] of aquifers, lakes and rivers used for the irrigation of biofuel plants is relevant for the water balance (Rösch and Skarka 2009). Besides, water quality can be affected by using fertilizers and pesticides to grow biofuel plants if these substances end up in surface or ground water.

20.4 Towards Environment Friendly Biofuels in Africa

Harmonized approaches to life-cycle analysis, greenhouse gas balances and criteria for sustainable production should be developed in order to ensure consistency. To achieve these goals, three strategies could be adopted: (1) research, (2) application of

[1] Blue-water is made of 38.8% of total precipitation and is equivalent to the natural water resources collected in rivers, lakes, wetlands and groundwater. This water is available for withdrawal (1.5% for direct human use) before it evaporates (1.3%) or reaches the ocean (36%).

tools to define and assess the suitability of areas and technologies, and (3) development and enforcement of policies and rules including criteria and indicators for sustainable biofuels production and sustainability standards.

20.4.1 Research

To date, there is insufficient localized data on land potential, growing conditions and crop suitability. In addition, policies to guide biofuels development are largely non-existent. As a result, most planning and decision making tends to be haphazard. African governments should consider a variety of measures including additional research in (1) development of harmonized product standards, (2) measures to limit the expansion of arable land into high-value natural ecosystems, (3) environmental suitability and agro-environmental zoning for biofuels production as recently undertaken in Kenya (Muok et al. 2010) to delineate sensitive ecosystems such as natural forests and national parks, which though suitable, would be unavailable for biofuels expansion, and (4) environmental performance of advanced generation biofuels such as those derived from wastes and sources such as switch grass and marine algae. To overcome negative side effects there is need to define sustainable agricultural practices in relation to biofuels.

20.4.2 Tools for Environmental Sustainability of Biofuels

A number of tools to assess and define areas and technologies suitable for sustainable bioenergy production exist. These tools include (1) the High Conservation Value (HCV) Areas concept originally developed by the Forest Stewardship Council (FSC) for use in forest management certification but now being expanded to all habitats, (2) the Key Biodiversity Areas (KBAs) methodology for identifying and mapping biologically critical sites at the scale of practical management units to inform protected area targets and identify gaps. KBAs include Alliance for Zero Extinction sites (AZEs) which safeguards key sites where species are in imminent danger of disappearing, and Important Bird Areas (IBAs) where key sites for bird conservation, small enough to be conserved in their entirety and often already part of a protected-area network, (3) Integrated Biodiversity Assessment Tool (IBAT) which facilitates siting and management decisions, (4) the ARtificial Intelligence for Ecosystem Services (ARIES) for rapid ecosystem service assessment and valuation at multiple scales, (5) Agro-Ecological Zoning (AEZ), (6) Environmental Impact Assessment (EIA), as well as appropriate policies, standards, and certification. Although regulatory approaches to standards and certification may not be the best option, they guarantee the sustainability of the production process and maximize the positive environmental and development benefits. For selected crops that can be used for energy purposes like soy, palm oil, and sugarcane, there are commodity roundtables and scorecards such as the Roundtable on Sustainable Biofuels (RSB, focusing on creating a meta-standard), Roundtable on Sustainable Palm Oil

(RSPO) and the Round Table on Responsible Soy (RTRS), the Better Sugarcane Initiative (BSI), the Council on Sustainable Biomass Production (CSBP, focusing on second-generation feedstock) (van Dam et al. 2008); and multilateral scorecards including IADB Biofuels Sustainability Scorecard for pre-project screening and the World Bank Biofuels Scorecard for both project screening and management.

In applying these tools in Africa, it should be clear that no single concept or tool provides all answers and that they define intrinsic/ecological boundaries based on specific procedures and standards to safeguard environmental integrity. Further, the tools have specific pros and cons which have to be evaluated on a case by case basis.

20.4.3 Integrated Energy-Environment Analysis

Lazarus et al. (1995) described three basic principles that underlie integrated energy-environment analysis. Firstly, the analysis considers all fuels and technologies, whether on the supply or demand side, on equal footing. Secondly, the ultimate goal is the provision of end-use services and amenities, rather than simply fuels or electricity, at the least social cost. Finally, the broader analysis seeks to incorporate the economic externalities (most notably environmental and equity impacts) absent from a traditional cost-benefit analysis based on market prices alone.

20.5 Conclusion

Environmental problems fall into specific domains, but a wide range of tools for environmental sustainability exist to guide development and deployment of biofuels in Africa. Since most land in Africa is highly degraded due to agricultural mismanagement, it becomes difficult to single out the potential impacts of biofuels production (especially at large-scale) and care should thus be taken not to associate all environmental degradation to biofuels.

There are areas of research that are vital for sustainable development of the biofuels sector in Africa. From a policy perspective, it will be important to (1) balance growth of feedstock supply against other existing and potential uses of land and (2) deploy an assessment of sustainability that encompasses the complete cycle from growth of the raw material to end use.

References

Becker K, Francis G (2003) Bio-diesel from Jatropha plantations on degraded land. Department of Aquaculture Systems and Animal Nutrition, University of Hohenheim, Stuttgart

Berndes G (2002) Bioenergy and water – the implications of large-scale bioenergy production for water use and supply. Glob Environ Change 12:253–271

Biofuelwatch (2007) Agrofuels – towards a reality check in nine key areas. Report prepared for the twelfth meeting of the Subsidiary Body on Scientific, Technical and Technological Advice (SBSTTA) of the convention on biological diversity, Paris, 2–6 July 2007

De Fraiture C, Giordano M, Yongsong L (2008) Biofuels and implications for agricultural water use: blue impacts of green energy. Water Policy 10(Suppl):67–81

Fargione J, Hill J, Tilman D, Polasky S, Hawthorne P (2008) Land clearing and the biofuel carbon debt. Science 319:1235–1238

GOK/GTZ (2008) A roadmap for biofuels in Kenya: opportunities and obstacles. Government of Kenya, Kenya & Ministry of Agriculture/Gesellschaft für Technische Zusammenarbeit (German Technical Cooperation – GTZ)

GRAIN (2007) The new scramble for Africa. http://www.grain.org/seedling/?id=481

Hoogeveen J, Faurès J, Van De Giessen J (2009) Increased biofuel production in the coming decade: to what extent will it affect global freshwater resources? Irrig Drain 58:S148–S160. http://repositories.cdlib.org/cgi/viewcontent.cgi?article=1000&context=its/tsrc. Accessed 5 Dec 2007

IFAD/FAO/UNF (2008) International consultation on pro-poor Jatropha development -consultation papers. www.ifad.org/events/jatropha

Kartha S (2006) Environmental effects of bioenergy. In: Hazell P, Pachauri RK (eds) Bioenergy and agriculture: promises and challenges. 2020 Vision Focus 14. International Food Policy Research Institute/The Energy and Resources Institute, Washington, DC/New Delhi

Lambrou Y, Laub R (2006) Gender, local knowledge, and lessons learnt in documenting and conserving agrobiodiversity. UNU-WIDER. Research paper 2006/69, 2

Lazarus M, Von Hippel D, Hill D, Margolis R (1995) A guide to environmental analysis for energy planners. Stockholm Environment Institute, Boston

Licht FO (2007) World biodiesel output. Planet Ark World Environmental News

Muok BO, Ouma BO, Esilaba AO, Nandokha T (2010) Environmental suitability and agro-environmental zoning of Kenya for biofuel production. ACTS/PISCES/UNEP, Nairobi

Righelatto R, Spraklen D (2007) Carbon mitigation by biofuels or by saving and restoring rain forests? Science 317:5840

Rösch C, Skarka J (2009) The European biofuels policy and sustainability. International Association for Energy Economics. Third Quarter 2009, pp 31–35

Tilman D, Hill J, Lehman C (2006) Carbon-negative biofuels from low-input high-diversity grassland biomass. Science 314:1598–1600

Turner B, Plevin R, O'Hare M, Farrell A (2007) Creating markets for biofuels: measuring and improving environmental performance. Institute for Transportation Studies, UC Berkeley Transportation Sustainability Research Center, UCB-ITS-TSRC-RR-2007-1

van Dam J, Junginger M, Faaij A, Jurgens I, Best G, Fritsche U (2008) Overview of recent developments in sustainable biomass certification. Biomass Bioenergy 32:749–780

Chapter 21
Potential Impacts of Biofuel Development on Biodiversity in Chobe District, Botswana

Donald L. Kgathi, Barbara N. Ngwenya, and Mogodisheng B.M. Sekhwela

Abstract There is a need to use cleared idle agricultural land for biofuel production in order to avoid adverse impacts on food security and biodiversity. This chapter examines the potential impacts of biofuel development on biodiversity in Chobe District, Botswana, using literature review and stakeholder interviews. The stakeholders interviewed confirmed that there are significant areas of idle agricultural land available in the district, but most of it is not cleared. Therefore, the production of biofuels in Chobe District may on the one hand negatively affect biodiversity through the clearing of new land, but on the other hand it may not adversely affect food security since idle agricultural land will be used. The use of marginal land for biofuel production may also harm biodiversity (plant and animal species). This chapter shows that the use of jatropha and sweet sorghum for biofuel production is likely to have a lower impact on biodiversity compared to corn. In conclusion, research on biology, chemistry as well as agronomic aspects of energy crops should be undertaken prior to large-scale biofuel development in Botswana.

Keywords Biofuels • Biodiversity • Land use change • Climate change • Botswana

D.L. Kgathi (✉) • B.N. Ngwenya
Okavango Research Institute, University of Botswana, Shorobe Road, Matlapana, Maun, Botswana
e-mail: kgathi@mopipi.ub.bw; bntombi@orc.ub.bw

M.B.M. Sekhwela
Office of Research and Development, University of Botswana, 4775 Notwane Rd, Gaborone, Botswana
e-mail: Sekhwela@mopipi.ub.bw

R. Janssen and D. Rutz (eds.), *Bioenergy for Sustainable Development in Africa*,
DOI 10.1007/978-94-007-2181-4_21,

21.1 Introduction

Biofuels are produced from biomass for a wide range of applications such as cooking, heating, cooling, and transport. Biofuels can be solid (e.g. fuel-wood), liquid (e.g. bioethanol, biodiesel) or gaseous (e.g. biogas) (FAO 2008). However, the term biofuel nowadays mainly refers to liquid biofuels such as bioethanol and biodiesel (UNEP 2008) and this chapter has adopted this definition. The major driving forces behind biofuel development are concerns about increasing energy prices, depletion of fossil energy sources, concerns about the environment and rural development (Dufey 2006; Goldemberg 2009). With respect to the environment the major issues of concern revolve around the following themes: greenhouse gas emissions, land-use change, water quality and quantity, air quality and biodiversity loss, the subject of this chapter (Dufey 2006).

The relationship between biodiversity and biofuel development in Botswana is still not clearly understood. Biodiversity is defined as the number, variety and variability of species of plants, animals, and micro-organisms as well as their ecosystems and ecological processes (Wallace 2007; Pearce et al. 1991, 1993). Biodiversity is essential for maintaining ecosystems which are important for the provision of ecosystem services essential for the well-being and survival of humans. The Millennium Ecosystem Assessment (MEA 2003) defines ecosystem services as "the benefits which people obtain from ecosystems". Ecosystem services are categorized by the MEA as follows: (1) provisioning services such as food, medicines and water, (2) regulating services such as water quality regulation and flood control, (3) cultural services such as recreation and tourism and aesthetic values and (4) supporting services such as soil formation and photosynthesis (Wallace 2007).

Not all the economic benefits associated with biodiversity conservation are known. Each species is unique and not necessarily substitutable, and it is not known which species will become resources in the future (Bishop 1978). There is currently an uncertainty about future losses in biodiversity and their irreversibility. Although biofuel development is associated with a number of benefits, it can also have negative impacts, particularly if not guided by appropriate practices and policies. It can, for instance, lead to loss in biodiversity, increase in greenhouse gas emissions, food insecurity and aggravation of poverty (UNEP 2008).

Compared to other countries in southern Africa, Botswana has high faunal and low floral biodiversity. The number of species in Botswana is estimated to be as follows: 150 mammals, 570 birds, 131 insects, 82 fish, and 2,150–3,000 plant species (MEWT 2007). Most of these species are concentrated in conservation areas of Chobe National Park, Moremi Game Reserve in the Okavango Delta, and Makgadikgadi Pans (MEWT 2007). In these areas, like in other parts of southern Africa, high rainfall is associated with high species diversity (O'Brien 1993). According to Sekhwela (2000), there is an increasing number of species with increasing rainfall. A feasibility study undertaken in 2007 on behalf of the Ministry of Minerals, Energy and Water Resources suggested that there is a potential for the production and use of liquid biofuels for transport in Botswana, with high potential for bioethanol in the northern part of the country and biodiesel in the Central District (EECG 2007).

This chapter presents a critical analysis of the potential impacts of the production of biofuels on biodiversity in Chobe District in Botswana. The specific objectives of the study are: (1) examine, through a state of the art literature review, how biofuel crops recommended for Botswana are likely to contribute to sustainable production and use of biofuels, (2) assess the potential impacts of using pristine land for biofuel production on biodiversity using existing literature, and (3) assess the perceptions of relevant stakeholders on the potential impacts of biofuel development on biodiversity.

21.2 Drivers of Loss in Biodiversity

Biofuel development may affect biodiversity in a negative or positive way, though most of the impacts are likely to be negative. According to Sala et al. (2009) and Omann et al. (2009), the major drivers of loss in biodiversity include habitat loss, increase in invasive species, pollution and climate change. Habitat loss, which is a major threat to biodiversity in Botswana (MEWT 2007), can be caused by direct and indirect impacts of biofuel production. Direct impacts occur when there is conversion of land from forest or grassland, whereas indirect impacts occur when conversion of agricultural land induces conversion of natural vegetation to agricultural production elsewhere (Smeets 2008). According to Fargione et al. (2008), the conversion of natural habitats such as "rainforests, peatlands, savannas, or grasslands" for biofuel production creates the so-called "biofuel carbon debt". This carbon debt is caused by CO_2 emissions which can be many times larger than the greenhouse gas (GHG) savings resulting from the substitution of fossil fuels by biofuels. It would thereby take decades to repay this debt from the benefits of carbon obtained from biofuel production. Furthermore, it is estimated that 10% substitution of petrol and diesel by biofuels will require 38% and 43% of the current cropland in Europe and USA, respectively (UNEP 2008).

If forests or grasslands are replaced by a single crop with low genetic diversity, the adverse impacts of habitat loss are even higher because monocultures are vulnerable to attacks by pests and diseases than diverse habitat patchworks (Royal Society 2008). The increase in pests and diseases is likely to lead to an increase in the use of pesticides/herbicides. In Botswana, the use of pesticides and herbicides has been found to affect non-target species and contribute to further loss in biodiversity (Arup Atkins 1990). The invasiveness of biofuel crops depends on the environment and type of the biofuel feedstock (Groom et al. 2008). According to the Royal Society (2008), some of the characteristics that make crops suitable for biofuel production (e.g. fast growth and high yields) also make them potential candidates for invasiveness. In addition, species which are not invasive in their native environment may become invasive in other environments.

Finally, climate change is one of the key drivers of loss in biodiversity, particularly in dry-land countries such as Botswana, where the global long-term predictions suggest that the climate is likely to become drier and hotter (MEWT 2007). According to Ravindranath et al. (2010) and FAO (2008), greenhouse gas emission savings of first generation biofuels are estimated to range from 20% to 60%. However, most of these studies exclude the impact of biofuels on land use change, which may be a major contributor to CO_2 emissions from biofuels (Ravindranath et al. 2010). Therefore, climate change and biodiversity are closely linked and policies for reducing climate change are also policies for conservation of biodiversity (UNEP 2008).

21.3 Principles for Sustainable Biodiversity

In recent years, there have been a number of initiatives worldwide concerned with the development of the sustainable production of biofuels. The Roundtable on Sustainable Biofuels (RSB) brings together different types of stakeholders concerned with the sustainability of biofuels (RSB 2009). The RSB has developed 12 principles for the production and use of biofuels. Biodiversity conservation, which is covered by Principle 7, states that "biofuel operations shall avoid negative impacts on biodiversity, ecosystems and other conservation values". The Principle states that in order to achieve this objective there is need to ensure that biofuel production takes place in areas with minimum risk to biodiversity loss. The RSB further states that production of biofuels should only take place in areas of high risk if strict observation of conservation values is ensured.

Further principles for production and use of biofuels have been developed by the Competence Platform on Energy Crop and Agro-forestry Systems (COMPETE) project. With regards to biodiversity, COMPETE recommends that good practices which do not harm biodiversity and the ecosystem should be adopted in the production and use of biofuels (Janssen et al. 2009). Apart from the general principles on production and use of biofuels, other scholars have developed principles specific to the impacts of biofuels on biodiversity. For instance, Groom et al. (2007, p. 608) have developed 12 principles for the promotion of "sustainably grown, biodiversity friendly biofuels". Some of the important issues raised by these principles include: (1) minimal use of land and agricultural inputs (fertilizer, pesticides and water), (2) promotion of restoration of degraded or marginal areas, (3) avoided use of invasive species, and (4) adoption of conservation tillage or other conservation oriented methods.

21.4 Research Methods

21.4.1 Study Area

According to the 2001 Census Report (CSO 2002), Chobe District has a land area of 22,052 km^2, and a human population of 18,258. The district is situated in northern Botswana and shares the border with Namibia in the north-west, Zambia in the

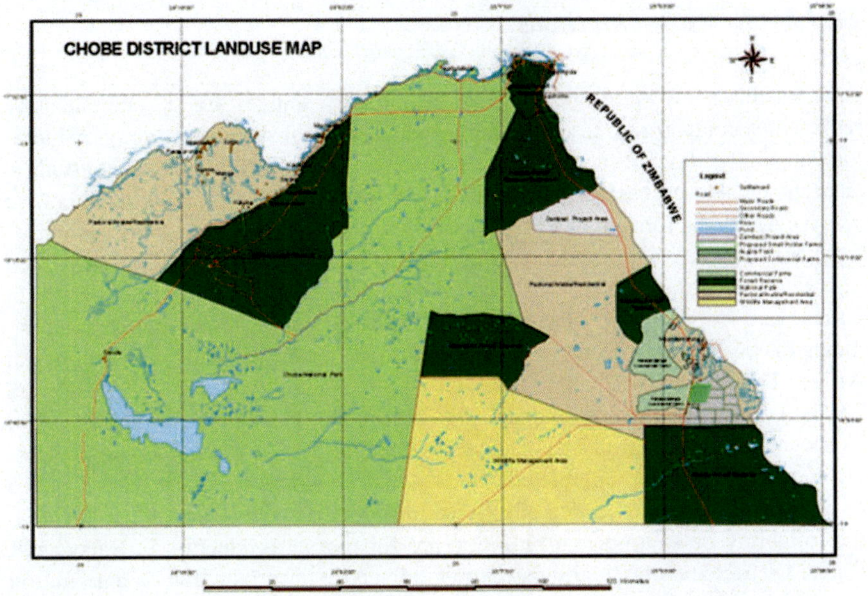

Fig. 21.1 Map of Chobe district, Botswana (Source: Ministry of Agriculture 2010)

north, and Zimbabwe in the east (Fig. 21.1). Chobe District has a low population density of one person per km^2, compared to the national average of three persons per km^2. The district has two land tenure categories, namely state land and tribal land. State land accounts for 69% of the district, mainly in the form of the protected areas of Chobe National Park and forest reserves which, as noted above, are areas of relatively high species diversity. Chobe National Park, covering an area of 10,566 km^2 (50% of the land in Chobe District) is the second largest park in Botswana with a very high diversity of wildlife, including a high number of elephants. There are six forest reserves in Chobe District covering an area of 4,096 km^2 or 19% of the land mass in the district, namely Kasane, Kasane Extension, Chobe, Kazuma, Maikaelelo and Sibuyu forest reserves (Kgathi and Sekhwela 2003). Tribal land accounts for 31% of the land area in Chobe District. Part of this land is designated as wildlife management areas (WMAs), which cover 10.7% of the land in the district. Eight villages of Pandamatenga, Lesoma, Kazungula, Mabele/Muchenje, Kavimba, Satau, and Parakarungu are settled on tribal land. About 80% of the district is therefore gazetted as conservation areas with free roaming wildlife. Thus, there is a shortage of land for livestock, crop production and other livelihood activities. Biomass is a major source of household energy in the district as 94% of households use fuelwood as their principal source of energy.

21.4.2 Research Methods

Data for this study were obtained through primary sources and a comprehensive review of secondary sources, both published and unpublished literature (grey literature, unpublished reports and planning documents). These included feasibility studies, environmental impact assessment studies and development plans. Published literature was mainly in the form of recent academic journal articles.

Qualitative data was collected through field observation and semi structured interviews were conducted with key stakeholders from relevant departments in Kasane and Gaborone (in March/April 2010 respectively). These included departments of Agricultural Research, Crop Production, Energy Affairs, Environmental Affairs, Forestry and Range Resources, Plant Resources, Ministry of Wildlife and Tourism, Physical Planning, Office of Research and Development (University of Botswana), commercial farmers and subsistence farmers. A total of 16 respondents (12 men and 4 women) completed a self-administered questionnaire that consisted of both closed and open ended questions. Respondents were asked to comment on the suitability of a number of biofuel crops for biofuel production in Botswana's Chobe District and potential impacts on biodiversity for production on agricultural land, marginal land, and forest reserve land.

21.5 Results

21.5.1 Biofuel Crops

The type of feedstock used for the production of biofuels is a major factor which determines the extent to which biofuels impact on sustainability, including biodiversity (Groom et al. 2008). The feasibility study for the production and use of biofuels in Botswana recommended sugarcane and sweet sorghum as suitable biofuel crops for production in Chobe district, whereas jatropha was recommended for growing in other parts of the country. These crops were selected because of their low costs of production and their suitability to the environmental conditions (EECG 2007). The production costs for sweet sorghum were estimated to range from Botswana Pula (BWP) 50/ton (8 US$) for small-scale production to BWP 60/ton (10 US$) for mechanized production. Those for jatropha were estimated to be BWP 610/ton/ha for plantations producing seed for 20 years and BWP 540 for those producing seeds for a longer period of 40 years (EECG 2007). The competitive advantage for these crops has also been analyzed. For instance, it was estimated that the price of jatropha based biodiesel in Botswana would be competitive with diesel at crude oil prices of 70 US$ per barrel (BWP 434). On the other hand, ethanol produced from sweet sorghum and sugarcane was found to be competitive with petrol at crude oil prices of 50 US$ (BWP 310) and 70 US$ (BWP 434) per barrel, respectively

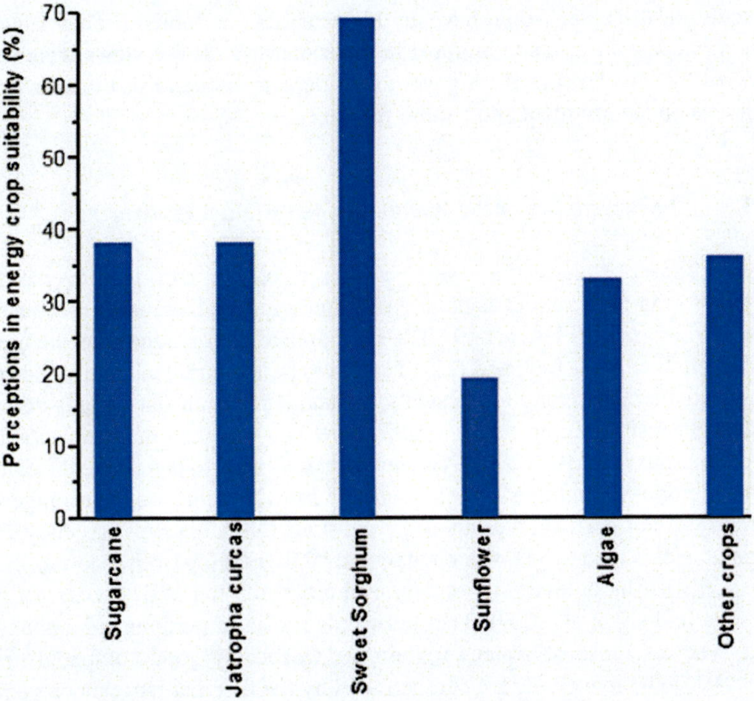

Fig. 21.2 Stakeholder perceptions on suitability of biofuel feedstocks in Botswana

(EECG 2007). This feasibility study included economic assessments only and did not address social and environmental costs (EECG 2007).

21.5.1.1 Perceptions About the Suitability of Feedstocks

The survey showed that sweet sorghum is considered the most suitable crop for biofuel production in the district by 69% of the stakeholders as it is already grown at subsistence level (Fig. 21.2). The stakeholders were aware that efficient first generation conversion technologies are available for producing bioethanol from sweet sorghum. Most of the stakeholders expressed reservations about the suitability of growing jatropha in the district and only 38% of them thought it should be encouraged. Most of them expressed the view that decisions on the selection of biofuel crops should be research-led rather than policy-led. Regarding sugarcane, the general concern is its high water needs, making it unsuitable for semi-arid conditions of Botswana. Few stakeholders are against growing of biofuel crops in the district, arguing that the district is already experiencing shortage of land, as 80% of the land is occupied by protected areas and forest reserves. In their view, Chobe District has a comparative advantage for tourism rather than biofuel development and the production of the latter would have an adverse effect on tourism. They suggested that it was better to produce biofuels in other districts. These views were mainly expressed by stakeholders from government departments and institutions dealing with issues on environment and natural resources.

21.5.1.2 Characteristics of the Recommended Biofuel Feedstocks

Literature sources suggest that sweet sorghum as biofuel crop may overcome the problem of food-fuel conflict as the crop produces fuel and animal feed from stalks and food from grains (FBAE 2009). The same area of land can therefore be used for the production of food, fuel, and feed. This saves agricultural land and avoids negative impacts on biodiversity associated with land conversion. Jatropha is described as a drought resistant crop which can be grown on marginal lands unsuitable for most other crops (UEMOA 2009). However, there is need to determine the extent to which the crop is drought resistant as evidence from Namibia seems to suggest that the crop does not perform well in arid conditions (Namibia Press Agency 2010). In general, yields tend to be lower on marginal lands, ranging from as low as 2–3 ton of dry seeds per ha on marginal or degraded areas to 5 ton of dry seeds per ha on good soils (Ndong et al. 2009). The fuel yields per ha of jatropha are amongst the lowest, whereas those of sweet sorghum are reasonably good compared to corn (Table 21.1). The low yields are compensated by the fact that jatropha can be integrated with food crop production as practised in Mali (see Chap. 22). The production of biofuels from jatropha is not land intensive compared to other energy crops as the same area of land may be used to produce fuel, household energy and even soap (Kumar and Sharma 2008). The foregoing suggests that biofuels produced from jatropha and sweet sorghum are likely to have low ecological footprints (in terms of the amount of land needed for biofuel production) associated with low impacts on biodiversity.

Table 21.1 also shows that production of these energy crops requires low amounts of water, fertilizer and pesticide compared to the production of corn. Low use of these inputs suggests that biodiversity may not be much affected. Table 21.1 also presents GHG emission savings ranging from 20% to 72% for jatropha biodiesel in West Africa and Thailand, based on life cycle assessments (FAO 2008). GHG emission savings of biofuels from jatropha thus seem higher than those of other first generation biofuels, probably due to low requirements on input and tillage (Francis et al. 2005). Finally, even larger GHG emission savings are expected for next generation biofuels as indicated in Table 21.1.

21.5.2 Use of Agricultural Land

Biofuel development is considered biodiversity friendly if it does not replace cropland needed for food production. Most stakeholders emphasized the need to avoid

Table 21.1 Comparison of efficiencies and input requirement for energy crops

Biofuel	Crops	Greenhouse gas savings (%)	Energy conversion efficiency	Fuel yield (l/ha)	Water use	Fertilizer use	Pesticides use
First generation ethanol	Sugarcane	87–97[a]	8–10.2[a]	5,000[b]	High	High	Medium
First generation ethanol	Corn	9.6–13.8[a]	1.1–1.3[a]	2,370[b]	High	High	High
First generation ethanol	Sweet sorghum	53	–	3,000[b]	Low	Medium	Medium
Biodiesel	Jatropha	20[b]–72[c]	4.72–6.03[d]	1,250[b]	Low	Low	Low
Next generation biodiesel	Microalgae	320[a]	–	49,700–108,800[a]	Medium	Low	Low

[a]Groom et al. (2008)
[b]Ravindranath et al. (2010)
[c]Ndong (2009)
[d]Prueksakorn and Gweewala (2008)

the trade-off between land for biofuel and food production. It is generally known that poverty is one of the main causes of biodiversity loss as it results in "forced over-use" of natural resources (MEWT 2007). More than 60% of the stakeholders said there is no idle, un-cleared agricultural land in Chobe District and that new land would have to be cleared for biofuel production. The impacts on biodiversity will depend on the type of land use which existed before the introduction of biofuel plantations. If natural habitats are replaced by monocultures, the impacts on biodiversity may be negative. The substitution of land already in use for food production would however have a negative impact on food production. This may have a negative impact on food security in Botswana, a country already with a high dependency on imports of cereals. According to Arup Atkins (1990), the conversion of natural habitats to arable agriculture in the Pandamatenga area in Chobe District was found to have adverse impacts on biodiversity, particularly with respect to certain antelope species. This suggests that further change of land use for biofuel development will negatively impact on the biodiversity of different land systems.

21.5.3 *Use of Marginal Land*

According to the Gallager Review, marginal land and idle land should be used for the production of feedstocks for biofuels (RFA 2008). Out of a total of 682,000 ha of arable land in eastern Botswana in 2009, 200,000 ha (29%) is marginal land (Fig. 21.2). 63% of the stakeholders said there is marginal land in Chobe District which is not used for agricultural production and could therefore be available for biofuel production. This marginal land refers to 5,000 ha of land in Pandamatenga considered not suitable for arable agriculture by the Ministry of Agriculture (Modise 2009). The soil in this area is mainly vertisolic clay not suitable for agriculture as it is not well drained and easily 'waterlogged'. However, the area is well endowed with 22 woody plant species, 15 grass species and 24 broadleaved/forbes species. It is not known whether these species are specific to this rather unusual habitat as a comprehensive plant survey has not been done in Chobe District. The area was found to be exclusive habitat of some rare antelopes (Arup Atkins 1990), such as Oribi (*Ourebia ourebi*), Reedbuck (*Redunca arundinum*), Roan (*Hippotragus equines*), Sable (*Hippotragus niger*), and Tsesebe (*Damaliscus lunatus*). These species were found to occur predominantly on these habitat types only found in the northern and southern plains in Pandamatenga area of Chobe District (Arup Atkins 1990).

Although energy crops such as jatropha are reported to perform well on marginal land, yields tend to be lower on such land. Jatropha yields on marginal land may be reduced by 10–20% compared to those in high rainfall areas (FAO 2008). In some cases, the development of biofuels on marginal lands, such as those of Pandamatenga, may negatively impact on biodiversity in the form of animal species, vegetation, and soil biota. In other cases, however, the growing of biofuel crops on marginal lands may have a positive impact on biodiversity since it may lead to the restoration of lands.

This suggests that the impact of biofuel development on marginal lands is context specific. In the context of marginal lands as defined in Chobe District (i.e. not suitable for arable agriculture), the development of biofuels may not necessarily lead to positive impacts on biodiversity. This suggests that there is need for internationally agreed definitions of the concepts of idle and marginal land as suggested by the Gallagher Review (RFA 2008).

21.5.4 Conversion of Forest Reserves

There is a global concern that biofuel production is one of the major factors that will contribute to loss of natural habitats in many parts of the world in the future. Most stakeholders (75%) were concerned that the production of biofuels could encroach on natural habitats of forest reserves which currently account for 19% of the land area in Chobe District. In their view, the replacement of pristine areas such as forest reserves by monoculture plantations could result in negative environmental impacts which could adversely affect biodiversity. They were also aware that the replacement of natural habitats with biofuel crops could also result in more carbon dioxide emissions into the atmosphere as carbon sequestration rates of natural habitats tend to be higher than those of biofuel crops. All stakeholders held the view that biofuel development will need to be controlled to avoid encroachment into conservation areas.

The forest surveys of 1992 found that the forest reserves hold a number of woody and herbaceous plant species that support and provide habitat to diverse fauna species (Norwegian Forestry Society 1992). The list of woody plant species recorded over forest reserves was over 90, whilst there were only seven shrub species and over 24 grass species, which increased when floodplain areas were included. Arup Atkins (1990) noted that a large number of wildlife species depend on the various vegetation and species types for their existence and hence such links are likely to be affected by any loss of habitat through land clearing. Therefore, the use of forest reserves for biofuel production will not only lead to loss of floral species, but also wildlife and soil biota species that are currently not well studied and documented in the forest areas of the Chobe District.

21.6 Conclusion

The use of recommended biofuel crops (jatropha and sweet sorghum) for biofuel production is likely to have a lower impact on biodiversity compared to corn. The results are based on analysis of factors likely to affect the ecological footprint including inputs (land, water, fertilizers, etc.), energy conversion efficiency and emission of greenhouse gases.

A number of studies have revealed that there is large potential for growing jatropha in semi-arid regions of Africa and Botswana in particular even though it is native to the coastal areas of South America and Mexico (Francis et al. 2005). However, Kumar and Sharma (2008) highlight the need for further information on biology, chemistry, and agronomic aspects before implementing industrial applications.

The study also suggests the need to use cleared idle agricultural land for biofuel production in order to avoid adverse impacts on food security and biodiversity. The stakeholders interviewed in this study considered that there was plenty of idle agricultural land in the district, but most of it was not cleared. This suggests that the production of biofuels in Chobe District may negatively affect biodiversity since it will result in the clearing of new land, but it may not adversely affect food security since idle agricultural land will be used.

Though marginal land exists in the agricultural area of Pandamatenga, its use is likely to adversely affect biodiversity since the area is richly endowed with plant and animal species. Furthermore, biofuel production on marginal lands will lead to low yields and low economic returns. The stakeholders also emphasised the need to avoid the replacement of pristine areas such as forest reserves by monoculture plantations as this will have a negative impact on biodiversity and climate change.

In conclusion, measures should be taken to ensure that the production of biofuels in Botswana is sustainable in social, economic and environmental terms. The failure to do so may result in adverse consequences on biodiversity. To achieve sustainability of biofuel development, it is necessary to develop criteria for their production and use in Botswana including criteria addressing biodiversity conservation. Other criteria should address greenhouse gas emissions, energy balances, other environmental aspects, as well as wider socio-economic and political issues. Whilst the production of first generation biofuels is generally encouraged in Botswana, slow and careful development is recommended. Furthermore, research on second generation biofuels produced from ligno-cellulosic biomass should be increased as they are likely to be more compatible with biodiversity conservation and sustainable development.

References

Arup Atkins (1990) Pandamatenga development study: annex P – environmental impact study. Final report. Ministry of Agriculture, Gaborone

Bishop RC (1978) Endangered species and uncertainity: the economics of a safe minimum standard'. Am J Agric Econ 60(2):10–18

Central Statistics Office (CSO) (2002) Report on the population census on 2001 population and housing census, Gaborone Ministry of Finance and Development Planning, Gaborone

Dufey A (2006) Biofuels production, trade and sustainable development: emerging issues. International Institute of Environment (IIED), London. www.earthprint.com

Energy Environment, Computer & Geophysical Application (EECG) (2007) The feasibility study for the production and use of biofuels in Botswana. Ministry of Minerals Energy and Water Resources, Gaborone

FAO (2008) Biofuels: prospects and opportunities. FAO, Rome

Fargione J, Hill J, Tilman D, Polasky S, Hawthorne P (2008) Land clearing and the biofuel carbon debt. Science 319(5867):1235–1238. doi:10.1126. http://www.sciencemag.org/cgi/content/abstract/152747

Foundation for biotechnology awareness and education (FBAE) (2009) Sweet sorghum: a new smart biofuels crop that ensures food security. http://fbae.org/2009/FBAE/website/biofuels_sweet-sorghum.html

Francis G, Edigner R, Becker K (2005) A concept for simultaneous wasteland reclamation, fuel production, and socio-economic development in degraded areas in India: need, potential and perspectives of *Jatropha* plantations. Nat Resour Forum 29:12–24

Goldemberg J, Guardabassi P (2009) Are biofuels a feasible option? Energy Policy 37(1): 10–14

Groom M, Gray EZ, Townsend P (2008) Biofuels and biodiversity: principles for creating better policies for biofuels production. Conserv Biol 22(3):602–609

Janssen R, Rutz D, Helm P, Woods J, Diaz-Chavez R (2009) Bioenergy for sustainable development in Africa: environmental and social aspects. http://www.compete-bioafrica.net/sustainability/OD9-2_paper_COMPETE_Janssen_090730.pdf

Kgathi DL, Sekhwela MBM (2003) Sustainability of the commercial exploitation and management of the Chobe forest reserves in Botswana. S Afr J Geogr 85(1):26–34

Kumar A, Sharma S (2008) An evaluation of multipurpose oil seed crop for industrial uses (jatropha curcas L): a review. Ind Crops Prod 28:1–10. doi:10.1016/j.indcrop.2008.01.001

Millennium Ecosystems Assessment (MEA) (2003) Ecosystems and human well-being. Island Press, London

Ministry of Agriculure (2010) Land-use Data Base, Ministry of Agriculure, Gaborone

Ministry of Environment, Wildlife and Tourism (2007) Botswana biodiversity strategy and action plan (Revised). Ministry of Environment, Wildlife and Tourism, Gaborone

Modise M (2009) Director of Crop Production, Ministry of Agriculture, Personal communication on use of agricultural land for biofuels, Ministry of Agriculture, Botswana

Namibia Press Agency (2010) Jatropha plants being uprooted in Shadikongoro www.nampa.org; 17/08/2010

Ndong R, Montrejaud-Vignoles M, Girons OS, Gabrielle B, Pirot R, Domergue M, Salblayrolles C (2009) Life cycle assessment of biofuels from *Jatropha curcas* in West Africa: a field study. GCB Bioenergy 1:197–210

Norwegian Forestry Society (1992) Chobe forests inventory and management plan: technical reports vol I and II. Division of Crop Production and Forestry, Ministry of Agriculture, Gaborone

O'Brien EM (1993) Climatic gradients in woody plant species richness: towards an explanation based on an analysis of Southern Africa's flora. J Biogeogr 20:181–198

Omann I, Stocker A, Jager J (2009) Climate change as a threat to biodiversity: an application of the DPSIR approach. Ecol Econ 69(1):24–31

Pearce D, Barbier E, Markandya A, Barrett S, Turner RK, Swanson T (1991) Blueprint 2: greening the world economy. Earthscan, London

Pearce D, Turner RK, Oriordan T, Adger N, Atkinson G, Brisson K, Brown K, Dubourg R, Fanhauser S, Jordan D, Moran D, Powel J (1993) Blueprint 3: measuring sustainable development. Earthscan, London

Prueksakorn K, Gheewala SH (2008) Full chain energy analysis of biodiesel from Jatropha curcas L in Thailand. Environ Sci Technol 42:3388–3393

Ravindranath NH, Lakshmi CT, Manuvie R, Balachandra P (2010) Biofuel production and implications for land use, food production and environment in India. Energy Policy. doi:10.1016.enpol2010.07.044

Renewable Fuels Agency (RFA) (2008) The Gallagher review of the indirect effects of biofuels production. RFA, East Essex

Round Table for Sustainable biofuels (RSB) (2009) RSB principles and criteria for sustainable biofuel production V1.0 Round table of sustainable biofuels. http://cgse.epfl.ch/biofuels

Sala OE, Sax D, Leslie H (2009) Biodiversity consequences of biofuel production. In: Howarth RW, Bringezu S (eds) Biofuels: environmental consequences and interactions with changing land use. Proceedings of the Scientific Committee on Problems of the Environment (SCOPE) International Biofuels Project Rapid Assessment, 22–25 Sept 2008, Gummersbach, Cornell University, Ithaca. (http://cip.cornell.edu/biofuels/)

Sekhwela MBM (2000) Woody biomass production ecology in Kalahari communal areas of Botswana. PhD thesis, Botany Department, University of Queensland, Brisbane
Smeets E (2008) Possibilities and limitations for sustainable bioenergy production systems. PhD thesis, Copernicus Institute for Sustainable Development and Innovation, Utrecht University, The Netherlands
The Royal Society (2008) Sustainable biofuels: prospects and challenges. Policy document 01/08 royalsociety.org
The West African Economic and Monetary Union (UEMOA) (2008) Sustainable bioenergy development in UEMOA and the hub for rural development in West and Central Africa, Dakar, Senegal
UNEP (2008) The potential impacts of biofuels on biodiversity. In: Conference of the parties to the convention on biological diversity, 9th meeting, Bonn, 19–30th May
Wallace KJ (2007) Classification of ecosystems services: problems and solutions. Biol Conserv 139:235–246

Chapter 22
Water Implications of Biofuel Development in Semi-Arid Sub-Saharan Africa: Case Studies of Four Countries

Donald L. Kgathi, Isaac Mazonde, and Michael Murray-Hudson

Abstract Biofuel production may have considerable impacts on water resources. To analyze the implications of biofuel development on water resources in the semi-arid parts of Botswana, Zambia, Tanzania and Mali, case studies were elaborated. In all four countries plans are under way to develop comprehensive biofuel policies. Botswana and Mali have similarities since they are water scarce countries with high dependency on imported food and energy. Whilst large areas of Mali and Botswana are semi-arid, the semi-arid parts in Zambia and Tanzania are of smaller size and have relatively suitable conditions for biofuel production. In Zambia and Tanzania, there are sugarcane and jatropha-based biofuel projects, whereas in Botswana and Mali biofuel production is mainly based on jatropha. It is shown that the expansion of biofuel projects in all four countries may adversely affect water resources. It is therefore recommended that water scarce countries such as Botswana and Mali should engage in biofuel projects which do not require much irrigation. The production model of integrating biofuel production with food crop production as practised in Mali provides useful lessons for Botswana and other countries. Whilst in Zambia and Tanzania large-scale projects based on contract farming could be expanded to reach the full potential of these countries, it is crucial to carefully monitor their impacts on water resources.

D.L. Kgathi (✉) • M. Murray-Hudson
Okavango Research Institute, University of Botswana, Shorobe Road, Matlapana, Maun, Botswana
e-mail: dlkgathi@orc.ub.bw; mmurray-hudson@orc.ub.bw

I. Mazonde
Office of Research and Development, University of Botswana, 4775 Notwane Rd, Gaborone, Botswana
e-mail: drd@mopipi.ub.bw

R. Janssen and D. Rutz (eds.), *Bioenergy for Sustainable Development in Africa*,
DOI 10.1007/978-94-007-2181-4_22, © Springer Science+Business Media B.V. 2012

Keywords Biofuels • Water use • Irrigation • Sub-Saharan Africa • Impacts • Botswana • Zambia • Tanzania • Mali

22.1 Introduction

In Sub-Saharan Africa, as in the rest of the developing world, biofuel development has started to attract growing attention because of concerns about energy security, climate change and rural development (Dufey 2006; Royal Society 2008). Biofuels are therefore likely to play a key role in the energy future of Sub-Saharan Africa. Globally, it is estimated that the amount of land under biofuels is about 11–12 million ha which is 1% of the total area under crop production (de Fraiture et al. 2008; Hoogeveen et al. 2009). In Brazil and USA, which are the leading producers of biofuels, the area under biofuel production is estimated at 2.5 million ha and 4 million ha (5% and 4% of the area under cropland), respectively (de Fraiture et al. 2008). The expansion of biofuel production will lead to an increase in the demand for land and water resources which are already constrained by the need for agricultural production. Water use efficiency varies according to the type of energy crop, climate, growing period, and agronomic practice, as well as the processing of biomass into biofuels and biomaterials (Demirbas 2009; Berndes 2008). It is also determined by whether or not the crop is cultivated under rain-fed or irrigated conditions.

Many countries of Sub-Saharan Africa have suitable conditions and very large areas of land which could be used for biofuel production. However, the production of biofuels on the scales necessary to meet national and global demand for these fuels could result in adverse impacts on water resources and contribute to the conflict between water for food and water for fuel. The large and fast growing economies of India and China are likely to experience serious water scarcities as they try to meet their demand for feed, food and biofuels (Berndes 2008; de Fraiture et al. 2008). According to de Fraiture et al. (2008), irrigation withdrawals for biofuel production globally amount to 44 km^3 or 2% of the total water withdrawals of 2,630 km^3. The successful implementation of biofuel policies on a global scale is likely to result in an additional use of cropland of 30 million ha and irrigation water withdrawals of 180 km^3 (de Fraiture et al. 2008). In fragile semi-arid regions of Sub-Saharan Africa already experiencing water scarcity, additional water use for biofuel production will add more pressure to water scarcity, and the problem may be worsened in the future by changes in water availability resulting from climate change (Bekunda et al. 2009; de Fraiture et al. 2007). In addition, international trade in biofuels will increase virtual water flows, and this may contribute to the deterioration of global water resources.

This chapter examines how water use is likely to be affected by the expansion of biofuel production aimed at meeting national goals in the semi-arid parts of the selected countries of Sub-Saharan Africa. The specific research objectives of this chapter are to: (1) examine the structure of water management systems in these countries, (2) assess the potential for irrigation of biofuels in key river basins, (3) identify the types of energy crops suitable for growing in these regions in terms of water availability, (4) assess if irrigation is required for producing biofuels, and (5) contribute to considerations for future biofuel policies. This chapter does not assess the implications of biofuel development on water quality. It is also crucial to note that no attempt has been made to discuss the details regarding the patterns of water use at a micro-level, as this would require field visits to study areas, a task beyond the scope of this study.

22.2 Sustainability of Biofuels and Water Resources

Today, there are growing concerns that the development of biofuels is associated with risks that result in adverse effects on sustainability (Naylor et al. 2007). The euphoric view that biofuels are mainly associated with positive aspects of sustainability has been proved to be incorrect by numerous scientific studies (Von Maltitz and Brent 2008). The major issues of concern about the sustainability of biofuel development revolve around the following themes: (1) greenhouse gas emissions, (2) biodiversity, (3) water resources, (4) land-use change, (5) impacts of increased nutrients and pesticides, (6) competition with food, and (7) human welfare and well-being (Smeets 2008; Yang et al. 2009; Royal Society 2008). More relevant to this chapter is the concern that the expansion of biofuels production may lead to the diversion of water resources from other development demands in Sub-Saharan Africa. Concerns related to biofuels sustainability have led to the development of certification schemes for biofuels, whereby the buyers of these fuels will ensure that their production complies with certain standards or sustainability criteria in order to avoid adverse environmental, economic and social impacts (Smeets 2008).

A number of techniques are used to measure the environmental sustainability of projects and policy programmes such as life cycle analysis, ecological footprint analysis, cost-benefit analysis, environmental assessment, and risk assessment (Paehlke 1999). The most important tools for measuring resource sustainability are life cycle analysis and ecological footprint analysis (Paehlke 1999). The concept of water footprint has been used in a number of studies to assess the impacts of biofuel production on water resources (Yang et al. 2009; Elena and Esther 2010). Hoekstra (2003) defines this concept as the "total volume of fresh water that is used to produce the goods and services consumed by the nation". It should not be confused with the concept of virtual water which refers to the volume of water used to produce a product "at the pace of production" (Elena and Esther 2010). Whilst the latter concept (virtual water) is defined from the perspective of production, the former is defined from the perspective of consumption (Elena and Esther 2010).

22.3 Study Area and Methods

22.3.1 Study Areas

The study sites for this particular research are the semi-arid (Mean Annual Rainfall – MAR < 600 mm) regions of Botswana, Zambia, Tanzania and Mali. This section provides an overview of the socio-economic and environmental background as well as the status of biofuel policies in these countries. These are four of the eight African countries which participated in the European Union co-funded project COMPETE (Competence Platform on Energy Crop and Agro-forestry Systems for Arid and Semi-arid Ecosystems in Africa).

Botswana is a semi-arid, landlocked country in the centre of southern Africa with a total area of 581,730 km^2. It receives a total mean rainfall of about 416 mm ranging from 250 mm in the south–west to a maximum of 650 mm per year in the extreme north (FAO 2008a). Rainfall distribution is extremely variable with lengthy dry spells, but with common dry years (Pallett 1997). Botswana is an upper middle income country with Gross National Income per capita of 5,900 US$ (2006) (World Bank 2008). Economic growth in Botswana has brought only a moderate success in human development. The 2007/2008 Human Development Report reveals that Botswana was ranked 124 out of 177 countries, with a human development index of 0.645 (where 0 is the minimum and 1 is the maximum value), as compared to the GDP index of 0.804 (UNDP 2007/2008). Thus, the GDP ranking is higher than the ranking of human development, suggesting that there has been moderate success in transforming economic growth into socio-economic development in Botswana. This country has relatively abundant land resources but limited water resources. It is currently in the process of finalizing a policy on biofuels after undertaking a detailed feasibility study on the production and use of these fuels (EECG 2007; Mabowe 2009).

Zambia has a population of 12 million (2006) and it covers an area of 752,610 km^2 (FAO 2007). The mean annual rainfall in Zambia varies from 600 mm in the south to 1,500 mm in the north. The areas with the highest rainfall are located in the north–west and north–east, whereas the driest areas are in the south–west and around the Luangwa and Zambezi river valleys and most of these parts are classified as semi–arid. According to the World Bank (2008), Zambia is a low income country with a gross national per capita income of 630 US$ (2006). Zambia is categorized as a low human development country with a rank of 165 out of 177 countries and a human development index of 0.434 (UNDP 2007/2008). This country has favourable land and water endowments for biofuel production (Janssen et al. 2009).

Like other African countries, Zambia is currently formulating a long-term Energy Strategy (2009–2030) which includes the development of a framework for sustainable production of biofuels.

The **United Republic of Tanzania** consists of mainland Tanzania and the island of Zanzibar. Its mean annual rainfall is 1,071 mm which varies from 500 to 3,000 mm, though most parts of the country have rainfall which varies from only 500 to 1,000 mm.

The highest rainfall of 1,000–3,000 mm occurs in the north–eastern part of the country in the Lake Tanganyika Basin, and in the Southern Highlands (FAO 2008b). It is a low income country with a Gross National income per capita of 350 US$ (2006). In the 2007/2008 Human Development Report the Republic of Tanzania was categorized as a low human development country with a rank of 159 out of 177 countries and a human development index of 0.467 (UNDP 2007/2008). The country has an area of 945,090 km^2, and its population in 2006 was 39 million (World Bank 2008). There are various biofuel initiatives in Tanzania for both local and international investors. However, the country does not as yet have a comprehensive biofuel policy for guiding these initiatives. According to Martin et al. (2009), the biofuel policy for this country is still under preparation and it is being facilitated by the National Biofuels Task Force.

Mali is a low income country with a Gross National income per capita of 440 US$ (2006) (World Bank 2008). Over 65% of the land area in Mali is desert or semi-desert (UEMOA 2009). The climatic groups of Mali vary from the arid desert in the northern part, arid to semi-arid in the centre, and savannah in the south. The Sahara region in the north-west tip of Mali covers 57% of the country and has an arid and semi-arid desert climate. The mean annual rainfall in Mali is 440 mm, and it ranges from 200 mm in the north to 1,200 mm in the south (Ouattara 2008). The amount of rainfall also varies across the three climatic zones of the Saharo-Sahelian with an annual rainfall of 100–200 mm, the Sahelo-Sudanian (200–400 mm), the Sudano-Guinean (400–800 mm) (Green 2009). In the 2007/2008 Human Development Report Mali is categorized as a low human development country with a rank of 173 out of 177 countries and a human development index of 0.380 (UNDP 2007/2008). Mali has a total surface land area of 1,241,000 km^2 and a population of 14 million (2006) of which 33.4% live in urban areas whilst 66.6% live in rural areas.

22.3.2 Methods

This study is a synthesis of information obtained from literature sources. The key sources of information included FAO reports on irrigation potential in Africa, water profiles in the African countries, consultancy reports on production and use of biofuels in Africa, and journal articles on biofuels and water in Africa. Through membership in the EU co-funded project COMPETE, literature on bioenergy relating to the study areas was obtained. A set of maps, produced on the study areas to serve as a template for areas available for biofuel production (COMPETE 2007), was used to guide the research and discussion on water implications of biofuel development in these areas. There is a general lack of recent data on the irrigation potential of the study areas. The only source of data on irrigation potential in most of the river basins in the study areas is the report published by FAO (1997), and more recent sources of data have not been identified.

22.4 Water and Biofuels in Botswana

According to FAO (2008a), Botswana acquires its water resources within the four major river basins, namely the Okavango in the north–western part of the country, the Limpopo in the extreme east, the Zambezi which forms the Chobe tributary in the far north of the country and the Orange River which originates from the Lesotho highlands with some tributaries in the southern part of Botswana. Botswana shares these major water resources with other SADC countries: Angola, Namibia, Zimbabwe, Zambia and South Africa. Apart from the perennial rivers and wetlands in the north and the over-utilized Limpopo and its tributaries in the east, Botswana suffers from a lack of surface water and therefore it is partly dependent on groundwater resources. Groundwater resources have an overall recharge rate of 1.7 km^3/year and they are found in small quantities almost everywhere in the country, being the main source of water for many rural households, livestock, and small-scale irrigation. However, there are some aquifers in current use which are not receiving any recharge. In many rural areas and remote towns, there is total dependence on groundwater resources (FAO 2008a). The major drainage systems of the Okavango, Chobe and Limpopo have irrigation potential, and abstraction of water for irrigation is limited to these rivers in Botswana (Masamba 2009).

22.4.1 *Limpopo Basin*

The Limpopo River and its tributaries are important sources of water in the eastern part of the country. The basin is estimated to account for 14% of the total area in the eastern part of the country (FAO 1997). The quantity of water produced by the Limpopo River within Botswana is estimated at 0.6 km^3/year, whilst the irrigation potential in the basin is estimated at 5×10^{-4} km^3 (FAO 1997), or less than 0.1%. The major towns of Botswana are located in the south–eastern part of the country, including the capital Gaborone situated on the Notwane River basin, one of the tributaries of the Limpopo River. This region has high water demand for residential and industrial use. The use of water for large-scale biofuel production in this area will therefore add pressure to the already existing water management challenges. However, this will depend on the kind of energy crops grown.

22.4.2 *The Okavango Basin*

The Okavango River, originating from the Angolan highlands, flows for over 500 km before it reaches the border between Angola and Namibia, and a further 113 km before reaching the Okavango Delta in Botswana. The Okavango Delta, a Ramsar Site of international importance, is an important attraction for tourism, the second most important economic activity after diamonds (Kgathi et al. 2007). The average

Table 22.1 Zambezi basin: irrigation potential, water requirements and areas under irrigation

Country	Irrigation potential ha	Gross potential irrigation water requirement m³/ha/year	Gross potential irrigation water requirement Total (km³/year)	Area under irrigation ha
Angola	700,000	13,500	9.450	2,000
Namibia	11,000	5,000–25,000	0.255	6,142
Botswana	1,080	5,500	0.006	0
Zimbabwe	165,400	10,500	1.737	49,327
Zambia	422,000	12,000	5.064	41,400
Tanzania	0	11,000	0.000	0
Malawi	160,900	13,000	2.092	28,000
Mozambique	1,700,000	11,000	18.700	20,000
Total	3,160,380		37.303	146,869

Source: FAO (1997)

rainfall in the Botswana part of the Okavango Basin is 495 mm, ranging from 415 to 570 mm (FAO 1997). Mean annual discharge of the Okavango River at Mohembo is 9 km³. According to FAO (1997), the maximum irrigation potential for the Okavango basin in Botswana is 9,060 ha of which 3,000 ha would require construction of infrastructure and storage. This suggests that the actual irrigation potential is 6,060 ha (FAO 1997). Currently, irrigation in the Botswana portion of the Okavango Basin is limited to 125 ha of vegetables in Shakawe in the upper part of the Okavango Delta called the Panhandle, and also around Maun in the form of small and medium-scale horticultural enterprises. Much of the land in the Okavango Delta is either protected or in the form of Wildlife Management Areas. The promotion of commercial irrigated agriculture for biofuels and other crops conflicts with national and district conservation objectives of land use zoning in the Okavango Delta (van der Heiden 1991) as it is associated with the use of agro-chemicals, which may have an adverse effect on the environmental sustainability of the ecosystem and the tourism industry.

22.4.3 *The Zambezi Basin*

The Chobe River, one of the tributaries of the Zambezi River, forms the border between Namibia and Botswana. Originating in western Zambia and Angola, this river crosses the Caprivi Strip, where it has an annual discharge of 1.3 km³. It flows back into the Zambezi River with an annual discharge of 4.1 km³ at a point where the four countries of Botswana, Zambia, Namibia and Zimbabwe meet. The irrigation potential for the Botswana part of the Zambezi basin is 80 ha without major water development works and 11,080 ha if major water development works are undertaken (FAO 1997). However, 10,000 ha of this total area is located in the Pandamatenga Plains outside the Zambezi basin in the north–eastern part of Botswana, suggesting that total irrigation potential in the Botswana part of the Zambezi basin is 1,080 ha (Table 22.1). There is a plan to transfer 495 million m³ of water per year (0.495 km^3/year) from the Chobe/Zambezi River for agricultural development in Pandamatenga area (Water Resources Consultants 2008). Currently, the land-uses of Chobe District include the Chobe National Park, forest reserves, commercial farms and traditional subsistence farming. The expansion of commercial farms in this area may also be in conflict with other forms of land use. In the Pandamatenga area, about 5,000 ha of land has been identified as marginal land and it is regarded as potential area for biofuel production by the Ministry of Agriculture as the soils are not suitable for food crops (Modise 2009).

22.4.4 Implications for Biofuel Production

Although the Government of Botswana is still in the process of finalizing a policy for biofuels, the available policy statements in various official documents suggest that it will recommend the use of jatropha and sweet sorghum as feedstocks for biofuel production. These crops were recommended by the Feasibility Study on Biofuel Production and Use in Botswana (EECG 2007). The study also revealed that there appears to be sufficient land in Botswana for the production of jatropha and sweet sorghum. It was estimated that only 30% of the 400,000 ha of land demarcated for arable agriculture during the period 2004–2007 was used and the rest of the land (70%) was lying fallow (EECG 2007). It was also found that only 28% of agricultural land was used in eastern Botswana in 2009, and the rest was found to be idle (Kgathi and Mfundisi 2009). The feasibility study also suggested that the production of biofuels in Botswana would not have adverse effects on water resources in the country (EECG 2007). The Botswana National Development Plan 10 (NDP 10), states that a plant of 50 million litres per year, using jatropha as a feedstock, will be constructed by the Botswana Government during this plan period and biodiesel production is targeted for 2012 (MFDP 2009). The plant, which is estimated to cost 100–150 million Botswana Pula (BWP) (15–24.8 million US$), will require 50,000–75,000 ha of land for the production of feedstock (MFDP 2009). According to the Principal Energy Officer in the Department of Energy, a plantation of 20,000 ha will be established by the Government of Botswana for experimental purposes (Mabowe 2009). Due to erratic and unreliable rainfall in semi-arid Botswana (Fig. 22.1), it is intended that groundwater will be used for the irrigation of the crops. It is estimated that six billion litres (600,000 m^3) of water will be required annually to irrigate a plantation of 20,000 ha since 30,000 l of water are required to irrigate one ha of jatropha (Mabowe 2009).

The recommended biofuel crops of jatropha and sweet sorghum have high water efficiencies as revealed by their low water footprints, and are therefore considered suitable to for Botswana with its unfavourable water endowment. It has been found that jatropha requires an annual rainfall of 500–600 mm to produce good yields. Using figures from China and other countries, Yang et al. (2009) revealed that the water footprint for sweet sorghum was 0.7 m^3/l (cubic metres of water/litres of biofuel) as

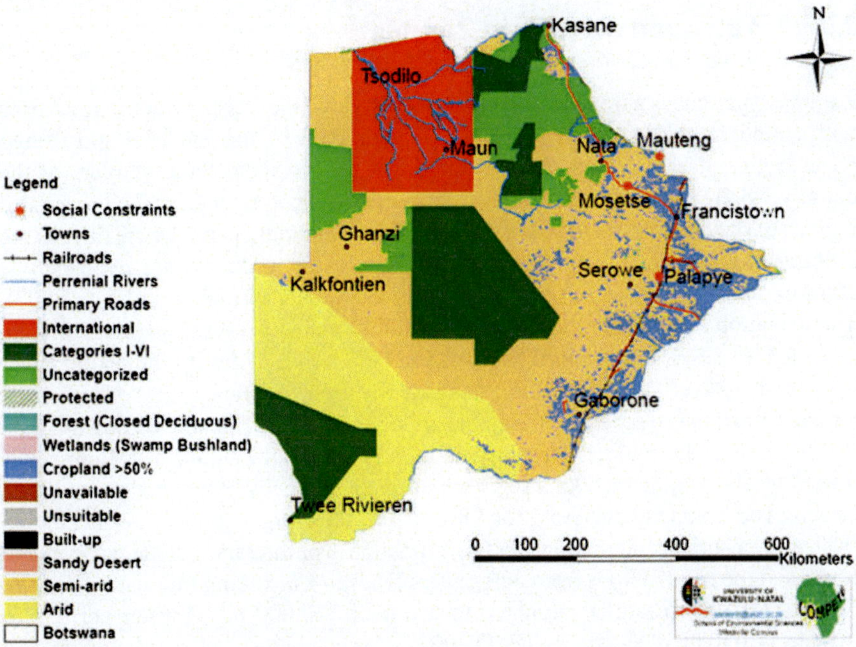

Fig. 22.1 Map of semi-aid areas in Botswana

compared to that of maize, sugarcane, and rapeseed of 1.47, 2.01 and 5.82 m^3/l respectively. On the basis of the amount of rainfall and available water resources, it can be concluded that there are possibilities for growing jatropha and sweet sorghum for biofuel production in river basins in the Chobe, Central, and Gaborone districts. In Chobe District, biofuels could be grown in the Pandamatenga area, where the Government is planning to transfer water from the Chobe/Zambezi River in order to boost agricultural development in the area. A recent survey undertaken in Chobe District revealed that most of the policy-makers held the view that there is a potential for biofuel development in Chobe District though some thought that it is likely to be in conflict with the tourism and forestry sectors. According to the SADC Water Protocol, Botswana is supposed to make a formal request to other member states to abstract water from this river, but such a request has not yet been made (Water Resources Consultants 2008). Given the pressure on water demand in the south–eastern part of the country, near the capital Gaborone, the use of drought resistant crops such as jatropha and sweet sorghum might be an appropriate water conservation strategy in this region. The possibility of using waste-water effluent and contaminated groundwater (e.g. around Ramotswa in the South East District) could also be explored.

22.5 Water and Biofuels in Zambia

Zambia experiences a tropical climate, and it is one of the largest countries in Africa with abundant surface water resources. It is drained by the Zambezi and Congo River basins. The Zambezi basin in the south covers about three quarters of the country, whereas the Congo basin in the north covers one quarter. The semi-arid regions of the country are found in the western and southern parts, where there is the Zambezi River (Fig. 22.2). Zambia's surface water resources are estimated at 105 km^3 and most of these resources (80%) are produced internally (FAO 2007). The irrigation potential in the Zambezi basin in Zambia is estimated at 422,000 ha of which 37.7% is located in the upper Zambezi basin, 47.9% in the Kafue River basin and the rest (14.5%) in the Luangwa River Basin (Table 22.2). There are a number of irrigation schemes in Zambia which include the Nakambala Sugar Estate (11,345 ha) in Mazabuka District in Southern Province of Zambia. The company is owned by Zambia Sugar Plc, a registered private company belonging to the Illovo Group. The company accounts for 98% of the total sugar production in Zambia (Kaizen Consulting International 2006). About 65% of the cane is obtained from the estate and the rest (35%) is from outgrowers (Kaizen Consulting International 2006; FAO 2007). The estate is estimated to use over 720,000 m^3 of water per day to irrigate 11,050 ha of sugarcane (FAO 2007).

In line with the national energy policy of ensuring sustainable exploitation of biomass resources, there are plans to expand the production of sugar at Nakambala Estate by 70% from the 2006/2007 production level of 260,000–440,000 ton. This expansion will also lead to an increase in the production of molasses, which will make it possible to produce bioethanol. The construction of a bioethanol plant, which will produce 25 million litres of bioethanol per year, is therefore anticipated (Kaizen Consulting International 2006). It is stated that the expansion will lead to an increase in water demand, but no figures are given. However, it is unlikely that the project will have adverse direct impacts on water supply as the area is endowed with abundant water resources and will not have exploited its full potential.

The semi-arid area of Southern Province of Zambia (Fig. 22.2) seems to have good conditions for biofuel production, particularly around Mazabuka District. The district has an annual rainfall of 708 mm. It is situated 125 km from the capital Lusaka and 360 km from the southern town of Livingstone. The area has sufficient surface and groundwater resources, the Kafue River and its tributaries being important sources of water and hydro-electricity. In addition to the planned agricultural production of sugarcane and its conversion to ethanol, the Southern Province is one area in semi-arid Zambia where additional biofuel projects could be established. Production of biofuels could be launched in other districts of the Southern Province. In addition to sugarcane, drought resistant energy crops such as jatropha and sweet sorghum could be established, particularly in areas which are not swampy.

In other parts of the country with abundant and reliable rainfall, there are several initiatives for biofuel development which occur under rain-fed agricultural conditions.

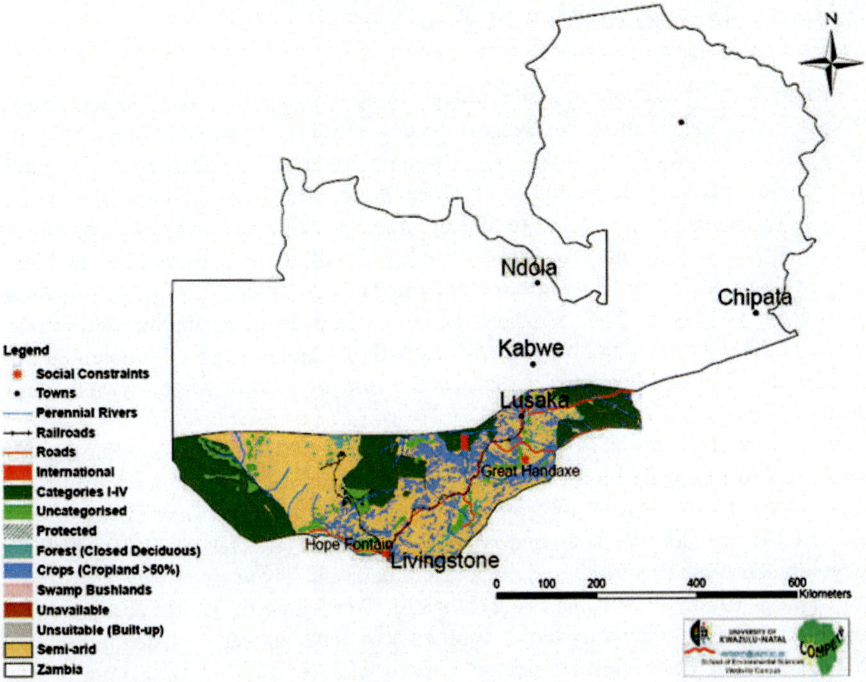

Fig. 22.2 Map of semi-arid areas in Zambia

Table 22.2 Irrigation potential in the different Zambezi sub-basins in Zambia

Types of irrigation	Upper Zambezi river basin (ha)	Kafue river basin (ha)	Luangwa river basin (ha)	Total for Zambezi river basin (ha)
Located	112,000	165,000	14,000	291,000
Groundwater	15,000	15,000	15,000	45,000
Commercial	2,000	2,000	2,000	6,000
Dambos	30,000	20,000	30,000	80,000
Total	159,000	202,000	61,000	422,000

Source: FAO (1997)

In most of these areas the common biofuel feedstock is jatropha. A Zambian Company known as Marli Investment has been involved in jatropha outgrower contract farming since 2003. According to Janssen et al. (2009), the company has distributed more than 12 million seedlings to outgrowers in different parts of the country. A total of 25,000 outgrowers have joined the scheme and their plantations are estimated at 12,000 ha. As long as these schemes continue to be mainly under rain-fed agriculture, it is unlikely that they will have a negative impact on water resources in the country.

22.6 Water and Biofuels in Tanzania

Tanzania lies within nine major drainage basin systems of Lake Victoria, Lake Nyasa, Lake Tanganyika, Lake Eyasi/Bubu depression, Lake Rukwa, Pangani River, Rufiji River, Ruvu/Wami River, and Ruvuma River (FAO 2008b). Of the total renewable surface water resources of 93 km^3/year, 90.3% are produced internally. Groundwater resources amount to 30 km^3/year. In 2002, the total water consumption in mainland Tanzania was estimated at 5,142 million m^3 (5.1 km^3), of which the largest share (90%) of 4.624 km^3 was consumed by agriculture. The Irrigation Water Master Plan Study of 1992 estimated the irrigation potential in mainland Tanzania at 2,123,700 ha (FAO 2008b). The regions with the highest irrigation potential are Mbeya and Iringa, Morogoro, Arusha and Kilimanjaro, and Mara, Mwanza and Kagera (FAO 2008b).

The semi-arid regions of Tanzania, in the central and southern parts, have good conditions for growing biofuels such as sufficient rainfall, abundant land resources, and potential for irrigation of crops, particularly in Morogoro District, where the potential is 376,000 ha. The country has identified several feedstocks for biofuel development such as jatropha, sugarcane and palm oil. There are already sugarcane and jatropha initiatives in mainland Tanzania. According to Martin et al. (2009), biofuel initiatives (including those of biogas) in Tanzania can be categorized into the following scales of development: (1) micro-scale (up to 200 ha), (2) small-scale (200–2,000 ha), (3) medium scale (2,000–50,000 ha), and (4) large–scale (over 50,000). Figure 22.3 shows the location of these initiatives. Micro-scale initiatives are mainly in the form of biogas (introduced in 1993) and jatropha oil production for electricity generation (introduced in 2006). Small- and medium-scale initiatives mainly focus on jatropha production (Martin et al. 2009). An example of a medium scale initiative is Diligent Tanzania Limited. The company intends to produce 1.5 billion litres per year of jatropha-based biodiesel for export to the European market (Martin et al. 2009). Although based in Arusha, the company covers other parts of the country, including those in the semi-arid parts (Martin et al. 2009). Large-scale initiatives are still in infancy as they were introduced in 2007, and they aim at producing sugarcane and jatropha based biofuels in the study areas as well as in other parts of Tanzania (Fig. 22.3).

Currently, the use of water for biofuel development does not seem to have adversely affected water resources in Tanzania as indicated in Table 22.3. The table compares water withdrawals associated with the production of sugarcane for biofuel production in Tanzania with those of other countries. It shows that the total water demand for ethanol production in Tanzania was estimated at 0.10 km^3 in 2008 and it is projected to increase by 50% to the level of 0.15 km^3 in 2017. The amount of water withdrawn for biofuel production in Tanzania in 2008 as a percentage of agricultural water withdrawals in 2000 was only 2% and this figure is projected to increase to 3% in 2017. By comparison, water withdrawals for biofuel production in Brazil, South Africa, Ethiopia, and Mozambique in 2008 as percentages of agricultural water withdrawals in 2000 were 14%, 11%, 5%, and 7%, respectively.

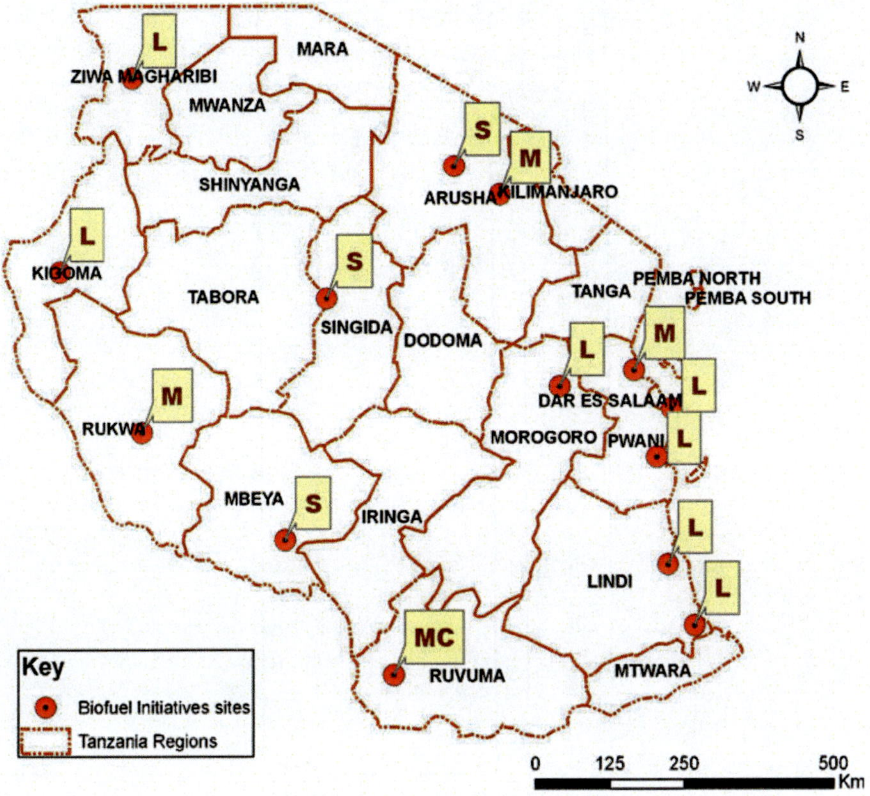

Fig. 22.3 Biofuel initiatives in Tanzania; Notes: *L* large-scale biofuel, *M* medium scale, *S* small-scale, *MC* micro scale (Source: Martin et al. 2009)

In addition, stress on water resources as indicated by the amount of water withdrawals as a percentage of the total agricultural water withdrawals in Tanzania was only 5%, compared to 16% in South Africa, 4% in Ethiopia, 29% in India and 15% in China.

22.7 Water and Biofuels in Mali

Mali generates its water resources within two major river basins of the Niger and the Senegal Rivers. The Niger River basin covers 7.5% of the continent and spreads over ten countries. In 2001, renewable water resources in Mali were estimated at 100 km³ (UEMOA 2009).

According to UEMOA (2009) and Ouattara (2008), Mali has a very limited potential to produce food as only 4% of the land area is arable and the rest is either used

Table 22.3 Implications of bioethanol production on water resources per country

	Main feedstock	% of production	Water demand for ethanol production (km³)		Total agricultural water withdrawals (km³)	Irrigation water for biofuel production as% of agricultural water withdrawals in 2000		Stress on water resources due to agriculture (%)
			2008	2017	2000	2008	2017	2000
USA	Maize	20	6.21	8.48	198	3	4	6
Brazil	Sugarcane	15	5.18	9.49	37	14	26	0
China	Maize	40	7.71	11.78	427	2	3	15
India	Sugarcane	80	4.18	7.82	558	1	1	29
South Africa	Sugarcane	40	0.85	1.58	7.8	11	20	16
Ethiopia	Sugarcane	100	0.25	0.48	5.2	5	9	4
Tanzania	Sugarcane	100	0.1	0.15	4.6	2	3	5
Mozambique	Sugarcane	100	0.04	0.05	0.6	7	8	0

Source: Hoogeveen et al. (2009)

for pasture (25%) or it is classified as other land used for other purposes (71%). Like Botswana, Mali has a high dependence on imported food and fossil energy sources and therefore its food and energy security situation is very fragile (Green 2009). In 2002, the total area under irrigation in Mali was 2,369 km^2. Irrigation is undertaken on 5% of the arable land, especially in the area around the inner Niger Delta (Ouattara 2008). Within the Niger River basin, the irrigation potential has been estimated at 556,000 ha with 200,000 ha (36%) fully controlled and the remaining 356,000 ha (64%) partially controlled. However, only 187,500 ha (34%) is already under irrigation and the irrigation potential in Mali within the basin is limited by the topography.

The majority of Mali's population has no access to electricity (UEMOA 2009). As a result, around 90% of the population uses firewood and charcoal as the main energy sources (Ouattara 2008). It is estimated that only 15% of the population has access to electricity in Mali and in rural areas the proportion is even lower (Ouattara 2008). The potential for jatropha development remains considerable in Mali as areas not suitable for crop production or which are not fully utilized for arable agriculture could be used for its cultivation (GTZ 2008). Figure 22.4 shows the potential areas for growing jatropha in southern Mali.

Bioenergy development in Mali is given strong support by national energy policy. Small-scale production of oil from jatropha for use in "agricultural machinery and rural electrification" has been going on for two decades (United Nations 2008; UEMOA 2009; GTZ 2008). Jatropha has been traditionally used as a hedge for fences and gardens in Mali, currently estimated at 20,000 km (Ouattara 2008). The potential use of the crop for oil production was recognized in the 1930s, but the activities remained on a small-scale until 1987, when Ouattara reinitiated jatropha activities and various opportunities for using it for rural and bioenergy development were explored. For instance, a local NGO, Mali Folkecenter (MCF), has started a project in southern Mali which aims at producing a 1,000 ha jatropha plantation via the small-scale outgrower farming approach. According to Quattara (2008) and Jumbe et al. (2009), communities living within a radius of 20 km are provided with mechanical power through multifunctional platforms for various purposes such as food processing, charging batteries, and rural electrification.

In 2008, the company Mali Biocarburant, established by the Royal Tropical Institute and financed by the Dutch Government, was established in order to produce biodiesel from jatropha already planted as hedges for the purpose of fencing fields or roads or from trees intercropped with other crops in existing farming systems. Bekunda et al. (2009) emphasize the distinct production model of this company whereby production of biodiesel is entirely based on seeds collected from jatropha trees already planted for fencing rather than from plantations. The company also addresses problems associated with intercropping by leaving a 5-m rather than a 2–3 m space between the crops. This is done in order to avoid the problem whereby the canopy becomes too large to allow the sunlight to reach other crops (Green 2009). The first site chosen for biodiesel production is Kalikoro, located in a semi-arid region. This is one of the areas identified by GTZ as a medium potential for jatropha production (Ouattara 2008). It is estimated that Mali Biocarburant will

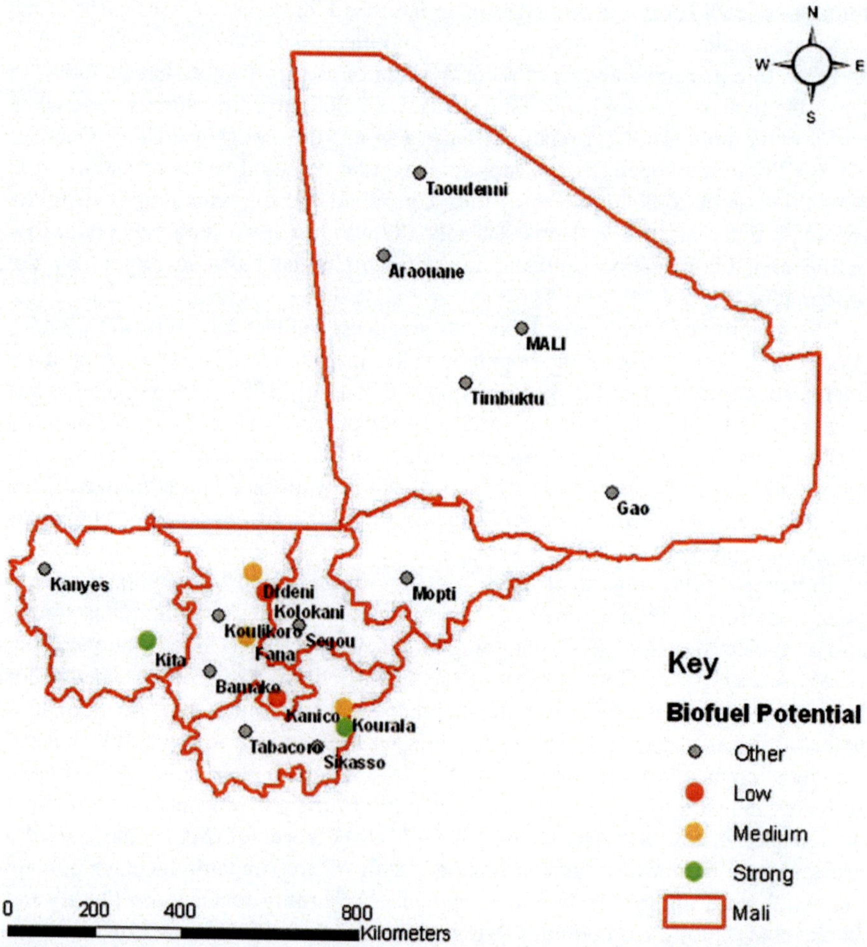

Fig. 22.4 Potential for growing jatropha curcas in southern Mali (Source: Mali Folkecenter 2008)

produce 750,000 litres of biodiesel per year (United Nations Economic Commission for Africa 2008). The company plans to expand to other parts of the country by establishing 20 similar units.

Given the fact that the average rainfall in Mali is 440 mm, the cultivation of jatropha may require irrigation in order to obtain higher crop yields, particularly in semi-arid areas. Studies undertaken in India suggest that if jatropha is grown on marginal land without irrigation, its yield will be 1.1–2.75 ton/ha (Green 2009). Whilst there is potential for irrigation of jatropha in Mali, it is crucial to limit biofuel production to modest levels and not to aim at exporting the fuel as the country is constrained by limited water resources. According to Green (2009, p. 17), "… Jatropha cultivation for biofuel will also be severely constrained by water in Mali and could put immense stress on water resources" and this will be made worse by climate change since various models have predicted drier and hotter conditions for Mali.

22.8 Conclusion

Botswana and Mali have similarities as water scarce countries with high dependency on imported food and energy. In Mali agricultural development is largely constrained by low rainfall and limited water resources. There are initiatives to produce jatropha-based biodiesel using trees already planted as hedges in various parts of the country. This approach is different from other farming systems as it does not lead to production of biofuels on new land. In Botswana, the Government is still finalizing a biofuels policy. However, available policy statements in various official documents suggest that it will recommend the use of jatropha and sweet sorghum as feedstocks for biofuel production. Botswana suffers from a lack of surface water and therefore the country is more dependent on groundwater resources. The major drainage systems of Okavango, Chobe and Limpopo have irrigation potential. On the basis of the amount of rainfall and available water resources, it can be concluded that there are possibilities for growing jatropha and sweet sorghum for biofuel production along river basins in the Chobe, Central, and Gaborone districts.

Whilst large proportions of Mali and Botswana are semi-arid, the semi-arid areas in Zambia and Tanzania are smaller. The semi-arid areas of **Zambia and Tanzania** have relatively suitable conditions for biofuel production. There are sugarcane and jatropha based biofuel projects in these countries, and possibilities for expanding production are being considered. The available land and water resources in these countries make them well positioned to undertake large-scale biofuel projects under rain-fed (e.g. *Jatropha curcas*) and irrigated conditions (e.g. sugarcane). In Zambia, there are plans to expand the production of sugar at Nakambala Estate by 70%. This expansion will lead to an increase in the production of molasses, which will make it possible to produce bioethanol. The construction of a bioethanol plant to produce 25 million litres of bioethanol is under consideration.

Expansion of biofuel projects in these countries has the potential to adversely affect water resources. In water scarce countries such as Botswana and Mali, it is better to engage in biofuel projects which do not require much irrigation. The use of irrigation water to produce fuel from marginal lands may face the primary limitation of water availability, rather than land suitability. The production model of integrating biofuel production with food crop production as practised in Mali can provide useful lessons for other countries. In Zambia and Tanzania, large-scale projects based on contract farming could be expanded to reach the full potential of these countries, but their impact on water resources should be carefully monitored.

In all these countries, detailed environmental assessments, including those of water footprints, should be made mandatory before introducing large-scale biofuel projects likely to deplete water resources. It is also necessary to estimate the theoretical virtual water content of biofuels. When biofuels are produced for export, the export includes virtual water content, which means that there is a net loss of environmental capital. In addition, research and development efforts on second generation biofuels should be continued as the production and use of these fuels is likely to lead to a reduction in water use.

References

Bekunda M, Palm CA, de Fraiture C, Leadley P, Maene L, Martinelli LA, McNelly J, Otto M, Ravindranath NH, Victoria RL, Watson H, Woods J (2009) Biofuels in developing countries. In: Howarth RW, Bringezu S (eds) Biofuels: environmental consequences and implications of Changing Land Use. Proceedings of the Scientific Committee on Problems of the Environment (SCOPE) International Biofuels Project Rapid Assessment, 22–25 Sept 2008, Gummersbach, Cornell University, Ithaca, pp 249–269. http://cip.cornell.edu/biofuels

Berndes G (2008) Water demand for global bioenergy production: trends, risks and opportunities. Report Commissioned by the German Advisory Council on Global Change (WBGU). Chalmers University of Technology, Göteborg, Sweden. http://www.wbgu.de/wbgu_jg2008.html

Competence Platform on Energy Crop and Agro-forestry Systems for Arid and Semi-arid Ecosystems-Africa (COMPETE) (2007) First periodic report, COMPETE, Munich

de Fraiture C, Smakhtin V, Bossio D, McCornick P, Hoanh C, Noble A, Molden D, Gichuki F, Giordano M, Finlayson M, Turral H (2007) Facing climate change by securing water for food, livelihoods and ecosystems. SAT e-J 4(1):1–21

de Fraiture C, Giordano M, Liao Y (2008) Biofuels and implications for agricultural water use: blue impacts of green energy. Water Policy 10(1):67–81

Demirbas MF (2009) Biorefineries for biofuel upgrading: a critical review. Appl Energy 86(1):S151–S161

Dufey A (2006) Biofuels production, trade and sustainable development: emerging issues. International Institute of Environment and Development (IIED), London

Economic Commission for Africa (2008) Biofuels: what strategies for developing the sector in West Africa? United Nations Economic Commission for Africa, Adis ababa, Ethiopia

Elena GC, Esther V (2010) Virtual water content and water footprint of biofuel consumption in Spain. Energy Policy 38:1345–1352

Energy Environment, Computer and Geophysical Application (EECG) (2007) The feasibility study for the production and use of biofuels in Botswana. Ministry of Minerals, Energy and Water Affairs, Gaborone

FAO (1997) Irrigation potential in Africa: a basin approach. Natural Resources Management and Environmental Department, UN, Rome. http://www.fao.org/docrep/W4347E/w4347e0h.htm

FAO (2007) Water profile in Zambia. In: Clevelend CJ (ed) Encyclopedia of Earth. Environmental Information Coalition, National Council for Science and Environment, Washington, DC

FAO (2008a) Water profile in Botswana. In: Clevelend CJ (ed) Encyclopedia of Earth. Environmental Information Coalition, National Council for Science and Environment, Washington, DC

FAO (2008b) Water profile in Tanzania. In: Clevelend CJ (ed) Encyclopedia of Earth. Environmental Information Coalition, National Council for Science and Environment, Washington, DC

Green L (2009) An analysis of possible implications for food security in Mali. Honours thesis, Dalhousie University, Halifax

Hoekstra AY, Chapagain AK (2007) Water footprints of nations: water use by people as a function of their consumption pattern. Water Resource Management 21:35–38

Hoogeveen J, Faures JM, Van de Giessen N (2009) Increased biofuel production in the coming decade: to what extent will it affect global freshwater resources. Irrig Drain 58:S148–S160

Janssen R, Rutz D, Helm P (2009) Bioenergy for sustainable development in Africa-Environmental and social aspects. A paper presented at the 17th European biomass conference, Hamburg, 29 June–3 July 2009

Jumbe C, Msiska FBM, Madjera M (2009) Biofuels development in sub-Saharan Africa: are the policies conducive. Water Policy 37(11):4880–4986

Kaizen Consulting International (2006) Zambia sugar PLC: expansion project at Nakambala estate. Environmental impact statement. Kaizen Consulting International, Lusaka

Kgathi DL, Mfundisi KB (2009) Potential impacts of the production of liquid biofuels on food security in Botswana. Report submitted to the COMPETE project as a contribution to the tasks of Work Package One (WP1) activities: current land use patterns and impacts. http;//www.Compete-Bioafrica.net/publications.html

Kgathi DL, Ngwenya BN, Wilk J (2007) Shocks and rural livelihoods in the Okavango Delta, Botswana. Dev S Afr 24(2):289–308

Mabowe BR (2009) A proposal for biodiesel development from Jatropha seeds. Ministry of Minerals, Energy and Water Resources, Gaborone

Mali Folkecenter (2008) MFC Jatropha activities. Mali Folkecenter, Bamako. http://www.malifolkecenter.org/lowersection/Dep3_NRM/jatropha_plantation_Map.html

Martin M, Mwakaje AG, Eklund M (2009) Biofuel development initiatives in Tanzania: development activities, scales of production and conditions for implementation and utilization. J Cleaner Prod 17(1):S69–S76

Masamba WR (2009) Transboundary diagnostic analysis of the Botswana portion of the Okavango River basin. Output 3: irrigation development. Harry Oppenheimer Okavango Research Centre, Maun

Ministry of Finance and Development Planning (2009) National development plan 10, April 2009 – March 2016. Government Printer, Gaborone

Modise M (2009) Director of Crop Production, Ministry of Agriculture, Personal communication on use of agricultural land for biofuels. Ministry of Agriculture, Botswana

Naylor RL, Liska AJ, Burke MB, Falcon WP, Gaskell JC, Roselle SD, Cassman KG (2007) The ripple effect: biofuels, food security, and the environment. Environment 49(9):30–43

Outtara O (2008) Mali case study: Jatropha for local energy supply & rehabilitation of degraded land. In: Duchrow A (ed) Biofuels development in drylands-panacea or empty promise. Deutsche Gesellschaft Fur Technische Zusammenarbeit (GTZ), Eschborn, Germany

Paehlke R (1999) Towards defining, measuring and achieving sustainability: tools and strategies for environmental valuation. In: Becker E, Jahn T (eds) Sustainability and social science. ZED Books, London

Pallett J (ed) (1997) Sharing water in Southern Africa. Desert Research Foundation of Namibia, Windhoek

Smeets E (2008) Possibilities and limitations for sustainable bioenergy production systems. PhD thesis, Copernicus Institute for Sustainable Development and Innovation, Utrecht University, Utrecht

Society R (2008) Sustainable biofuels: prospects and challenges. Royal Society, London

UEMOA (2009) Sustainable bioenergy development in UEMOA member countries. UEMOA and the Hub for Rural Development in West and Central Africa, Dakar

United Nations Development Programme (UNDP) (2007/2008) Human development report, UNDP, New York. 2007/2008

van der Heiden LJ (1991) Land use and development plan Kwando and Okavango wildlife management areas. First draft. District Land Use Planning Unit, North-West District Council, Maun

Von Maltitz GP, Brent A (2008) Assessing the biofuel options for southern Africa. Science real and relevant. In: 2nd CSIR biennial conference, CSIR International Convention Centre. Pretoria, 17 and 18 Nov 2008, pp 16. http://hdl.handle.net/10204/2579

Water Resources Consultants (2008) Detailed environmental impacts assessment study for a prefeasibility/feasibility on utilization of water resources of the Chobe/Zambezi river-draft environmental impact assessment report. Water Resources Consultants, Gaborone

World Bank (2008) World development report. World Bank, Washington, DC

Yang H, Zhou Y, Liu J (2010) Land water requirements of biofuel and implications for food supply and the environment in China. Energy Policy 37:1876–1885

Chapter 23
Implications of Climate Change on Sustainable Biofuel Production in Africa

Sumedha Malaviya and Nijavalli H. Ravindranath

Abstract The global demand and production of biofuels is likely to increase in the future due to their ability to mitigate climate change and reduce reliance on expensive oil imports. Many African nations have formulated biofuel policies and have set ambitious biofuel targets. In Africa biofuel production from first generation biofuel feedstock like sugarcane and maize is being supplemented by crops like cassava and sorghum. Jatropha is also being promoted for biodiesel production. Biofuel crops are likely to be impacted by climate change similar to food crops and forests. Limited studies on projected changes in the distribution and productivity of biofuel feedstock in Africa are available. This chapter reviews and highlights some key climate change impacts on African biofuel crops like sugarcane, maize, sorghum, cassava and jatropha. Africa is projected to experience climate change on a significant scale with rise in temperature, decline in rainfall in many regions, and increased occurrence of droughts and floods. The projected climate change is likely to lead to land degradation, water stress, increased pest occurrence and ultimately reduction in yields of annual as well as perennial plantation crops. If biofuels are to be cultivated on a large-scale in Africa, it is very important to develop and adopt adaptation practices and strategies for sustainable crop yields.

Keywords Climate change • Africa • Mitigation • Adaptation • Climate change projections • Climate variability

---

S. Malaviya (✉) • N.H. Ravindranath
Center for Sustainable Technologies, Indian Institute of Science, Malleshwaram, Bangalore 560 012, India
e-mail: sumedhamalaviya@gmail.com; ravi@ces.iisc.ernet.in

R. Janssen and D. Rutz (eds.), *Bioenergy for Sustainable Development in Africa*,
DOI 10.1007/978-94-007-2181-4_23,

23.1 Introduction

Climate change is one of the most important global environmental concerns, which is likely to impact natural ecosystems such as forests, wetlands and grasslands and economic systems such as food production and coastal settlements. Africa is one of the regions which are more likely to be adversely impacted by climate change than any other region, due to a large dependence of the population on natural resources. Furthermore, the adaptive capacity of the population is low. Mitigation and adaptation are the two measures to address climate change. The priority in Africa should be on adaptation, since the contribution of Africa to the global GHG emission is insignificant. In the global efforts to mitigate climate change, biofuels which substitute fossil fuels, are amongst the critical mitigation options (IPCC 2007a).

Worldwide the use of biomass energy will increase in the future. World biofuel production has surpassed 100 billion litres of annual production in 2009. Biofuels are gaining additional importance in the context of unprecedented increase in price of petroleum fuels. In the recent years a number of countries, both developed and developing, have set targets for substituting diesel and gasoline by biofuels, with proportions ranging from 5% to 20% to be met at various timelines within the period 2012–2030. It has been estimated that global consumption of biofuels is expected to increase from 75 Mtoe in 2015 to nearly 125 Mtoe in 2030 (ELOBIO 2010).

In the global efforts to address climate change and oil security, biofuels are likely to play an important role. Africa has significant unused land resources which can be used sustainably to reduce food insecurity and also contribute to domestic, regional and international biofuel markets. However, environmental goals need to be carefully managed (SCOPE 2009). Africa has vast potential land area of around 870 million ha suitable for agricultural production. Currently in Africa only about 224 million ha are under crops. There is large interest in growing biofuel crops in Africa due to easily available land, labour and climate. Many countries in Sub-Saharan Africa like Kenya, Uganda, Sudan and Zimbabwe have programmes and policies supporting biofuel expansions. Countries like Mozambique and Tanzania have progressive and ambitious biofuel policies.

If annual or perennial biofuel crops have to be grown to meet the biofuel demands on a large-scale, it is necessary to consider the likely impacts of climate change on sustainable production of biofuel crops. In this chapter, an attempt is made to assess the likely implications of the projected climate change on biofuel crop production and possible adaptation measures to cope with the projected climate change. The current area under biofuels is negligible but is likely to increase in the coming few decades (Table 23.1).

Table 23.1 Land requirements for biofuel production globally

	2004[a]		2030 reference scenario[b]		2030 alternative policy scenario[c]		2030 second-generation biofuels case[d]	
	Million ha	% arable	Million ha	% arable	Million ha	% arable	Million ha	% arable
United States and Canada	8.4	1.9	12.0	5.4	20.4	9.2	22.6	10.2
European Union	2.6	1.2	12.6	11.6	15.7	14.5	17.1	15.7
OECD Pacific	Neg	Neg	0.3	0.7	1.0	2.1	1.0	2.0
Transition economies	Neg	Neg	0.1	0.1	0.2	0.1	0.2	0.1
Developing Asia	Neg	Neg	5.0	1.2	10.2	2.5	11.5	2.8
Latin America	2.7	0.9	3.5	2.4	4.3	2.9	5.0	3.4
Africa and Middle East	Neg	Neg	0.8	0.3	0.9	0.3	1.1	0.4
World	13.8	1.0	34.5	2.5	52.8	3.8	58.5	4.2

Source: FAO (2008)

[a]Data of 2004

[b]Reference Scenario refers to an outcome, on given assumptions about economic growth, population, energy prices and technology, if nothing more is done by governments to change underlying energy trends and takes account government policies and measures that had already been adopted by mid-2007

[c]Alternative Policy Scenario, takes into account those policies and measures that countries are currently considering and are assumed to adopt and implement the share is projected to increase to 3.3% in 2015 and 5.9% in 2030, corresponding to an increase in total volume to 78 Mtoe in 2015 and 164 Mtoe in 2030

[d]Second-generation liquid biofuels become widely commercialized before 2030

23.2 Current Climate Variability in Africa

The African continent is characterized by a large variety of climate systems ranging from humid equatorial systems, through seasonally-arid tropical, to sub-tropical Mediterranean type climates (UNFCCC 2006). The temporal variability of rainfall is high throughout these different climates. The African Sahel therefore provides the most dramatic example worldwide of climate variability that has been directly and quantitatively measured (Hulme 2001). Climate sensitivity of the major livelihood providing sectors like agriculture and water resources makes climate an important determinant of economic development in Africa. Climate studies about Africa have been limited by lack of accurate and sparse baseline weather data (Collins 2010). There are 1,152 World Weather Watch (WWW) stations in Africa giving a station density of one per 26,000 km^2, eight times lower than the World Meteorological Organization (WMO) minimum recommended level (UNFCCC 2006).

23.2.1 Temperature Variability

Based on historical records, a warming of approximately 0.7°C over most of the continent during the twentieth century is reported by the IPCC (2007a) with a decadal warming rate of 0.05°C and slightly higher warming in June–November seasons than December–May. The IPCC WG II report (2007b) highlighted decadal warming rates of 0.29°C in the African tropical forests and 0.1–0.3°C for South Africa, with minimum temperatures increasing more than mean and maximum temperatures in South Africa and Ethiopia, increasing number of warm spells over southern and West Africa for 1961–2000 period and decreasing number of extremely cold days, whilst in eastern Africa decreasing temperatures have been reported from weather stations located close to coasts or major inland lakes. In Sahelian Africa a clear trend towards an increase in mean annual temperatures has been observed (Hulme 2001) Another study reported low inter-annual temperature variability (less than 5%) for Nigeria in Sub-Saharan Africa and a wide range of coefficient of variation for rainfall (0–600%) (AIACC 2006). A recent study by Collins (2010) investigated the variability of near surface air temperatures in Africa over a 50 year period (1949–2009) using the National Centers for Environmental Research/National Center for Atmospheric Research (NCEP/NCAR) Reanalysis. The trend in mean temperature was determined for three time periods: 1948–1977, 1978–2009 and 2000–2009 and reported an increase in temperatures for 1948–1978 and 1978–2009 time periods all over Africa for the DJF months (December, January and February) with cooling reported for Saharan regions of Nigeria/Cameroon and Senegal/Mauritania in the same season. Large warming during JJA (June, July, August) with a difference of as much as 3°C between periods 1948–1978 and the recent 1978–2009. This is especially true for the Sahara region with warming being attributed not to the El Niño Southern Oscillation (ENSO) but natural climate variability or human activity.

23.2.2 Rainfall Variability

Rainfall variability is much more widely studied and reported for Africa than temperature variability. Data report a high degree of spatial and temporal variability (IPCC 2007b), large inter-annual variability for most parts of Africa specially post 1970s, substantial multi-decadal variability and higher rainfall anomalies. In West Africa, annual rainfall has declined since the 1960s with a decrease of 20–40% observed between the periods 1931–1960 and 1968–1990, with a decline in mean annual precipitation of around 4% in West Africa, 3% in north Congo and 2% in south Congo for the period 1960–1998, a 10% increase in annual rainfall along the Guinean coast in the last 30 years (IPCC 2007b). In different parts of southern Africa (Angola, Namibia, Mozambique, Malawi, Zambia), a significant increase in heavy rainfall events has also been observed. Fauchereau et al. (2003) studied precipitation changes in southern Africa and found no trend towards wetter or drier conditions, but a marked increase in rainfall variability. In the Sahelian zone a clear trend towards declining rainfall can be observed (Collier et al. 2008).

23.2.3 Droughts and Rainfall Extremes

Various parts of North Africa have seen severe droughts in the past 30 years and even back in history in the sixteenth, eighteenth and twentieth centuries (Touchan et al. 2010). Beginning in the 1960s, peaking in the mid 1970s, Burkina Faso and other West African countries of Sahel have witnessed extreme droughts and an average rainfall as low as 381 mm for 1985 (Kasei et al. 2010). The number of drought events per country has been high for countries in Sahelian Africa and West Africa (Fig. 23.1) with number of drought events greater than 10 in Niger, Chad, Sudan and Ethiopia. More recently the 1999–2002 droughts in North Africa appear to be the worst in history (Touchan et al. 2010). Variations in sea surface temperature have been linked to persistent droughts in West Africa with drought severity being attributed to natural variations in sea surface temperatures like Atlantic temperatures (Shanahan et al. 2009) as one of the factors controlling West African Monsoon (WAM). Time series analysis of the Standard Precipitation Index (1961–2005) for the Volta basin in West Africa revealed a drought intensity of less than −2 over nearly 75% of the region indicating that a major part of the region was under extreme drought conditions during this period with an areal extent of 90% (Kasei et al. 2010). Floods have been recurrent in some cases with ENSO events

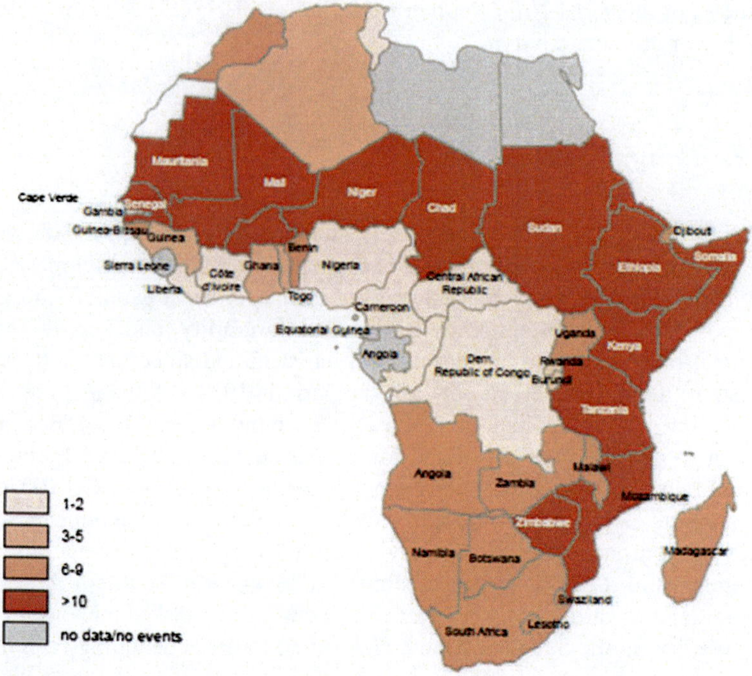

Fig. 23.1 Drought events per country from 1970 to 2004 within Sub-Saharan Africa (Source: UNECA 2008)

(e.g. in Mozambique, Algeria, Tunisia, Somalia). The years 2000 and 2001 witnessed a huge flooding in Mozambique, particularly along the Limpopo, Save and Zambezi valleys (IUFRO 2010).

23.3 Climate Change Projections for Africa

The Fourth Assessment Report of the IPCC projects the increase in average temperatures in Africa to be substantially higher than the global average change (IPCC 2007b) with drier subtropical regions warming more than the moister tropics. Some of the key limitations in knowledge regarding future African climate have been identified by Hulme (2001) as the mainly poor representation of El Niño in General Circulation Models (GCM) and the absence of any representation of regional changes in land cover, dust and biomass aerosol loadings in these models. The general consensus of global climate models is that climate in Africa is likely to become more variable. IPCC (2007a) also predicts a likely decrease in rainfall in much of Mediterranean Africa, northern Sahara and South Africa as well as an increase in mean annual rainfall in East Africa. The future behaviour of rainfall in Sahel and the Guinean coast and Sahara remains unclear.

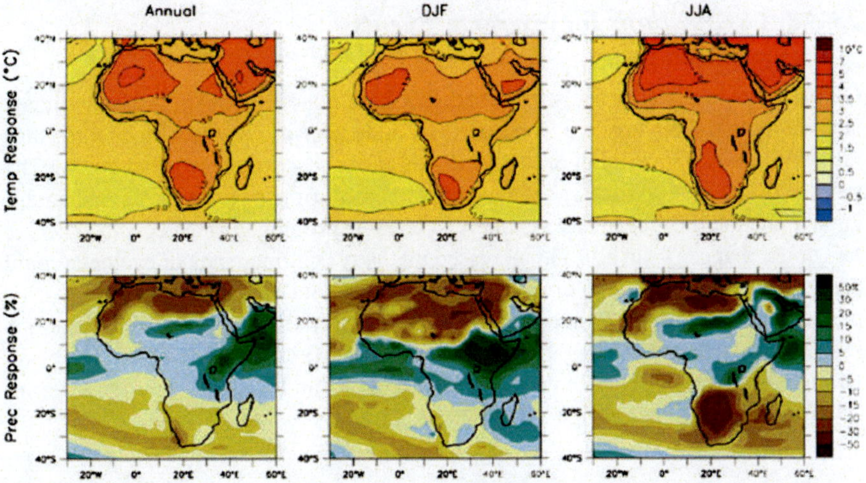

Fig. 23.2 Temperature and rainfall projections for Africa; change from 1980–1999 to 2080–2099 for IPCC scenario A1B, averaged over 21 atmosphere-ocean general circulation models (annual average (*left*), December–February (*centre*), June–August (*right*)) (Source: IPCC 2007b)

23.3.1 Temperature Projections

In all the African regions, namely West Africa, East Africa, South Africa and Sahara, the mean temperature increase has been predicted to be between 3°C and 4°C, i.e. roughly 1.5 times the global mean temperature change. The eastern and western regions of the continent are expected to experience higher temperatures. The largest temperature responses in North Africa are projected to occur in JJA, in southern Africa during September, October and November (SON) (Fig. 23.2).

23.3.2 Rainfall Projections

The eastern and western regions of the continent are expected to experience more rainfall. Further, the rainfall is likely to decrease in much of the Mediterranean Africa, northern Sahara and southern Africa, and some parts of East Africa may receive increased rainfall (Fig. 23.2). Some models of future climate change suggest precipitation changes that may ultimately lead to a more humid regime in the Sahel and parts of the Sahara and a potential increase of vegetation cover of up to 10% of the Saharan area per decade as a result of increased CO_2 concentrations (IUFRO 2010). There is some uncertainty about whether additional rainfall will lead to greater availability of water resources for consumption and food production.

23.3.3 *Projections for Extreme Events*

Global warming is also likely to impact the three major drivers of African climate (Inter Tropical Convergence Zone, ENSO, and the West African Monsoon), thus affecting also the severity of droughts, floods and other extreme weather events. The risks of droughts in Africa are connected to the occurrence of the El Niño phenomenon in the Pacific Ocean and are likely to become prolonged and frequent. There is less doubt that sea level rise and increased intensity and frequency of cyclones could cause problems for coastal cities and major river delta areas. Li et al. (2009) estimated global drought frequencies for Special Report on Emissions Scenarios (SRES) using Palmer Drought Severity Index (PDSI) and calculated the Cropland Drought Risk Index (DRI) with the highest DRI value for Africa that ranges from 95.77 in baseline to 205.46 in 2100.

23.4 Impacts of Climate Change

IPCC (2007a) report declares "Africa to be the most vulnerable region to climate change and variability, a situation further aggravated by the interaction of multiple stresses at various levels and a low adaptive capacity". Assessments of climate change impacts on crop production must consider uncertainties in both future climate projections and the response of crops to these changes (Li et al. 2009).

Agriculture is likely to be impacted the most, severely compromising the already existing food shortage. Rain fed farming occupying 97% of crop area in Sub-Saharan Africa makes it more vulnerable to rainfall changes in the future (Calzadilla et al. 2009). The impacts of climate change on food are likely to be different in different geographical locations for e.g. moderate warming (increases of 1–3°C) is expected to benefit crop and pasture yields in temperate regions, whilst in African tropical and seasonally dry regions, it is likely to have negative impacts, particularly for cereal crops (FAO 2010). It is estimated that by 2100, parts of Saharan Africa are likely to emerge as most vulnerable with losses amounting to 2–7% of GDP. Impacts in western and central Africa also range from 2% to 4% whilst somewhat lower losses are estimated for northern and southern Africa (0.4–1.3%) (IPCC 2007a). Marginal areas (e.g. semi-arid areas) are likely to become more marginal with the area under arid and semi-arid lands expected to increase by 5–8% by 2080. Impacts on agriculture will be further exacerbated by increased soil erosion losses, changes in the onset of rainy days and variability of dry spells. Agriculture losses are shown to be possibly severe for several areas (e.g. the Sahel, East Africa and southern Africa) accompanied by changes in length in growing periods impacting mixed rain-fed, arid and semi-arid systems under certain climate projections. At the local level, many people are likely to suffer additional losses to their livelihood when climate change and variability occur together with other stressors. However, some positive impacts are projected for certain areas in Africa like parts of Ethiopian highlands and parts of southern Africa like Mozambique, with a lengthening of the growing season due to increased temperature and rainfall, with further benefits proposed for irrigated croplands and dry land farms. A study by Ward et al. (2010) reported some positive implications of climate change on agriculture. It estimated that a 1°C increase in temperature would, on average, result in only a 6% decrease in yields of cereals, and even for areas without irrigation a 1°C increase in mean temperature would only be expected to reduce yields by 8%. It also indicated that a 1% increase in precipitation is expected to result in a 7% reduction in cereal yields.

Climate change and variability are likely to result in species loss and extinctions and also constrain the 'climate spaces' and range of many plants and animals in Africa. Lack of access to safe water, arising from multiple factors, is a key vulnerability in many parts of Africa. This situation is likely to be further exacerbated by climate change. Some assessments show severe increased water stress and increased drought risk for parts of north and southern Africa and increase in runoff in East Africa. Africa is characterized by low coping and adaptive capacity. This is due to the extreme poverty of many African countries, frequent natural disasters such as drought and floods, and agriculture heavily dependent on rainfall. The implications of climate change that are projected to lead to land degradation, water stress, decline in the crop yields, threat to forest biodiversity and biomass production will have implications for biofuel crops.

23.5 Biofuel Options and Crops for Africa

Many countries in Africa with vast tracts of arable lands and fertile soils have favourable climate for growing biofuel crops. Amongst the biofuel crops ethanol is the most promising product that can be produced from different raw materials by African countries (Jumbe et al. 2009). Many countries in Africa can produce ethanol from molasses using sugarcane, whilst jatropha and oilseeds are the main source for producing biodiesel. Jatropha is cultivated for example in Sahel, Togo, Ghana, Niger, Tanzania and Mozambique. There are variations in yields of different energy crops across countries, possibly due to differences in varieties, soil conditions, and technologies used in oil production. Palm trees are naturally found in Angola, Democratic Republic of Congo, Nigeria, Ghana, Tanzania, and Malawi. Other crops being used for production of bioenergy include soybean (e.g. Malawi, Tanzania, Ghana), coconut (e.g. Senegal, Mozambique, Nigeria), sunflower (e.g. Angola, South Africa, Namibia), groundnut (e.g. Zambia, Gambia), avocado (e.g. South Africa, Nigeria, Ghana), cashew nuts (e.g. Angola) and castor beans (e.g. Democratic Republic of Congo) (Jumbe et al. 2009).

The use of both arid and semi-arid land for production of biofuels can be an option for countries like South Africa and Kenya. As shown in Table 23.2, the potential for utilization of semi-arid and arid lands is least for Burkina Faso. Most of the arid regions in Botswana and Senegal have high potential for biofuels on arid

Table 23.2 Percentage of the arid and semi-arid regions in eight Sub-Saharan African countries available and suitable for biofuel crop production

Country	% arid	% semi-arid
Burkina Faso	0	15
Senegal	72	6
Mali	31	29
Kenya	91	75
Tanzania	–	46
Zambia	–	42
Botswana	80	42
South Africa	94	70

Source: SCOPE (2009)

Table 23.3 Potential rainfall, land and cultivation practices for bioenergy crops

Bioenergy crops	Rainfall (cm)	Land and climate	Cultivation practice
Sugarcane	150–250	Tropical and sub-tropical	Intensive irrigation and fertilization required at all times
Oil palm	180–500	Humid tropical, moderate to high quality land	No irrigation required in humid areas; Fertilization for high yields
Maize (grain)	70–150	Semi-arid, moderate to high quality land;	Intensive irrigation and fertilization for high yields
Jatropha (seeds)	60–120	Tropical, sub-tropical and semi-arid poor to moderate land	No irrigation and fertilizer application, drought resistant
Woody biomass (ligneous biomass)	50–500	Arid, semi-arid, humid Tropical to temperate Poor quality land, drought resistant	Less intensive fertilization and no irrigation required
Sweet sorghum	50–100	Arid and semi-arid	More drought tolerant than sugarcane; Suitable for wider areas whilst still achieving high yields

lands (80% and 72%, respectively). Only a third of Mali's arid and semi-arid region and only 6% of semi-arid region in Senegal can be considered for biofuel production.

The broad categories of biofuels, climate and land requirement as well as intensity of cultivation are provided in Table 23.3.

23.5.1 Bioethanol Crops

Sugarcane and maize are the principal ethanol yielding crops in Africa. Besides, crops like cassava and sweet sorghum are also being considered both of which can provide high yields per hectare.

Table 23.4 Areas suitable for sugarcane production, given in 1,000 ha as a percentage of land area of the country

	Malawi		Mozambique		Zambia	
	1,000 ha	%	1,000 ha	%	1,000 ha	%
Country area	9,408		78,409		74,339	
Potentially suitable for sugarcane	742	7.9	4,966	6.3	3,546	4.8
Protected areas filtered out	595	6.3	4,622	5.9	2,433	3.3
Slopes>16% filtered out	580	6.2	4,350	5.8	2,427	3.3
Crops and wetlands filtered out	316	3.4	3,733	4.8	1,726	2.3
Existing sugarcane filtered out	314	3.3	3,771	4.8	1,726	2.3
Areas<500 ha filtered out	256	2.7	3,470	4.4	1,485	2.0
Unsuitable soils and rainfall filtered out	206	2.2	2,338	3.0	1,178	1.6

Source: Watson (2008)

Sugarcane is grown mainly in tropical and sub-tropical climates and needs intensive irrigation and fertilizer application. Africa has a much lower ethanol yield from sugarcane of 4,000 l/ha compared to Brazil and is cultivated in 37 countries till 2006. The large producers are Sudan, Kenya, Swaziland, Zimbabwe and Mauritius. Malawi is amongst the few countries that started producing bioethanol from sugarcane molasses since the early 1980s (Jumbe et al. 2009). According to the global agro ecological assessment for agriculture study done by IIASA (2002), 45.6 million ha of land in East Africa is considered very suitable (80–100% of the maximum yield attainable) for cultivation of rain fed and irrigated sugarcane, whilst parts of Cameroon, Nigeria, Ghana and Sudan are moderately to marginally suitable (20–60% of the maximum yield attainable). Sugarcane potential under rain-fed conditions is limited. However, irrigation could change this significantly. In Tanzania, ethanol from sugarcane is a competitive option that requires large-scale industrial set up (FAO 2010) whilst in Ethiopia the total irrigable land for sugarcane production for ethanol is about 700,000 hectares (IRCEED 2009). Based on evaluated yields, percent sugar recovery, length of vegetation cycle in southern Africa, Malawi, Mozambique, and Zambia are best suited for sugarcane cultivation (FAO 2010) (Table 23.4).

Maize is an annual crop that often needs irrigation for high yields. Suitability analysis for rain fed Maize shows that parts of Sudan and Chad are highly suitable whilst other parts of Africa like Nigeria, Ghana, Angola, Tanzania, and South Africa are only moderately suitable for rain fed maize (IIASA 2002). However, the biggest producers of maize ethanol are South Africa, Nigeria, Ethiopia, Kenya and Tanzania.

Sorghum, which has drought adaptation capability, is a preferred crop in tropical, warmer and semi-arid regions of the world with high temperature and water stress. Sorghum is the most important food crop in many African nations. It requires minimal water and nutrient inputs. With a yield of 3,000–6,000 litres per hectare with two harvests a year, sweet sorghum is being increasingly cultivated as a feedstock for ethanol. Sweet sorghum presents a possible alternative to sugarcane because of its lower water requirements, but as new crop it would need further investigation. Nigeria is the biggest producer of Sorghum (Wetland International 2008). Cultivation of rain fed sweet sorghum has been estimated to be highly suitable for parts of Sudan and Egypt with good to medium suitability in parts of Nigeria, Congo and South Africa (IIASA 2002).

Cassava is a perennial shrub that can grow under conditions of low rainfall up to elevations of about 1,000 m. It is a staple food for poor farmers in Africa. As a bio-ethanol feedstock, cassava has some advantages over other feedstock like the ability to remain in ground for months without any deterioration. It can be cultivated at a considerable distance from the processing plant as it can be processed within 2–3 days after harvesting. Cassava is not as bulky as sorghum or sugarcane stalks, therefore it is much easier and cheaper to transport. Cassava has large production potential throughout Tanzania (FAO 2010). Parts of the Democratic Republic of Congo have been estimated to be highly suitable whilst parts of Nigeria, Chad, Sudan and South Africa are declared to be moderately suitable (IIASA 2002). Yields of ethanol from cassava in Nigeria are up to 1,480 l/ha (SCOPE 2009). Cassava can be a promising feedstock for ethanol production owing to its high starch content as it is widely grown in Africa and it can be cultivated with minimal inputs.

23.5.2 *Biodiesel Crops*

Both oil palm and jatropha are being used in Africa for the production of biodiesel. Other oil crops under both experimental and commercial cultivation are cotton seed, groundnut, soybean, castor bean and sunflower.

Cultivation of jatropha is being promoted in many African countries because of its ability to grow relatively well in marginal areas, reclaim degraded lands and contain soil erosion. Besides, jatropha can also be cultivated with other annual crops in agro forestry systems. It is perennial and drought resistant and can be grown on nutrient poor soils. Although many countries grow jatropha, Togo, Ghana and Niger have large jatropha farms, whilst Mali and Tanzania have used it for production of electricity using jatropha oil to facilitate rural economic development (Jumbe et al. 2009).

Amongst the oil crops used for producing biodiesel, oil palm has the highest yield potential per hectare (5,950 l/ha) with large areas under oil palm cultivation in Angola, Democratic Republic of Congo, Nigeria, Ghana and Tanzania (Jumbe et al. 2009). Suitability studies for rain fed oil palm indicate only Democratic Republic of Congo to be the most suitable.

23.6 Impacts of Climate Change on Biofuels in Africa

There are no dedicated studies on the likely impacts of projected climate change on different biofuel crops. There are annual and perennial biofuel crops as well as irrigated and rain-fed crops. The IPCC (2007b) predicts that yields of grains and other crops could decrease substantially across the African continent because of increased frequency of droughts despite potential production increases due to increases in CO_2 concentrations.

23.6.1 IPCC Assessment

The IPCC summarizes the impacts of climate change on important relevant crops as follows:

- The impacts on perennial crops are likely to be more severe than those on annual crops as damages due to temperature stress, pest outbreaks, increased damage from climate extremes as well as benefits (extension of latitudinal optimal growing ranges) may accumulate with time.
- Moderate local increases in temperature can have small beneficial impacts on rain fed maize. But, in dry and tropical areas yield reductions are expected. Even a 1–2°C rise in temperatures is expected to cause a decline in maize yields.
- There can be an increase in the yearly variability of yields of all crops, especially of annuals.
- Potential yield increases for lignocellulosic crops like switch grass and other C4 crops are projected.
- In the absence of studies on the impact of climate change on other biofuel crops such as jatropha, oil palm and pongamia, their response to climate change could be similar to other food crops grown in the region.

23.6.2 Other Studies on Impacts of Climate Change on Biofuel Feedstock

Impacts of climate change on important biofuel feedstock can be divided into impacts on annuals like maize and sorghum and impacts on perennial feedstock like cassava, sugarcane, jatropha and oil palm.

Impacts on Annual Feedstock: Using the South African core climate change scenario for Botswana, Chipanshi et al. (2003) estimated a decline in yields by 36% for maize due to shorter growing seasons. However, recent studies have demonstrated a change in yield up to –22% for un-irrigated maize in Sub-Saharan Africa (Schlenker and Lobell 2010). Burke et al. (2009) examined likely shifts in crop climates in Sub-Saharan Africa under climate change to 2050, and show that the overlap between historical and projected future growing season temperatures for maize will be 58% by 2025, 14% by 2050 and 3% by 2075. Walker and Schulze (2008) used the CERES-Maize model for South Africa and concluded that an increase in temperature has negative effect on maize in yields in drier parts of South Africa. A rise in temperature is projected to have negative effects on maize yields in wetter parts of South Africa. Maize yields are predicted to decline by about 10–15% with a 1°C increase in growing season temperature, a temperature change lower than most published estimates (around 3°C) (Ward et al. 2010). Changes in rainfall patterns and increased pest incidences were blamed for a recent failure of sorghum in Tanzania (Page et al. 2009). Shifts in distribution of sorghum under climate change have been predicted with a 54% overlap between historical and 2025 growing seasons (Burke et al. 2009). In Botswana, a reduction in crop yields by 31% has been predicted (Chipanshi et al. 2003). Besides, Schlenker and Lobell (2010) estimate a change in yields by −17% for un-irrigated sorghum in Africa.

Impacts of Climate Change on Perennial Feedstock: A preliminary assessment of the impacts of climate change on sugarcane in Swaziland indicated that with climate change relatively minor increases in productivity are expected due to increased radiation levels and higher temperatures (1–6% and 10–29% above baseline). However, a decline in the water use efficiency and a reduction in sucrose yield with no noticeable increase in biomass with increased CO_2 is also reported. An earlier study had reported non linear impacts of climate change on net revenue per hectare of sugarcane in South Africa with sensitivity to precipitation (Deressa et al. 2005) and indicated a decline in yields up to 27% with a 2°C rise in temperature.

Limited knowledge exists about impacts of climate change on other perennial crops like cassava, oil palm and jatropha. A decline in yields by about −8% has been predicted for cassava (Schlenker and Lobell 2010). A general conclusion about the negative effects of increased temperature on pestilence under climate change can be drawn.

23.7 Potential Adaptation Strategy for Sustainable Biofuel Production

In the context of projected climate change and the likely adverse impacts, it is necessary to develop and implement adaptation strategies for sustained biofuel production. Currently, there is limited research on potential adaptation practices and

Table 23.5 Potential adaptation strategies to climate change for biofuels

Crops	Adaptation practices	Adaptation policies
Sugarcane	Assured irrigation Increased irrigation efficiency Efficient water management	Insurance Area regulation for crops
Maize, Sweet sorghum	Soil and water conservation Drought resistant varieties Irrigation	Insurance
Jatropha	Soil and water conservation Breeding for drought and pest resistance	Cropping only on degraded lands
Wood-fuel for combustion and gasification	Mixed species plantation Anticipatory planting of crops Fire protection and management	Energy plantations only on degraded lands

strategies even for food crops. Thus, only potential win-win adaptation strategies could be considered. It is important to note that the majority of the adaptation strategies required to cope with current environmental stresses are also required to cope with climate change impacts. An illustrative list of potential adaptation strategies is given in the Table 23.5.

23.8 Conclusions

Africa is projected to experience climate change on a significant scale with rise in temperature, decline in rainfall in many regions, and increased occurrence of droughts and floods. The projected climate change is likely to lead to land degradation, water stress, increased pest occurrence and ultimately reduction in yields of annual as well as perennial plantation crops. If biofuels will be cultivated on a large-scale in Africa to produce ethanol and biodiesel, it is very important to develop and adopt adaptation practices and strategies for sustainable crop yields. However, it is important to state that there is limited research and knowledge on the impacts of climate change on biofuel crops or potential adaptation strategies, particularly at the regional level for Africa. Thus, it is very important to recognize the importance of climate change and initiate research on assessment of impacts of climate change and for developing adaptation strategies for different biofuel crops in different regions of Africa.

References

AIACC (2006) Food security, climate variability and climate change in Sub Saharan West Africa. Available online at: www.aiaccproject.org/Final%20Reports/.../FinalRept_AIACC_AF23.pdf

Burke MB, Lobell DB, Guarino L (2009) Shifts in African crop climates by 2050, and the implications for crop improvement and genetic resources conservation. Glob Environ Change 19:317–325

Calzadilla A, Tingju Z, Rehdanz K, Tol RSJ, Ringler C (2009) Economy wide impacts of climate change on agriculture in Sub-Saharan Africa. International Food Policy Research Institute (IFPRI) discussion papers 873

Chipanshi AC, Chanda R, Totolo O (2003) Vulnerability assessment of the maize and sorghum crops to climate change in Botswana. Climatic Change 61:339–360

Collier P, Conway G, Venables T (2008) Climate change in Africa. Oxford Rev Econ Policy 24(2):337–353

Collins JM (2010) Temperature variability over Africa. Clivar Exchanges 15(4):20–22

Deressa T, Hassan R, Poonyth D (2005) Measuring the impact of climate change on South African agriculture: the case of sugarcane growing regions. Agrekon 444(4):524–552

ELOBIO (2010) Impacts of biofuel expansion on world food systems and the environment. Available online at: http://www.elobio.eu/fileadmin/elobio/user/docs/WP2-PolicyAnalysis_v20080912.pdf

FAO (2008) The state of food and agriculture, biofuels: prospects, risks and opportunities. Chapter 4. Available online at: ftp://ftp.fao.org/docrep/fao/011/i0100e/i0100e04.pdf

FAO (2010) The BEFS analysis for Tanzania. Available online at: www.fao.org/docrep/012/i1544e/i1544e01.pdf

FARA (2010) Mapping food and bioenergy in Africa. Available online: http://fara-africa.org/library/browse/fara_publications/Mapping_Food_and_bioenergy_Africa__final_June_2010.pdf

Fauchereau N, Trzaska S, Richard Y, Roucouand P, Camberlin P (2003) Sea-surface temperature co-variability in the southern Atlantic and Indian Oceans and its connections atmospheric circulation in the southern hemisphere. Int J Climatol 23:663–677

Fresco LO (2006) Biomass for food or fuel: is there a dilemma? The Duisenberg Lecture, Singapore, 17 Sept 2006. Rabobank, The Netherlands

Hulme M (2001) Climate perspectives on Sahelian dessication: 1973–1998. Glob Environ Change 11:19–29

IIASA (2002) Global agro-ecological assessment for agriculture in the 21st century. Available online at: http://www.iiasa.ac.at/Research/LUC/SAEZ/index.html

IPCC (2007a) Climate change 2007: synthesis report, summary for policy makers. Intergovernmental Panel on Climate Change, Geneva

IPCC (2007b) Summary for policy makers. Working Group II, Intergovernmental Panel on Climate Change, Geneva

IRCEED (2009) Biofuels in Africa: a pathway to development. Available online at: http://ccb.colorado.edu/pdf/Biofuels_in_Africa_ICEED_No.43_2009.pdf

IUFRO (2010) Climate change impacts on African forests and people. IUFRO occasional paper no. 24

Jones PG, Thornton PK (2003) The potential impacts of climate change on maize production in Africa and Latin America in 2055. Glob Environ Change 13:51–59

Jongschaap REE, Corre WJ, Bindraban PS, Bandenburg WA (2007) Claims and facts on Jatropha curcas L.: global Jatropha curcas evaluation, breeding and propagation programme. Plant Research International, Wageningen

Jumbe CB, Msiska FB, Madjera M (2009) Biofuels development in Sub-Saharan Africa: are the policies conducive? Energy Policy 37:4980–4986

Kasei R, Diekkruger B, Leemhius C (2010) Drought frequency in the Volta Basin of West Africa. Sustain Sci 5:89–97

Li Y, Ye W, Wang M, Xiaodong Y (2009) Climate change and drought: a risk assessment of crop yield impacts. Climate Res 39:31–46

Mielke (2007) Oil world annual 2007. ISTA Mielke GmbH, Hamburg, Germany

Page SLJ, Karanja DK, Mbwaga AM, Letayo EAS, Nsemwa LTH (2009) The underlying cause of the 2009 sorghum failure in Kongwa district and its implications for Tanzania's vulnerability to climate change. Food Secur 2:157–167

Schlenker W, Lobell DB (2010) Robust negative impacts of climate change on African agriculture. Environ Res Lett 5:014010

SCOPE (2009) What are the final land limits. Available online at http://cip.cornell.edu/DPubS/Repository/1.0/Disseminate?view=body&id=pdf_1&handle=scope/1245782016

Shanahan TM, Overpeck JT, Anchukaitis KJ, Beck JW, Cole JE, Dettman DL, Peck JA, Scholz CA, Kings JW (2009) Atlantic forcing of persistent drought in West Africa. Science 324:377–380

Sims R, Taylor M, Saddler J, Mabee W (2008) From 1st to 2nd generation biofuel technologies: an overview of current industry and RTD activities – extended executive summary. OECD/IEA, Paris

Thornton PK, Jones PG, Alagarswamy G, Andersen J (2009) Spatial variation of crop yield response to climate change in East Africa. Glob Environ Change 19:54–65

Thow A, Warhurst A (2007) Biofuels and sustainable development. Maplecroft report, UK

Touchan R, Anchikaitis KJ, Meko DM, Sabir M, Attlah S, Aloui A (2010) Spatiotemporal drought variability in northwestern Africa over the last nine centuries. Climate Dyn 5:1–16

UNECA, United Nations Economic Commission for Africa (2008) Africa review report on drought and desertification. Available online at: www.un.org/esa/sustdev/csd/csd16/rim/eca_bg3.pdf

UNFCCC (2006) Impacts, vulnerability and adaptation to climate change in Africa. Available online at: unfccc.int/…/adaptation/…/200609_background_african_wkshp.pdf

Walker NJ, Schulze RE (2008) Climate change impacts on agro-ecosystem sustainability across three climate regions in the maize belt of South Africa. Agric Ecosyst Environ 124:114–124

Ward P, Florax JGM, Flores-Lagunes A (2010) Agricultural productivity and anticipated climate change in Sub-Saharan Africa: a spatial sample selection model, agricultural and applied economics association>2010. Annual Meeting, 25–27 July 2010, Denver. Available online at: http://ageconsearch.umn.edu/bitstream/61635/2/Ward%20et%20al%202010%20AAEA%20Paper.pdf

Watson H (2008) WP1 first year report. COMPETE. Competence platform on energy crop and agroforestry systems for arid and semi-arid ecosystems – Africa, Kwazulu Natal University

Wetlands International (2008) Biofuels in Africa; an assessment of risks and benefits for African Wetlands. Wetlands International, Wageningen

Winkler H (2006) Energy policies for sustainable development in South Africa. Energy Research Centre University of Cape Town. Available online at: www.iaea.org/…/ST/…/South_Africa_Report_May06.pdf

Chapter 24
Bioenergy and Sustainable Adaptation to Climate Change in Africa

Kirsten Ulsrud

Abstract In this chapter the role of bioenergy for sustainable adaptation to climate change is discussed. The concept of sustainable adaptation is promoted to address the critical need for radical changes in society due to at least three global problems: (1) The huge poverty problems in the world, which should not be accepted as a natural part of society, (2) The consequences of climatic variability and change for individuals, communities and the society as a whole, (3) The critical need to reduce greenhouse gas emissions drastically, and develop a low-emission-society, as well as a green society that maintains and repairs ecosystems. Production and use of bioenergy are amongst the emerging socio-technical systems that have the potential to address such global challenges. Energy access and strengthened livelihoods for rural populations, in particular, may be important for reducing vulnerability to climate change amongst the poor. However, the variability, contingency and risks in the bioenergy field create uncertainty about its sustainability and opportunities for contributing to climate resilient social and economic development. This chapter will demonstrate how sustainable adaptation to climate change can be applied, and discuss whether and how bioenergy developments can contribute to sustainable adaptation by reducing climate risk and vulnerability or increasing adaptive capacity and resilience, in green and equitable ways.

Keywords Sustainable adaptation • Adaptation to climate variability and change • Poverty reduction • Climate risk • Vulnerability • Adaptive capacity • Local energy supply

K. Ulsrud (✉)
Department of sociology and human geography, University of Oslo,
Molkte Moes vei 31, P.O. Box 1096, Blindern 0317, Oslo, Norway
e-mail: kirsten.ulsrud@sosgeo.uio.no

R. Janssen and D. Rutz (eds.), *Bioenergy for Sustainable Development in Africa*,
DOI 10.1007/978-94-007-2181-4_24,

24.1 Introduction

In every effort to introduce technical and social changes, like activities aimed at introducing and developing new energy systems, climatic factors need to be taken into account. Different types of weather and climate conditions can influence the success of the activities themselves, and/or those people and communities who are meant to benefit from them and their ability to cope with and adapt to climatic variability and change. Climatic stressors can seriously influence people's lives, as shown in the climate change adaptation literature, and more and more so with the ongoing climatic changes (IPPC 2007). Climate adaptation should therefore be an integrated part of every development measure, including attempts to increase the activities in the production and use of bioenergy in developing countries.

In this chapter the role of bioenergy will be discussed for what has been called sustainable adaptation to climate change. The concept of sustainable adaptation is promoted because there is critical need for radical changes in society due to not only climate change and climatic stressors in general, but also due to the huge poverty problems in the world, which should not be accepted as a natural part of society. This problem interacts with the impacts of climatic variability and change on individuals, communities and the society as a whole. Furthermore, the concept is developed to also keep in mind the critical need to reduce greenhouse gas emissions drastically, and develop a low emission society, as well as a green society that maintains and repairs ecosystems.

Production and use of bioenergy are amongst the emerging socio-technical systems that have the potential to address such global challenges. Energy access and strengthened livelihoods for rural populations, in particular, may be important for reducing vulnerability to climate change amongst the poor. However, the variability, contingency and risks in the bioenergy field create uncertainty about its sustainability and opportunities for contributing to climate resilient social and economic development.

24.2 The Sustainable Adaptation Framework

The sustainable adaptation framework developed by Eriksen et al. (2007), Ulsrud et al. (2008) and Eriksen and O'Brien (2007) can be used for the systematic analysis of bioenergy activities and their effects on the ways in which poor people are influenced by climatic variability and change and their opportunities to deal with such climatic stressors. The concept is also a part of attempts to address climate adaptation within larger transformations of society and thus considers the role of social and technological innovations in transitions to sustainability.

The background for the concept for sustainable adaptation is social science research on the linkages between climate adaptation and poverty reduction. The central concerns in this research has been the ways in which contextual factors and multiple stessors at different societal levels hinder or facilitate the efforts of people on the ground to cope with and adapt to climatic variability and change. A central message from this literature is the way in which climatic factors constantly interact with non-climatic factors to shape people's situation on the ground (Eriksen et al. 2007; Leichenko and O'Brien 2008). Thus the first and central idea behind the concept of sustainable adaptation is that responses to environmental changes should be done in consideration of how they influence society, livelihoods and the social equity and wellbeing of people, including the opportunities for climate adaptation on the ground in local communities.

The focus is here on social sustainability, the social context at different societal levels and the diverse reasons for people's vulnerability (Eriksen et al. 2007). Another idea behind the concept of sustainable adaptation is that environmental concerns should be an integrated part of the work for social and economic sustainability of climate adaptation and poverty reduction. Research has shown that environmental degradation is one of the important causes of vulnerability and insecurity for poor people (Ulsrud et al. 2008). The loss of ecosystem services means loss of livelihood assets such as clean water, biomass for energy, raw materials, soil, forest products, animal grazing, game meat and so on. In the next sections at the potential role of bioenergy activities for sustainable adaptation will be assessed, including the potential risks of bioenergy activities to work in the opposite direction of sustainable adaptation.

24.3 Bioenergy, Climate Change and Poverty

A central reason for the promotion of bioenergy, including liquid biofuels for transport energy, is the aim of reducing CO_2 emissions. The use of bioenergy can integrate the production and burning of fuel in the earth's natural CO_2 circulation, although the carbon account varies between different ways of producing and using bioenergy. It has also been argued that the production of bioenergy could have positive effects on poverty because it could benefit poor countries through export revenues on biofuel crops or processed fuel, and thus benefit people through increased opportunities for work and income (EuropaBio 2008).

Some types of bioenergy activities have been recommended also before the strong focus on biofuels as replacement for fossil fuels came as response to climate change, for example through the work on improved cooking stoves and biogas systems for improvements in provision of cooking energy in developing countries. Furthermore, there is also mention of bioenergy in connection with climate adaptation by Ulsrud and Eriksen (2007), in their discussion on the role of local and innovative ecosystem management for climate adaptation. Some types of bioenergy activities could have the potential to contribute to reduced risk to climatic challenges or reduced vulnerability, or even increased capacity for poor people to cope with and adapt to climatic variability and change. However, emphasis is on the importance of efforts to manage, maintain and strengthen ecosystems in ways that enhance ecosystem services and improve access to these services for the poor. Such access to biomass for sustainable energy supply or other livelihood activities could be important for people's ability to cope with and adapt to stressors. Therefore, some types of bioenergy activities could be regarded as sustainable adaptation measures because of their potential for environmentally, socially and economically sustainable contributions to people's livelihoods and people's ability to deal with climatic stressors. The discussion below will shed light on different types of factors that need to be taken into account when assessing the contingent sustainability of biofuel production and other types of bioenergy production and use.

24.4 Systematic Analysis of the Role of Bioenergy for People Affected by Climatic Variability and Change

Bioenergy is a wide and diverse category, as demonstrated in other chapters in this book, since it consists of a range of different fuels that can be used in different types of technologies to run engines or produce heat or electricity. These fuels are produced from a range of different plants and types of biomass, including residues such as rice husks, organic waste and straw (ESMAP 2005). They can be produced in large-scale farms and factories for marketing nationally or internationally or they can be grown or collected in smaller scale, either for local provision of energy or for supply to industry for large-scale processing. Some examples of bioenergy crops are jatropha, sunflower, cotton, sugarcane, sorghum, and maize. Oil seeds are used as pure plant oil and for biodiesel production for engines of different kinds. Sugarcane and cereal crops are used for the production of bioethanol for engines, replacing gasoline. Many types of biomass can be used for the production of briquettes for cooking energy, drying of agricultural crops, etc. Traditional and modern production of charcoal is also amongst the bioenergy technologies.

Because of this large diversity in crops, technologies and products, the role of bioenergy for social and economic development and climate adaptation highly depends on the type of bioenergy in question, and the way it is produced, regulated and used (Diaz-Chavez and Woods 2007). Some bioenergy activities involve risks for environmental degradation, exploitation of workers, acquisition of land from people who have used the land for their livelihoods, competition with food production, and other negative effects that may increase vulnerability and poverty, and at the same time give no improvement in the energy supply for poor people. Charcoal, for example, is produced, transported and sold under different social, political and regulatory circumstances that influence the opportunities for poor people to earn a fair income or to use it for their own energy needs (ESDA 2005).

24.5 Linkages Between Biofuels/Bioenergy and Sustainable Adaptation

The following section describes how the sustainable adaptation framework (Eriksen et al. 2007; Ulsrud et al. 2008; Eriksen and O'Brien 2007) can be applied in the field of bioenergy. It also briefly discusses whether and how bioenergy technologies and developments can contribute to sustainable adaptation by reducing climate risk and vulnerability or increasing adaptive capacity and resilience, in green and equitable ways.

Since the sustainability of bioenergy activities depends fully on how they are carried out, it can be difficult to consider the different ways that such activities can influence the different aspects of climate adaptation. According to Eriksen et al. (2007) social and technological changes for adaptation to climatic variability and change should not only address the risk of damages from climatic stressors on crops and property, but also the vulnerability and adaptive capacity/resilience of people and communities in the face of climatic stressors. These are aspects that are influenced by political, economic and institutional factors at different levels. This approach, looking at risk, vulnerability and adaptive capacity, together with the concern about environmental sustainability, can provide a systematic way of considering whether a given bioenergy activity can represent sustainable adaptation or not. Similar systematic considerations have been done in a comprehensive way in the "More than Rain" report (Ulsrud et al. 2008), where different types of development measures were analyzed. Here, only a few examples will be addressed how bioenergy activities may influence climate risk, vulnerability and the capacity to cope with and adapt to climatic stressors.

24.5.1 Bioenergy and Climate Risk

Climate risks – the direct consequences of climatic challenges – can be such as animals killed by drought, suffering and starvation amongst people, flash rains, floods and landslides washing away soil, homes, animals and crops, changed sowing and harvesting times and changed pests. Bioenergy activities can influence climate risk positively or negatively, depending on its type. If the use of water for irrigation of a large biofuel field reduces the access to water for other purposes, or the growing of biofuel feedstock leads to a land use change that gives increased risk for erosion, the risk for damages from climatic stressors increases. Such effects depend on the type of crops, ways of growing them, the used agricultural techniques (e.g. manure, soil management), the soil type, and the natural conditions in general, etc. Some bioenergy activities can have the potential to reduce risk for damage from climate events, for example if they include the use of perennial instead of annual plants, or the use of trees and bushes instead of cereal crops, thus preventing soil erosion.

24.5.2 Bioenergy and People's Vulnerability to Climatic Stressors

People's vulnerability to climatic stressors is shaped by factors such as unemployment, degraded environment, remoteness, disease, lack of public services, import competition to their agricultural crops, difficult housing conditions, crime and conflict. Bioenergy activities can influence vulnerability either positively or negatively depending on their type and how they are carried out. Vulnerability can be increased for example if the production of specific bioenergy crops or processing of biofuels harm the natural environment and disturb the functioning of ecosystems on which people depend for their livelihoods. Importantly, large-scale bioenergy projects can threaten poor people's weak rights to land, which is already shown in different cases. The risk of such threats to people rights must be seriously considered because of the very unsettled situation for distribution of land in many countries, for example in South Africa, Tanzania, Kenya, Zambia, Senegal, Mali, Burkina Faso and Botswana (Clover and Eriksen 2009). At the same time, governments and investors in many of these countries are looking at the opportunities for production of bioenergy crops. On the other hand, if bioenergy production gives employment with a decent salary and working conditions, without negatively affecting livelihoods or ecosystems, it could reduce some people's vulnerability to climatic stressors as long as climate change doesn't undermine this agricultural production. The energy supply in a country can also become less dependent on world oil markets.

24.5.3 Bioenergy and the Capacity to Adapt and Create Resilient Livelihoods

People's and communities' capacity to cope with and adapt to climatic conditions is constituted by different factors such as access to knowledge, information, access to lessons learned in other places, technology, funding for investments, skills, education, health, physical strength, local labour force. This capacity is influenced by a whole range of societal factors, including the support or hindrances created by governments and established social, political and socio-economic structures. Also different types of bioenergy activities may influence such capacity positively or negatively. Some types of bioenergy technologies can improve the energy supply in households and income generation options in ways that are climate resilient and sustainable (WBGU 2008). For example, smart and environmentally conscious ways of producing and using firewood, charcoal, biogas, briquettes and other types of bioenergy types that can be produced in sustainable ways, can reduce the workload for women, increase incomes, improve working conditions and reduce health problems from smoke in houses. As shown by ESDA (2005) production of charcoal, an activity that is often considered to be unsustainable, in fact can be done in ways that make it a sustainable, efficient and practical way of using biomass. Moreover, local production of oil seeds for generation of electricity in local generators, replacing diesel, can also be a sustainable activity that positively influences livelihoods and quality of life.

24.6 Uncertainty Because of Variability in the Bioenergy Field

A characteristic of this discussion on bioenergy and sustainable adaptation is the use of words like "may", "potential", "could", etc. This illustrates the variability, contingency and uncertainty in the bioenergy field, and the problems of ensuring that the development will go in sustainable directions. The policy recommendations in the area suggest a large range of measures to ensure the environmental and social sustainability of bioenergy activities (COMPETE 2009; UNCTAD 2008; DG TREN – European Commission 2009; Schlegel and Kaphengst 2007), and it may seem relatively unrealistic to implement all of these. Sustainability considerations may be overlooked if demand in the biofuel sector in Europe becomes an increasingly strong driving force for production of biofuel crops in Africa. This means that some of the risk for unsustainable production of bioenergy may only be avoided by looking at other alternatives for energy production and use, including other systems and technologies for transport in the world, such as electric vehicles using electricity from a diversity of renewable energy sources.

24.7 Wider Approach than Bioenergy

Furthermore, it would be important to open up the bioenergy discussion to look at people's opportunities for taking advantage of ecosystem services both in traditional and new ways, in order to go in the direction of an economically, socially and environmentally sustainable society where communities and livelihoods are more resilient to a variable and changing climate than today. Furthermore, there is a need to think more broadly about the facilitation of the use of environmental technologies and the harvesting of different types of local resources such as solar power, wind and geothermal energy. The solar energy potential for example, is much larger than the potential for the use of bioenergy, and has fewer contingencies when it comes to sustainability problems than bioenergy in terms of being less diverse and less interlinked with the functioning of ecosystems both inside and outside agriculture. It is also less dependent on cheap labour and less land intensive. Thus there are huge variations between different socio-technical systems/knowledge systems and technologies in terms of their potential to contribute to a fruitful human – nature interaction which is the prerequisite for sustainable adaptation and a sustainable future. Clearly, the diversity of new renewable energy technologies makes it possible to choose energy sources which are resilient to changes in local climate conditions, whilst it should be remembered that many large-scale hydro projects, despite their renewable nature, increase vulnerability by flooding pasture land and forest resources that people depend on during droughts for example (Ulsrud and Eriksen 2007). The opportunities for using technologies for harvesting local renewable energy resources in sustainable ways depend on the creation of new socio-technical systems that give people access to such technologies and related services. In order to be able to start-up local energy supply, using biomass or other renewable resources, it is important for people to get access to information, equipment and financing, and there would be need for local institutions to handle the system for energy supply. Some facilitation by governmental or NGO actors would be necessary, not least through the removal of barriers constituted by established legal and political structures.

24.8 Recommendations

In the report "More than Rain" by Ulsrud et al. (2008) looking at what sustainable adaptation can mean in practice, some guidelines were suggested that could contribute to ensure that development activities would be sustainable both environmentally and socially. These guidelines were developed to help addressing one or more of the three aspects of climate adaptation mentioned above: reduction of climate risk, reduction of vulnerability towards climatic variability and change, and increase in the capacity to adapt to climatic and other changes. Increased resilience of livelihoods and communities could then be a possible result. The guidelines highlighted the necessity of having sufficient understanding of the local livelihood strategies and vulnerabilities. Without such understanding of people's realities and their efforts for getting food and income it is not possible for involved governmental actors or organizations to understand how adaptation and poverty reduction measures should be planned and how one can be able to address issues that are important for poor people's livelihoods, and build on the strengths and capacities of people themselves.

Understanding of a diversity of knowledge systems is also a central part of the knowledge that is needed in order to promote activities that can work as sustainable adaptation measures (Ulsrud et al. 2008). Such knowledge systems can be about efficient, locally adapted energy supply, conscious implementation processes for such energy supply, organic/sustainable agriculture in combination with traditional techniques, and ecological sanitation, just to mention a few examples.

Identification and understanding of political, economical and institutional barriers for change is also one of the recommendations for how to facilitate sustainable adaptation (Ulsrud et al. 2008). The role of social capital and the role of different actors at different geographical levels in hindering or facilitating activities on the ground can also be crucial for the opportunities for social and technological changes in local communities. Furthermore, identification and spread of good examples, including south-south learning, was also pointed out as an important element in the facilitation of useful changes in communities.

24.9 Conclusion

The concept of sustainable adaptation and its different aspects can be used to show different ways in which bioenergy activities can influence society. This includes the climate risk, vulnerability and adaptive capacity of individuals and communities, especially in relation to the resilience of the livelihoods of poor people in the face of climate variability and change. The examples of potential effects of bioenergy activities on the different aspects of sustainable adaptation that have been presented in this chapter indicate the diverse ways in which bioenergy activities can influence climate risk, vulnerability and adaptive capacity either in positive or negative ways. This chapter thus argues for the need for systematic analysis of bioenergy activity in different geographical and social contexts. Thus, the uncertainty and contingency of the desired sustainability of bioenergy activities was highlighted.

One way of addressing this uncertainty and contingency, is to apply the shown tool, which can help to identify many different potential effects of the activities. Another way of addressing the uncertainty of the role of bioenergy production for sustainable adaptation is to open up the perspective from bioenergy technologies, to different types of renewable energy technologies. These can also sometimes be used in combination with bioenergy such as in solar power supply, where biofuel generators or small-scale gasifiers can complement the solar system in ways that address the need for battery storage of electricity in the solar system, whilst not depending fully on the continuous supply of feedstock for the biomass systems. The discussion can also be opened up in the sense that it includes poor people's opportunities to benefit from ecosystem services in different and innovative ways.

Importantly, the potential negative effects of large-scale biofuel production constitutes a serious risk to the opportunities for poor people to cope with and adapt to climatic stressors, especially if it threatens their rights to land, which is very weak already in many African countries.

References

Clover J, Eriksen SEH (2009) The effects of land tenure change on sustainability: human security and environmental change in southern African savannas. Environ Sci Policy 12(1):53–70

Compete (2009). Third periodic activity report (01.01.2009–31.12.2009) Jan 2010 ANNEX 3-3-3: good practice guidelines to project implementers. Competence platform on energy crop and agroforestry systems for arid and semi-arid ecosystems – Africa (COMPETE)

Diaz-Chavez R, Woods J (2007) The environmental certification of biofuels. Discussion paper no. 2007–6 Dec 2007, OECD, International Transport Forum. Imperial College, London

Eriksen SEH, O'Brien KL (2007) Vulnerability, poverty and the need for sustainable adaptation measures. Climate Policy 7:337–352

Eriksen SEH, Klein RJT, Ulsrud K, Næss LO, O'Brien KL (2007) Climate change adaptation and poverty reduction: key interactions and critical measures. Report prepared for the Norwegian Agency for Development Cooperation (NORAD). University of Oslo, Oslo

ESDA (2005) National charcoal survey report. Energy for Sustainable Development Africa (ESDA), Nairobi. www.esda.co.ke

ESMAP (2005) Advancing bioenergy for sustainable development. Guideline for policymakers and investors, vols. I, II and III. Energy Sector Management Assistance Programme, ESMAP, Stockholm Environment Institute, Stockholm

EuropaBio (2008) Biofuels and developing countries. EuropaBio Fact Sheet. The European Association for Bioindustries (EuropaBio), Brussels. http://www.worldbiofuelsmarkets.com/reports/Developing_countries_March08.pdf

European Commission (2009) Directive 2009/28/EC of the European Parliament and of the Council of 23 April 2009 on the promotion of the use of energy from renewable sources and amending and subsequently repealing Directives 2001/77/EC and 2003/30/EC

IPCC (2007) Impacts, adaptation and vulnerability. Contribution of working group II to the fourth assessment report of the intergovernmental panel on climate change. In: Parry ML, Canziani OF, Palutikof JP, van der Linden PJ, Hanson CE (eds) Climate change (2007). Cambridge University Press, Cambridge, p 976

Leichenko R, O'Brien KL (2008) Environmental change and globalization: double exposures. Oxford University Press, New York

Schlegel S, Kaphengst T (2007) European union policy on bioenergy and the role of sustainability criteria and certification systems. J Agric Food Ind Organ 5(2). doi:10.2202/1542-0485.1193, Article 7

Ulsrud K, Eriksen SEH (2007) Vulnerability and adaptation to climate change. New challenges for poverty eradication. In: Norwegian church aid and church of Sweden, climate change threatens the fight against poverty – understanding the issue 1/2007. Norwegian Church Aid and Church of Sweden, Oslo, pp 23–51

Ulsrud K, Sygna L, O'Brien KL (2008) More than rain: identifying sustainable pathways for climate adaptation and poverty reduction. Report prepared for the Development Fund in Norway, University of Oslo, Oslo

UNCTAD (2008) Making certification work for sustainable development: the case of biofuels. UNCTAD/DITC/TED/2008/1. http://www.unctad.org/en/docs/ditcted20081_en.pdf

WBGU (2008) World in transition – future bioenergy and sustainable land use. German Advisory Council on Global Change. Earthscan, London

Part V
Financing and Socio-Economic Issues

Chapter 25
Keynote Introduction: Socio-Economic Impacts of Different Scales of Biofuel Production in Africa

Dominik Rutz and Rainer Janssen

Abstract Two thirds of the population in the developing world, where poverty is at its peak, derive their incomes from agriculture. Thus, the cultivation of energy crops as feedstock for biofuels and bioproducts is considered as large opportunity to improve their economic situation. At the community level farmers who produce dedicated energy crops can increase their incomes and grow their own supply of affordable and reliable energy for their internal needs. At the national level, cultivating biofuel crops may generate new industries, technologies, jobs and markets. At the same time, producing more biofuels will reduce energy expenditures and allow developing countries to put more of their resources into health, education and other services for their neediest citizens. On the other hand, the cultivation of large energy crop monocultures for industrial use will most likely be dominated by international large-scale companies and investors. This may cause negative socio-economic impacts, especially on land tenure issues. Many large-scale economic models discourage pro-development practices. In order to prevent negative social, economic, and environmental impacts, suitable policies, enforcement of legislation and sustainability schemes are needed. The present chapter shows risks and opportunities of biofuel production in Africa related to different production scales and related to the type of end use of the biofuel.

Keywords Socio-economic impacts • Domestic use • Export • Biofuels • Africa • Agriculture • Farming • Monocultures • Subsistence farming • Industrial farming • Smallholders farming • Super-farms • Small-scale • Large-scale

D. Rutz (✉) • R. Janssen
WIP Renewable Energies, Sylvensteinstr. 2, 81369 Munich, Germany
e-mail: dominik.rutz@wip-munich.de; rainer.janssen@wip-munich.de

R. Janssen and D. Rutz (eds.), *Bioenergy for Sustainable Development in Africa*,
DOI 10.1007/978-94-007-2181-4_25,

25.1 Introduction

Recent developments of bioenergy have high impacts on two main sectors: agriculture and energy.

In agriculture-based countries, such as in most African countries the **agricultural sector** generates on average 29% of the gross domestic product (GDP) and employs 65% of the labour force (The World Bank 2007). Thus, it is one of the most important sectors for economic growth in Africa. Thereby a large portion of agriculture is related to subsistence agriculture and small-scale farming. The debate about sustainable and suitable sizes and forms of agriculture for Africa has already started long time ago during colonization of Africa. This debate includes discussions about preferences for local consumption or export of agricultural commodities. Originally mainly focusing on traditional commodities like food, luxury goods (e.g. tobacco, coffee, tea) and materials (e.g. cotton), the debate today is enlarged by another dimension: bioenergy.

The second sector which is affected by the introduction of bioenergy is the **energy sector** which was so far dominated by fossil energy carriers (crude oil, natural gas, coal) and by traditional bioenergy (firewood, charcoal). Similar to the export of commodities in the agricultural sector, also fossil energy carriers were exported very often without leaving revenues in the African countries.

The market for fossil energy carriers is dominated by multinational large-scale companies, whereas the agricultural sector usually involves a mix of small to multinational large-scale companies.

If not properly managed, there is a risk of unsustainable exploitation of biofuels, as it happened in the energy and agricultural sectors. This is especially relevant for liquid biofuels since they can be easily exported and globally transported due to their high energy density.

The sustainability debate in Africa is mainly influenced by socio-economic aspects, whereas the main driver for biofuel development in Europe is environmental aspects.

The following sections describe the socio-economic impacts of biofuel production related to different agricultural scales and biofuels with emphasis on liquid biofuels.

25.2 Overview of Socio-Economic Impacts

One of the main arguments in favour of biofuel production in Africa is opportunities for African people and economies. On the other hand, there is also serious concern about the socio-economic sustainability of biofuel production. Biofuel production has impacts on different levels: single persons, households, farmers, communities, regions and countries. Thereby, impacts can be positive or negative. This section gives an overview on the main socio-economic impacts of biofuels in Africa which are influenced by the scale of the agricultural system and of the biofuel production concept.

Food security refers to food availability and access to food. A household is considered food-secure when its occupants do not live in hunger or fear of starvation. Food security is often related to biofuel production. In many cases no differences between food availability and access are made. Food availability means that sufficient quantities of appropriate, necessary types of food from domestic production, commercial imports or donors are continuously available to the individuals or are within reasonable proximity to them or are within their reach. Food access means that individuals have adequate incomes or other resources to obtain levels of appropriate food needed to maintain consumption of an adequate nutrition level. Biofuels may contribute to increased or reduced food security, depending on policies, agricultural systems, markets, prices and income level of the poorest. For instance an increase of food prices may be an opportunity for farmers due to increased income and thus to increased food access. On the other hand increased food prices may not be affordable for the poorest. Increased food prices especially affect food importing economies which are many African countries (Rossi and Lambrou 2009). The current impact of biofuels on food security is still low, but may increase with increasing biofuel demand.

Employment opportunities are associated with the growing global demand for bioenergy, thus leading to increased income generation and rural development. Employment increase is generally related to all steps of the value chain, from agricultural feedstock production, to the conversion process, and to the end use. In comparison to fossil fuels, the employment rate of biofuel production is much higher. However, the positive or negative impacts on employment largely depend on the scale of the production systems. With increasing mechanization of agricultural production in many countries, and substitution of traditional agricultural systems, the number of agricultural jobs associated with the production of liquid biofuels is likely to decrease over time.

Rural development is seen as key for poverty reduction in Africa (Anríquez and Stamoulis 2007). Biofuel production can contribute positively to rural development, but this depends on the scale of the production and on the feedstock type.

Land use is one of the most controversial subjects in developing countries since any land use change has positive and/or negative impacts. It is important to recognize in biofuel projects that very often also so-called marginal land has value for the local population in Africa (Scott 2009). In general, increasing biofuel production will increase land use competition. An option to avoid land use conflicts especially in Africa is the improvement of agricultural productivity.

Land ownership systems in Africa are often associated with uncertainties, since land property is often not officially secured, and cadastral registries are often non-existent. Land is frequently leased from the state or held communally and is not based on private property. Therefore, land rights are regularly in dispute. Large-scale production of feedstock is often criticized for depriving small farmers of their properties. Unclear land rights and poorly regulated land acquisition – conditions which often prevail in developing countries – may lead to displacement of local farmers to non-arable regions or urban centres. These concerns are basically the same if dedicated energy crops are grown for first or second-generation bioenergy production (Eisentraut 2010).

Agricultural transition from small-scale to large-scale farming is increasing and related to economies of scale. Usually the production of energy crops is more cost efficient on large-scale. This may lead to an agricultural transition from small to large-scale agriculture with extensive monocultures especially in industrial scale biofuel projects. Insight is needed on the effects of this transition, especially on social impacts (Rutz and Janssen 2009).

Risk mitigation of food production and income failure is an important issue for farmers, especially for smallholders and subsistence farmers. The risk consists on the one hand on market risks (e.g. sudden price fluctuations) and on the other hand on losses of harvests due to unfavourable climatic conditions or due to failure of agricultural practices. The main objective of smallholders and subsistence farmers is to produce enough food for their family's daily nutrition. Only if this is ensured, farmers will take care of other crops for income generation. As the example of cotton and peanut cultivation in Mali shows, it is very risky for farmers to concentrate only on one cash crop. The decrease of cotton and peanut prices pushed many farmers in exodus. The diversification of crops can contribute to reduce the agricultural risk and shall be considered in small to medium-scale biofuel projects. Crops that have the ability to serve as food and fuel also reduce risks, since harvests can be sold either to the food or to the energy markets. Jatropha cultivation does not have this advantage since it is a dedicated energy crop and not usable for nutrition.

Risk of land devastation has to be considered for large-scale projects. In the case of failure of large biofuel feedstock monocultures, land can be devastated leaving it unusable for many years.

Apart from these impacts which are closely related to the scale of the agricultural system and of the biofuel production concept, other impacts have to be considered, as for instance **change of traditional agricultural knowledge**, **gender issues**, **working conditions**, and other issues.

25.3 Scale of Agricultural Systems

The scale of the agricultural system in which feedstock for biofuels is produced highly influences the sustainability of biofuels. It seems that feedstock production on a smaller scale is currently performing better in comparison to large-scale systems with regard to socio-economic impacts. However, also large-scale concepts can be beneficial for the local population as well as for national economies. The different scales of agriculture are described in the following chapters. Independent from the scale, agriculture in Africa has to improve in order to be efficient, competitive and to satisfy demand for food, fibre, feed and fuels.

25.3.1 Subsistence and Smallholder Agriculture

The majority of farmers in Africa are subsistence and smallholder farmers with farms ranging from 0.5 to about 4 ha (Aina 2007).

Subsistence agriculture is self-sufficiency farming in which farmers produce enough food to feed their families. This traditional agriculture is based on accumulated knowledge which is forwarded from one generation to the other. Production is usually done without entering any markets. Due to the lack of market involvement and income, farmers are prone to several risks, including hunger in case of harvest failure, lack of medical care, and lack of information/education.

Farmers who are supporting a single family with a mixture of cash crops and subsistence farming are called smallholders.

The involvement of subsistence farmers and smallholders in the biofuel value chain is an opportunity to increase the living standard by generating income and supply of modern energy. However access to markets, technologies, knowledge, and financing is very difficult for this group. An opportunity to participate in the biofuels market would be to establish local biofuel markets for local consumption (for rural electrification, multi-functional platforms, etc.) and to set-up outgrower models as described in Sect. 25.3.3.

25.3.2 Large-Scale Industrial Agriculture

Industrial agriculture is a form of modern farming that refers to the industrialized production of agricultural commodities, such as to feedstock production for biofuels. Industrial agriculture usually involves large investments and a high level of further processing of the agricultural product into high-value products. The objective of industrial agriculture in Africa depends on the actors and it includes economic, political, and strategic motivations. Industrial agriculture has access to innovative agricultural practices, biotechnology, global commodity markets and global trade, financing, patents, and other supportive tools.

An increasing number of international large-scale agro-industrial companies are interested in feedstock production for biofuels in Africa. The aim is usually to export biofuels to global markets with high prices for biofuels.

Apart from the commercial industrial agriculture and according to Collier and Dercon (2009) so-called "super-farms" include investments in vast tracks of land of thousands of hectares for food crop agriculture focused on exports. It is argued that these super-farms are fundamentally geopolitical rather than commercial and not an appropriate vehicle for encouraging growth in African societies. This applies not only to food crop agriculture, but especially also to biofuel feedstock production, since increased efforts are made by several international actors to get access to huge land areas for feedstock cultivation. These large geopolitical and strategic projects have the highest risk of being unsustainable due to the creation of large dependencies, monopolies, and often rigorous methods, harming the local population.

25.3.3 Outgrower System and Contract Farming

There exist several types of agricultural models between subsistence, smallholder, and large-scale industrial agriculture.

One model combining smallholders and industrial companies are outgrower or contract farming schemes. Both concepts are often used synonymously to describe ways of vertical integration between small farmers and agro processors or traders. They do not have precise definitions but enclose a wide variety of institutional arrangements of vertical integration (Brüntrup 2006).

Typically, the farmer agrees to provide established quantities of a specific agricultural product, meeting the quality standards and delivery schedule set by the purchaser. In turn, the buyer commits to purchase the product, often at a pre-determined price. In some cases the buyer also commits to support production through, for example, supplying farm inputs, training, land preparation, providing technical advice and arranging transport of products to the buyer's premises (FAO 2010). In these schemes, the agri-business company has often considerable control over the smallholder production process.

Contractual arrangements are expected to gain importance in the globalizing world, since they offer, for both firms and producers, a number of advantages: From the side of the agro industry, better availability of primary products and therefore better use of industrial capacities and the possibility/security for delivering "just in time", better control of quality from the field to the shelf for an ever growing consumer demand for quality and food safety, establishment of (private) labels often in an environment of fierce competition between huge agribusinesses and retailers, emergence of many niche markets in saturated food markets with respective requirements throughout the chain. From the farmer's viewpoint, in addition to maintaining or gaining access to modern global market chains through such agribusiness contracts, outgrower schemes often provide the only feasible access to technology, inputs, credit and/or information. Through cooperatives or the condensed demand of agro industries, economies of scale are possible. Not at least, farmer organizations gain in professionalism and standing in political debates (Brüntrup 2006).

As with any other form of contractual relationship, there are also potential disadvantages and risks associated with contract farming. If the terms of the contract are not respected by one of the contracting parties, then the affected party will be disadvantaged. Common contractual problems include farmer sales to a different buyer (side selling or extra-contractual marketing), a company's refusal to buy products at the agreed prices, or the downgrading of product quality by the company. A frequent criticism of contract farming arrangements is the uneven nature of the business relationship between farmers and their buyers. Companies who buy the products from the smallholders are invariably more powerful than farmers. They may use their bargaining power to their short-term financial advantage, although in the long-term farmers would cease to supply them, if there are other markets willing to buy their products. Nevertheless, the balance between advantages and disadvantages seems to be on the positive side for both companies and farmers: contractual arrangements are more and more frequently being used in agriculture worldwide (FAO 2010).

25.4 Impacts of Different Biofuel Systems

Besides the scale of the agricultural system the other important factor which influences the socio-economic impact is the type of the biofuel. Some biofuels are suitable for the production and use on small-scale, others are more suitable for large-scale. This is due to different properties of the fuel, agricultural practices of crop production, applicability of the fuel, production technologies and production costs. Different energetic uses of liquid biofuels in Africa include domestic transport, cooking, heating, lighting, export commodity for foreign biofuel markets, and local electricity generation. The following chapters describe different biofuels which have different socio-economic impacts in Africa.

25.4.1 Bioethanol

Conventional ethanol production from sugar or starch is usually made at large to medium size. Potential feedstock for ethanol production in Africa are sugarcane, cassava, sweet sorghum, corn, sugarbeet and other starch or sugar bearing plants.

Sugarcane (*Saccharum* sp.) is the main feedstock for ethanol production in several African countries. However, the potential of ethanol production from sugarcane in Africa is limited since sugarcane is mainly grown for sugar production in the food market. Production of bioethanol from sugarcane is typically considered commercially viable if conducted on large-scale. Production from sugarcane raises various environmental concerns, primarily related to fertilizer and fuel use. Pesticides and other pollutants can cause negative impacts. Smoke from burning fields also needs to be taken into account, as well as the use of water for irrigation. Expanding ethanol production has also affected biodiversity by clearing natural forests. All of these sustainable development concerns need to be addressed where bioethanol is to be produced on a large-scale (UN DESA 2007). Sustainable cultivation of sugarcane for biofuel production requires that small-scale farmers and village communities adequately share the benefits. This can be done by outgrower schemes.

Since the sugar market is volatile, combined sugar-ethanol biorefineries have the opportunity to sell either sugar or ethanol, depending on market prices for sugar and ethanol. This approach is widely implemented in the Brazilian sugar-ethanol market. However, in many cases, waste products from the sugarcane industry are not used, but just dumped.

Apart from direct conversion of sugar from sugarcane into biofuels, often molasses are used for bioethanol production. Molasses are viscous by-products of the processing of sugarcane (or other sugar crops) into sugar. This is implemented for example in South Africa.

Sweet sorghum (*Sorghum* sp.) is seen as potential ethanol crop in areas which are not suitable for sugarcane cultivation due to several advantages such as less water requirements as well as due to tolerance to soil acidity and drought. However, ethanol production from sweet sorghum faces similar challenges to sugarcane such as seasonality and instability characteristics of its fermentable sugars that require high investments in processing facilities which must be large enough to process the harvest within weeks. Unless other feedstock are available, ethanol production facilities will be underutilized or idle for many months each year (UN DESA 2007).

Although theoretically possible, the production of ethanol on small-scale is currently difficult due to the high costs, lack of suitable small-scale applications (in comparison to e.g. pure plant oil), and due to technical knowledge barriers.

In conclusion, most bioethanol production chains are targeting large-scale production. This creates an opportunity for stimulating national economies and for reducing dependency from fossil energy imports (if used in the national market), but there is risk of harming smallholders, landless people and the poorest if projects are not implemented in a sustainable way. For small-scale production in villages, ethanol is currently not suitable.

Bioethanol is not only produced at different scales, but also used for different applications. Thus, ethanol is used for chemical industries and as biofuels (mixtures) at large-scale.

At small-scale level, bioethanol can be also used for household applications, combining several advantages of both traditional and new fuels. It can be for example used in ethanol cooking stoves, as it is described by Robinson (2006) for Malawi, or promoted by the Gaia project (http://www.projectgaia.com/) in Ethiopia and Nigeria. The Gaia project is promoting clean-cooking fuels, particularly alcohol cooking stoves, to the poorest part of the population. A relatively new development is also to use ethanol-gel (gelfuel) instead of liquid ethanol for cooking stoves. However, Lloyd and Visagie (2007) showed that the use of gelfuel in practical applications is difficult, mainly due to pollutants and increased cooking time, since gelfuel has lower energy content than other (liquid) fuels.

25.4.2 *Pure Plant Oil and Biodiesel*

The large advantage of the production and use of pure plant oil is that it can be made at any scale. Thus, it could on the one hand contribute to rural development and access to modern energy at village level, but it could also meet industrial requirements and thus contribute to national economic growth. However, a major barrier is still the high costs of oil production. Due to very different characteristics of oil crops, each crop has to be investigated individually. The main oil crops which are currently discussed for biofuel production in Africa are jatropha, oil palm, castor, coconut, cottonseed, croton, rapeseed, and sunflower.

Any vegetable oil can be converted by transesterification to biodiesel. However, this requires sophisticated technology and high investment capital. Therefore, the conversion of vegetable oil into biodiesel at village level is difficult. However, the use of pure plant oil is sufficient for the use in small-scale applications.

Jatropha (*Jatropha curcas*) is currently one of the most discussed energy crops in Africa. Several demonstration and pilot projects on the cultivation and use of jatropha were implemented in several countries, such as for instance in Mali, Tanzania, Ghana, and Zambia. The projects include small-scale rural electrification projects with gen-sets and multifunctional platforms (MFP), outgrower schemes with centralized and decentralized pressing facilities, and large-scale jatropha plantations. Small-scale projects usually use pure plant oil directly, whereas large-scale projects usually target towards further processing of the oil into biodiesel. Although jatropha initially seemed a very promising crop suitable for any scale, discussions with involved stakeholders show more and more problems associated with jatropha cultivation. Jatropha was considered initially as crop for marginal land with low water and fertilizer requirements, and tolerant to many diseases. It is furthermore harvested only manually, thus it was considered as driver for creating new job opportunities for harvesting people. In reality it is shown, that jatropha needs water, fertilizer, and pesticide to ensure economic viability in large-scale projects (e.g. in monocultures). Bottlenecks for small-scale applications were low agronomic extension support which has led to poor production and unpredictable, scattered market for selling the seeds (GTZ 2009). This is due to the fact that jatropha is inedible and thus only an energy crop which cannot be sold to alternative markets such as to the food market. As described by GTZ (2009) for Kenya, jatropha currently does not appear to be economically viable for smallholder farming when grown either within a monoculture or in an intercrop model. The only model for growing jatropha that makes economic sense for smallholders is the cultivation as living fence with very few inputs and it is suggested "that all public and private sector actors for the time being cease promoting the crop amongst smallholder farmers for any plantation other then a fence" (GTZ 2009). In summary, jatropha cultivation currently seems to be only advantageous for small-scale production and use at local level. Successful small-scale projects for rural electrification with jatropha oil were for instance made in Mali and Tanzania. A promising outgrower scheme is implemented by Diligent in Tanzania.

Palm oil (*Elaeis* sp.) is similar to jatropha regarding the large labour needs for harvesting. The primary unit of production of the palm oil industry is the farm where the oil palm tree is cultivated to produce palm fruits. There are also wild groves of oil palm. The farm units are of different sizes and may be classified as small, medium, and large-scale estates (FAO 2002). Palm oil is currently mainly used as food commodity and production is usually organized by large-scale food industries.

Castor (*Ricinus communis*) is similar to jatropha since it is toxic, grows on marginal land and is often used as fence plantation. According to GTZ (2009), in Kenya it could present a positive investment opportunity for both, monocultures and fence plantations, thus being suitable for different scales. Nevertheless, similar to jatropha, cultivation of castor also requires inputs, is prone to pests in monocultures and only grows under certain climatic conditions. Therefore, the use of castor only in small-scale applications may be the main suitable approach.

Croton (*Croton megalocarpus*) is indigenous to East Africa. GTZ (2009) has shown in Kenya that croton nuts are a more economical source of biofuel than jatropha. However, production is currently small, but could be scaled up significantly due to the large availability of the seeds in eastern Africa (Kenya, Tanzania). However, it is not yet clear if it will be suitable for large-scale plantations or for small-scale applications only.

Sunflower (*Helianthus annuus*) oil is generally also suitable for biodiesel production. However, it requires good soils and water, thus biofuel production would compete for this land directly with sunflower cultivation for food. This applies for both large and small-scale production.

25.4.3 Biofuels from Ligno-Cellulosic Biomass

As described in previous chapters in this book, ligno-cellulosic biomass in Africa is currently mainly used as firewood in household applications. In Europe and North America, ligno-cellulosic material is considered as potential source for second generation liquid biofuels production, including biofuel production by thermo-chemical processes (BtL fuels, pyrolysis oil) and by bio-chemical processes (second generation bioethanol).

Although latest advancements of ligno-cellulosic ethanol production are promising, the step towards commercialization is still not made. It seems that even more effort is needed for the thermo-chemical processes to reach the commercial status.

However, if one of these technologies will be technologically and commercially mature, it could be also introduced at commercial scale in Africa. If properly applied, this could contribute to the general economic growth in Africa. Due to economic reasons of these technologies, fuel production will be implemented only on large and industrial scale. Potential socio-economic impacts could be positive and negative, depending on the type of feedstock which can be agricultural and industrial waste, dedicated woody energy crops, forest residues, but also wood from naturally managed forests. In theory, peaceful coexistence between smallholders and ligno-cellulosic ethanol production facilities could be achieved and participation of smallholders in this sector could be enabled. Nevertheless, this will largely depend on the introduction of suitable policies, enforcement of the legislation, and consideration of sustainability standards. As for any other energy crop, sustainability of second generation biofuel technologies will largely depend on the type and methods to gain feedstock material.

25.5 Conclusion

The introduction and support of biofuels in Africa, especially on small-scale, has the potential to increase national economies, contribute to rural development and to mitigate poverty. This depends on several factors including the introduction of suitable policies as well as the enforcement of legislation of sustainability measures.

Irrespective of the size of the agricultural and biofuel processing system, improved feedstock production and thus improved agriculture in general is key to the overall success of biofuels in Africa and also to meet the Millennium Development Goals.

The suitable size of feedstock production and conversion into biofuels depends on the characteristics of the feedstock as well as on the economy and complexity of the conversion process. In general there are various opportunities for introducing different scales of biofuel production in Africa. A mixture of different systems will be needed in the future to meet the different demands.

It seems for instance that the cultivation of jatropha and castor as living fences is suitable for small-scale production of vegetable oil which is used locally in multifunctional platforms or for electrification. On the other hand large-scale jatropha cultivation in monocultures, mainly targeting the export of biodiesel, does seem to be neither attractive nor sustainable, unless considerable efforts are undertaken to achieve improved varieties.

The production of ethanol seems to be mainly suitable for large-scale production. Smallholders for feedstock production can theoretically be included in this production by outgrower models. However, this has been only implemented to a limited extent so far. Future biofuel projects need to increasingly involve smallholders.

The production of second generation biofuels in Africa could offer a large opportunity for economic growth, but is far away from commercialization and thus does not need to be considered today.

The answer to the question whether local use or export of biofuels should be supported is influenced by several factors. It firstly depends on the produced fuel which has to meet certain standards to be internationally traded. Some fuels, especially those which are produced on small-scale and in decentralized production systems will not be able to meet these technical specifications.

Another factor is the market price of the fuel. In the long-term and unless no trade barriers will be introduced, biofuels will be sold to those international markets which offer the highest price. Depending on the nationality of the producer and on the national (tax) policies, revenues to the participation in this international market could contribute to economic growth in Africa.

Finally, an important factor is the national energy policies of the African countries. Many African governments maintain direct or indirect subsidies for fossil fuels. This can be a serious barrier for the introduction of biofuels. Incentives for local biofuel production and use instead of incentives for fossil fuels could contribute to local poverty mitigation, national security of energy supply, and to avoid the loss of own revenues due to the import of expensive fossil fuels.

In general, more insight into the various and complex socio-economic aspects of biofuel production in Africa has to be gained. This is currently done in the EU funded project Global-Bio-Pact (Rutz et al. 2010).

References

Aina LO (2007) Globalisation and small-scale farming in Africa: what role for information centres? World library and information congress. In: 73rd IFLA general conference and council, Durban, 19–23 Aug 2007

Anríquez G, Stamoulis K (2007) Rural development and poverty reduction: is agriculture still the key? ESA working paper no. 07–02. The Food and Agriculture Organization of the United Nations, Rome

Brüntrup M (2006) Perspectives of contract farming in this reader. In: Brüntrup M, Peltzer R, Outgrowers (eds) A key to the development of rural areas in Sub-Saharan Africa and to poverty reduction. Report of the DEG/DIE Workshop, Bonn, 18 Aug 2006, pp 1–3

Collier P, Dercon S (2009) Smallholders in a rapidly changing world? FAO expert meeting on how to feed the world in 2050, June. Food and Agriculture Organization, Rome. ftp://ftp.fao.org/docrep/fao/012/ak983e/ak983e00.pdf

Eisentraut A (2010) Sustainable production of second generation biofuels. Potential and perspectives in major economies and developing countries. International Energy Agency (IEA), Paris

FAO (2002) Small-scale palm oil processing in Africa. FAO agricultural services bulletin 148

FAO (2010) Food and Agriculture Organisation of the United Nations, Contract Farming Resource Centre (on-line service), Rome. http://www.fao.org/ag/ags/contract-farming/en/

GTZ (2009) Jatropha Reality Check. A field assessment of the agronomic and economic viability of Jatropha and other oilseed crops in Kenya. GTZ – Regional Energy Advisory Platform (East Africa). Nairobi, Dec 2009

Lloyd PJD, Visagie EM (2007) A comparison of gelfuels with alternative cooking fuels. J Energy S Afr 18(3), Aug 2007, p 26–31

Robinson J (2006) Bio-Ethanol as a household cooking fuel: A mini pilot study of the SuperBlu stove in peri-urban malawi. report elaborated with funding from: The Royal Academy of Engineering, Bluwave Ltd, Eco Ltd and GTZ ProBEC. Oct 2006

Rossi A, Lambrou Y (2009) Making sustainable biofuels work for smallholder farmers and rural households: issues and perspectives. Food and Agriculture Organization of the United Nations, Rome

Rutz D, Janssen R (2009) Intercultural aspects of bioethanol and biodiesel sustainability in Latin America. In: Proceedings of the 17th European biomass conference and exhibition, Hamburg, pp 2356–2363. ISBN 978-88-89407-57-3

Rutz D, Van Dam J, Janssen R, Hiegl W, Diaz Chavez RA, Woods J, Faaij APC, Smeets EMW, Vos J, Vis M, Bottriell K, Rettenmaier N, Reinhardt G, Nussbaum R, Gozzi A, Haye A, da Silva Walter AC, Hilbert JA, Coto C, Fallot A, Sawe EN, Togola I, Ouattare O, Jaax R (2010) Global socio-economic impact assessment of biofuel and bioproduct chains. In: Proceedings of the 18th European biomass conference and exhibition, Lyon. ISBN 978-88-89407-56-5

Scott A (2009) Large-scale biofuels programmes in Africa – Who benefits? Practical action, United Kingdom. In: Paper presented at the VENRO and German NGO forum on environment and development international symposium: rethinking biomass energy in Sub-Saharan Africa, Bonn, on 25 Aug 2009

The World Bank (2007) World development report 2008: agriculture for development. The International Bank for Reconstruction and Development/The World Bank, Washington, DC

UN DESA (2007) Small-scale production and use of liquid biofuels in Sub-Saharan Africa: perspectives for sustainable development. United Nations Department of Economic and Social Affairs; Background paper no 2, New York

Chapter 26
Social Impacts of Biofuel Production in Africa

Stanford Mwakasonda and Francesca Farioli

Abstract Within the last decade there have been significant discussions and initiatives on use of biofuels as alternatives to fossil fuels. Primarily, the interest in biofuels as alternative fuels has been prompted by their perceived ease of availability and increasing world oil prices. In Africa biofuel programmes have been seen to have the added advantage of catalyzing rural development. Brazil has frequently been cited as an example of what a biofuel programme can add to the development of struggling and infant economies in Africa. At the same time, a cautious approach for the implementation of biofuel programmes in Africa has been advocated by a number of civil societies and international organizations. Countries have been warned of the existence of risks that come hand in hand with the perceived benefits of biofuel programmes. Such concerns have spurred governments and institutions in Africa and elsewhere to come up with biofuel strategies for ensuring implementable and sustainable biofuel programmes. This chapter discusses some of the major social impacts of implementation of biofuel programmes in Africa, and makes recommendations on issues that need to be addressed for biofuel programmes to be sustainable in Africa.

Keywords Social impacts • Biofuels • Fossil fuels • Greenhouse gases • Energy poverty • Energy access • Clean Development Mechanism • Agro-forestry • Poverty alleviation • Multifunction platforms • Marginal lands • Migrant labourers • Food security • Labour laws • Millennium Development Goals

---

S. Mwakasonda (✉)
Engen Petroleum Ltd, Engen Court, Thibault Square, PO Box 35,
Cape Town 8000, South Africa
e-mail: Stanford.mwakasonda@engenoil.com

F. Farioli
Interuniversity Research Centre on Sustainable Development,
(CIRPS)- SAPIENZA Università di Roma, Piazza San Pietro
in Vincoli 10, 00184, Rome, Italy
e-mail: francesca.farioli@uniroma1.it

R. Janssen and D. Rutz (eds.), *Bioenergy for Sustainable Development in Africa*,
DOI 10.1007/978-94-007-2181-4_26,

26.1 Evolution of Biofuel Programmes and Related Social Issues

The last 10 years have witnessed significant discussions on alternatives to fossil fuels. Two major reasons have been behind this interest spike in moving away from fossil oil, a source of energy that has played, and continues to play, a major part in world primary energy supply now for more than 150 years. One of the reasons for the new interest in alternative fuels, particularly biofuels, is their perceived ease of availability in almost all regions of the world. Biofuels gained a particularly heightened interest in the 2006–2007 period when world oil prices reached a record breaking 147 US$ a barrel (Nicholls 2008). With such a world oil price scenario, many countries, especially in Africa where the majority of countries use up to 50% of export earnings for importing petroleum products, saw biofuels as the means to getting the required respite from the pressure of having to import ever increasingly expensive oil. Biofuels were seen as the solution to offsetting some of the cost of oil imports, and at the same time providing an opportunity for moving along an energy supply path that offers sustainable development. Success stories in countries like Brazil provided the required decisive factor for headlong jumping into a biofuel programme for some of the African countries. The other major reason for diversifying the energy sector has been the need to cut down greenhouse gases (GHG) known to cause global warming and climate change.

Factors mentioned above have provided much of the current momentum for Africa to start biofuel initiatives. During the opening speeches of the First African Union (AU) – UNIDO High Level Seminar in Addis Ababa, Ethiopia, in 2007 the UNIDO Director-General Kandeh Yumkella mentioned that there was an ongoing "biofuels revolution" that Africa could not afford not to be a part of, and underscored that the global fight against poverty needs to address the "energy poverty" question (COMPETE 2007).

Caution has been advocated by a number of civil societies and international organizations to African countries that want to implement biofuel programmes. The main message relayed is that there are significant risks that come hand in hand with the perceived benefits of biofuel programmes. There are pleas that African countries need to learn from implemented biofuel programmes and prepare appropriate strategies before embarking on this route. Whilst Brazil is cited as one of the best examples of successful biofuel initiatives, one needs to note that the country embarked on the biofuel blending initiative whilst already a leading sugar producer from sugarcane. It is acknowledged that Africa still has just a few well established biofuel programmes and as such there are only a few cases in Africa that can provide input on the social impacts of biofuel programmes (ENDA 2008). However, examples can be found in Asia and Latin America that provide lessons relevant to Africa. Other forms of commercial agricultural production in Africa can also provide a good insight of what could be the social impact of large-scale biofuel programmes in Africa (Cotula et al. 2008).

It is obvious in the context of sustainability of biofuel programmes in Africa that world oil prices have to be comparatively high for the interest in biofuels to be sustained.

Reduction of GHG emissions is not the main force that can propel biofuel programmes in Africa, nor any other perceived benefits of alternatives to fossil fuels. Once world oil prices become affordable the interest in biofuels is more than likely to disappear from the African energy supply scenery.

Further potential benefits that most literatures associate with the implementation of biofuels include the following (UN-Energy 2007; Cotula et al. 2008):

- Opportunity for meeting growing needs of energy, especially raising modern energy access in rural areas
- Reduction and offsetting the high cost of importing fossil fuels
- Enhancing employment opportunities
- Providing productive energy in rural areas and thus creating opportunities for entrepreneurship
- Providing an opportunity for clean development projects that can be undertaken by most countries on the continent and hence alleviate global warming and indoor air pollution
- Reclamation of waste land and improving food productivity

Africa is seen as a continent that has a lot of potential for biofuel growth in tapping the biofuel opportunities and benefits. Compared to Brazil, which produces about 34% of the world's bioethanol, Africa produces only 1.2% despite having perfect weather conditions for the large-scale production of most of the biofuel crops. Africa is also regarded as having significant availability of land and manpower necessary for successful biofuel programmes (Greenpower 2010).

Whilst there are many schools of thought that acknowledge biofuel programmes would offer some level of economic development opportunities for Africa, concerns are expressed on the impact of biofuel programmes on a number of socio-economic facets. Some of the major social concerns of biofuel programmes are mentioned to include the following (UN-Energy 2007; Malony and Smith 2010; Rossi and Lambrou 2008):

- Food security problems due to most efforts being directed towards energy crops production and sale
- Biofuel programmes becoming the sole domain of the rich
- Land use conflicts between energy crops and other economic activities like pastoralism
- Poor working conditions and child labour on energy farms
- Reduction in food crop land

At the same time, some biofuels investors have expressed concerns on the implementation of biofuel programmes in Africa. In addition to the generic foreign investment concerns around infrastructure and political stability, the concern on biofuel programmes is around the additional investment benefits from carbon trading. It was expected that the Clean Development Mechanism (CDM) of the Kyoto Protocol in the United Nations Framework Convention on Climate Change (UNFCCC) would provide the necessary take-off base for a fast implementation of biofuel programmes in Africa. This has not happened, partly because potential biofuel investors in Africa have been concerned by the unproven record of return on biofuel investments in Africa. There have been mixed signals from the CDM Executive Board on the issue of Life Cycle Analysis (LCA) for biofuel CDM projects. Other challenges faced by investors include failure to get enough land for economic viability of biofuel farming.

Such concerns have spurred governments and institutions in Africa and elsewhere to come up with biofuel strategies for ensuring implementable and sustainable biofuel programmes. Such effort has been spearheaded by organizations and institutions like United Nations agencies, international non-governmental organizations, regional and international energy networks, etc. In 2004 for example, the Southern African Development Community (SADC), through one of its agencies, Food, Agriculture and Natural Resources (FANR), released a study on the feasibility of the production of biofuels in the SADC region, recognizing that the SADC region has the largest untapped potential of biofuels in the world. The SADC Biofuel Strategy, completed in 2005, acknowledged the need for considerable investment for successful implementation of biofuel initiatives and called for all SADC nations to develop biofuel strategies that would make the sub-region more self reliant in energy production. South Africa, one of the first member countries to respond to this call, initiated a national biofuel strategy. The South African biofuel strategy aimed at re-energizing the agricultural sector and paving the way for a home-grown biofuels industry. The biofuel initiative was envisioned to create some 55,000 jobs and contribute to economic growth (SADC 2005; Sugrue 2007).

At the AU – UNIDO High Level Seminar in 2007, the UNIDO Director highlighted the importance of sustainability in biofuels development, and noted the need to create a better policy climate and add value to African commodities in international markets (COMPETE 2007). These remarks were in line with activities in several other African countries since 2005 to promote biofuel strategies. One common aspect in most strategies, whether nationally or internationally developed, is the critical importance for countries to have in place measures that address, not only economic benefits, but also concerns on both social and environmental issues. Due to the nature of the farming systems in Africa, need for appropriate strategies/policies for sustainable energy crop and agro-forestry systems in Africa, adapted to national circumstances, is critically important. Some of the issues recommended to be taken into consideration in developing biofuel strategies in Africa include the following (UN-Energy 2007; SADC 2005; Cotula et al. 2008):

- Biofuel strategies need to support sustainable energy, environment, forestry, water and land use/agriculture, etc.
- Focus needs to be placed on crops with competitive advantages in different countries
- Biofuels strategies need to be developed in consistency with existing policies and legislation on water resources, land use, forestry, agriculture, etc.
- Benefits from biofuel initiatives need to sustainable development

Key areas of social impacts of biofuel programmes in Africa are discussed in the following sections.

26.2 Availability of Land for Biofuel Programmes

One of the most important issues here is the assurance that biofuel programmes will not encroach on land used for subsistence farming by rural communities and result in deeper poverty of such communities. It is argued that large-scale energy crop farming operations are likely to have impact on the available productive land and may lead to disadvantage subsistence farmers. The proponents of large-scale biofuel programmes usually come with the counter argument that only degraded land would be used for such large-scale initiatives, with the advantage of rehabilitation of soils. Opponents of large-scale biofuel farming, however, claim that there is evidence that commercially viable biofuel yields can only be obtained on fertile soil and with enough water availability (Cotula et al. 2008; Rossi and Lambrou 2008).

In some countries where large-scale biofuel farms have been opened, community members argue that the projects were not participatory and they are not beneficial to them. Allegations are made that some of the farms were acquired through land grabbing from peasants. The following extract in Box 26.1 from Newscientist (2007) depicts the problems that implementation of large-scale biofuel programmes can cause:

Box 26.1 Potential Problems of the Implementation of Large-Scale Biofuel Programmes

"A row over the conversion of rainforests into biofuel plantations is creating a grave political crisis for a country until now seen as a beacon for democracy in Africa. The issue has brought to a head the simmering conflicts between short-term economic gains and the conservation of vital natural resources in the continent" (Newscientist 2007).

"The president of Uganda, Yoweri Museveni, is this week pressing ahead with plans to give a large chunk of one of the country's last protected forests to a sugarcane company so it can expand its operations. The Sugar Corporation of Uganda, which is owned by Ugandan Asians, wants to expand production to cash in on the booming global market in sugar for biofuels."

"The crisis reached boiling point last week when a demonstration against the plan in the capital Kampala turned into an ugly race riot. Asian shops were ransacked, an Asian was stoned to death and police killed two demonstrators."

"The demonstrations have resumed this week, with hundreds of defenders of the forest beaten up by squads of vigilantes known as kiboko, which local media claim are backed by the government…."

The decision to clear the green forests for a biofuel programme was later reversed by the Ugandan government (Guardian 2007).

Needless to say, it is necessary to perform an appropriate classification of national lands before embarking on large-scale biofuel programmes, and this would be a challenge for most Sub-Sahara African countries. Of importance would be to make sure that the cultivation of land for biofuels does not significantly encroach upon food production or creates pressure on arable land, resulting in the change of equilibrium of food production in a particular community, especially in rural areas.

Reports from Tanzania indicate that the land area which commercial biofuel investors in Tanzania have requested has ranged from 5,000 to two million hectares (Sulle and Nelson 2009). A company seeking to invest in an ethanol energy project found it quite hard to secure a land large enough for the cultivation of sugarcane. Legal procedures the company had to follow to acquire land for such an investment were seen to make the biofuel investment process quite difficult. The Tanzania government's position was that there would be no short-cut and the company could not start without following the required procedures, including compensating and relocating local residents. This government stance was seen as a significant investment hurdle by the potential foreign investor, and probably a subject of complaints to the local investment promotion centre (IPP Media 2010).

26.3 Biofuel Programmes and Issues on Land Ownership

Closely related to land availability is land ownership. Land ownership in Africa has traditionally been a source of many individual and community conflicts. Many wars have been fought in Africa because of land ownership issues. Most of the sung heroes in the history of fight against African colonization were chiefs and kings that opposed the grabbing of lands by white settlers. During the UNFCCC Kyoto Protocol negotiations, land ownership issues were often cited as some of the reasons why a number of African countries were opposed to inclusion of "sinks" in CDM projects. Such CDM projects were felt synonymous to re-colonization of African countries, as they would pave way for gradual ownership of national land by foreigners from developed Annex I countries, indicating sensitivities around land ownership issues in the context of implementing biofuel programmes in Africa.

The truth remains, that most of the foreign biofuel investors in Africa are particularly interested in the continent mainly because of the low cost of land and labour, especially in rural areas. The key question is whether land for large-scale biofuel programmes in Africa can be provided to investors without having to displace local communities. In Colombia, for example, it is reported that between 1.87 and 3.83 million people have been displaced by violence such as murder or massacres, threats, kidnapping and torture to pave way for biofuel programmes (Mingorance 2006). It is reported that their lands are seized and planted with oil palms. Are the authorities credible enough to prevent such atrocities from happening in Africa? For a continent saddled with the reputation of corruption, abuse of human rights and lack of governance transparency, providing assurance that biofuel programmes processes would adhere to the highest ethical standards is quite a challenge. It is widely recognized that the energy sector is one of the areas in Africa highly ridden with irregularities and corruption.

There are some countries in Africa that have recognized the sensitivities around land ownership. Such countries are in favour of proposals that encourage, rather than biofuel production resulting in the concentration of land ownership and land access into fewer hands, agro-forestry initiatives involving current owners of land in rural areas. In line with this position, advocates of sustainable development in Africa have pointed out that African governments could enhance rural development by favouring small-scale livelihood-oriented bioenergy initiatives as these are more likely to have higher socio-economic returns than large-scale biofuel initiatives. It has further been explained that development of integrated food and energy activities (IFES)[1] could result in win-win situation whereby both cash/food and energy crop production would be promoted simultaneously (FAO 2010). Agro-forestry and intercropping are both ways to combine the production of food and energy on the same plot. In some countries, energy crop cultivation has taken the form of contractual arrangement between local small-scale farmers and biofuel companies. This is a model borrowed from cash crop production like tea and sugarcane. Under this model of production, the local farmers maintain ownership of their land.

In reality, however, land ownership systems in Africa are quite diverse, ranging from traditional customary land ownership to centralized state ownership systems. This implies that no single solution can be prescribed that will work in all African countries. Land in Africa constitutes the basis of economic livelihood of indigenous people and small-scale rural farmers and landowners. If bioenergy initiatives entail dislocation of local communities, it is likely to cause limited access to lands that the communities used on the basis of customary or formal mechanisms.

Land ownership systems in Africa can thus be said to present both opportunities and difficulties to biofuel investors. Opportunities exist as there are significant tracts of land with no formal "owners", whereas the difficulties may arise where communities feel they have ancestral access right to some tracts of land. The latter could prompt serious conflicts between investors and the local communities. Whilst such conflicts could be mediated through a range of policies and processes, robust framework, good practices and innovative business approaches towards more equitable and sustainable land management, such processes could last for a significantly long time. In certain cases, change in the local political scenario can imply a shift in government stance on land ownership issues.

---

[1] Integrated Food Energy Systems (IFES) are "designed to integrate, intensify, and thus increase the simultaneous production of food and energy in two ways: (1) by combining the production of food and fuel feedstock on the same land, through mixed cropping and/or agro-silvo-pastoral systems, or (2) by transforming the by-products of one system into the feedstocks for the other, through the adoption of farming systems and agro-industrial technology that allow maximum utilization of by-products, diversification of raw materials, waste production on a smaller scale, and encouraging recycling and economic utilization of residues, for harmonization of energy and food production" (FAO 2010).

26.4 Biofuel Programmes and Poverty Alleviation

The potential of biofuel programmes for poverty alleviation is large, whether through employment, income opportunities, wider growth multipliers, and increased access to energy services. Important factors for such programmes to effectively contribute to poverty alleviation include type of energy crop, scale and business model used as well as regulatory or investment framework in place. As a matter of principle, it is generally agreed that biofuels initiatives should provide opportunities for income generation, especially to rural communities. Such income generation opportunities should lead to economic empowerment and improved living standards.

Biofuel programmes associated with poverty alleviation are mostly based on business models that do not involve large-scale land acquisitions, and that are characterized by collaborative arrangements between investors and local small-scale farmers and communities. They include a wide range of arrangements such as contract farming schemes, joint ventures, management contracts and supply chain relationships (Vermeulen and Cotula 2010). Understandably, such biofuel models are not easy to manage, as companies need to balance local development objectives with profits and interests of shareholders.

In the context of rural development in Africa, biofuels initiatives are seen as an important avenue for providing productive energy in rural areas and thus create opportunities for entrepreneurship. Commonly cited examples include the multi-functional platforms in some West African countries like Mali. If properly managed, rural biofuel programmes can be an effective means to create a source of income for the vast rural population and a more favourable economic environment for investment in rural infrastructure, health, and education. This would in turn have an impact on the problem of migration of rural populations to urban and semi-urban areas. The inferred conclusion is that biofuel programmes in rural areas have to be implemented in such a way that the biofuel processing facilities are near to farms, and the rural communities have a stake in value-added parts of the production chain, such as transportation and processing, and not only in the feedstock production.

Community owned biofuel multifunctional platforms could contribute to improve the living standard of rural populations, including improved access to modern energy and a reduced burden of rural women in searching for wood. The other area of poverty alleviation is the development of small cottage industries which can run on locally available biofuels.

On the other hand, it has been reported for some places in Africa that rural farmers rushed into jatropha farming on the promise that it would get them out of poverty. A few years later the farmers realized they were not getting the income they had been promised and could not afford food for their families, having locked themselves into a medium term non-food crop production with little uptake from the envisaged readily available market. This is a typical example of an improperly managed implementation of a biofuel programme, which can result in aggravating the poverty problem of rural communities.

Another aspect of rural biofuel programmes that can contribute to worsening the rural poverty problem is inherent in the use of marginal lands for biofuel crops.

Marginal, "idle" or "abandoned" lands, which are usually proposed to be the most appropriate for biofuel programmes, are reported to be particularly important to women as well as vulnerable groups that rely on this land as basis for their livelihoods, through crop farming, herding and gathering of wild products. In the context of small-scale biofuel programmes, studies show that in several Sub-Saharan African countries, women are usually allocated low quality lands so that they can undertake their horticulture and other small-scale agricultural activities. It is very unlikely that men will give up their lands in favour of having a small family energy crop farm. The first option thus would be the marginal lands used by women, and income from the sale of energy crops will likely end up in the hands of the head of the family unit, usually a man. In this context, small-scale energy crops in rural areas might not necessarily provide the envisaged benefits, especially in a setting where the husband determines how family income should be used. This is likely to result in a significantly hindered economic ability for women to be able to meet household obligations that require some form of income, including payment of school fees in families where the husband does not take his due responsibility (Rossi and Lambrou 2008).

In the context of large-scale biofuel programmes, use of marginal lands could result in the exacerbation of the burden of women in searching for firewood. There are places in Africa where marginal lands serve as a source of firewood for cooking. Use of such marginal lands for biofuel programmes would result in women having to travel longer distances to search for firewood. This would in turn result in reducing the time available to women to participate in income generating activities previously undertaken.

In many cases, lands perceived to be "idle", "under-utilized", "marginal" or "abandoned" provide a vital basis for the livelihoods of poorer and vulnerable groups, through grazing, bee keeping, gathering of wild products and other cultural activities that are part of customs and traditions of rural communities (Rossi and Lambrou 2008). These factors need to be assessed and given due consideration before implementing biofuel programmes, whether on small or large-scale.

26.5 Biofuel Programmes and Employment Opportunities

Creating enough employment opportunities, mainly for young people, is a challenge that most African countries face. Rural areas in Africa seem to be the hardest hit by the employment opportunity challenge. This usually results in an influx of the youth to urban areas in search of jobs. It is for this reason that biofuel development in rural Africa is seen as a significant opportunity to move development from cities to rural areas.

Generally, for a country that imports oil, a barrel of biofuel produced will have higher potential for job creation than a barrel of crude oil refined. A large number of employments in the biofuel sector would probably come with large-scale biofuel programmes and investments. However, investment in large-scale biofuel plants would imply a high density of energy crop planting, and this is likely to require large-scale energy crop farming. It has been argued that if this is not done, feedstock logistic costs will reduce returns, a scenario that is seen as not being attractive to investors. This special case would require careful planning and strategizing in order to minimize social impacts to vulnerable communities (Singh 2006; DME 2006).

The estimated employment generation effect through biofuel programmes is variable according to the type of energy crop and technology used. For example, soy cultivation for biofuels is usually highly mechanized and needs large land areas with low job generation potential. Non-mechanized sugarcane plantations can have a high potential for unskilled jobs, mainly sugarcane cutting. In most cases discussion around employment effects does not reflect the quality of jobs, which is misleading.

It is therefore necessary that discussions about the employment potential are accompanied by discussions about the nature and quality of work, as well as by programmes for workers qualification. In cases where a biofuel project uses modern techniques of handling and mechanization, it is necessary to assess the loss of jobs initially created and the quality and quantity of jobs remaining or to be created. It is important to ensure the availability of jobs for local people and an equal distribution of economic revenues from the biofuel project.

26.6 Labour Conditions, Gender Issues and Rights of Children

The discussion of labour conditions addresses wages, overtime, child labour and slavery. It is generally known that in most rural areas in Africa labour rights are poorly understood or not recognized completely. Some of the existing plantation projects in these areas comprise of appalling working conditions of almost slavery dimensions. Wages paid can be as little as one dollar in a 12 h-shift. In some cases it is reported that women and children work without payment in order to help their bread earner meet the daily quota. It is further reported that in many cases, women are discriminated in terms of employment benefits and exposure to occupational safety and health risks (Rossi and Lambrou 2008). On some farms in South Africa, it is common practice for employees to be paid in kind rather than in monetary wages. Use of illegal migrant labourers is common, taking advantage of their desperation and the fact that they would not be able to complain to authorities. This has been linked to xenophobic attacks when local indigenous people feel foreigners are given employment preference. In addition, farm owners in Africa often do not provide appropriate working tools and safety equipment to the workers.

These are issues that need to be considered carefully and included in a national regulatory framework before embarking on biofuel programmes that include large-scale plantations. Whilst most of the above mentioned labour problems are not unique to energy crop plantations, it is important that biofuel initiatives should not be used as a loophole for disregarding international labour laws, especially in rural areas. National employment regulations in line with international labour laws need to be enforced and adhered to.

26.7 Biofuel Programmes and Food Security

Food security has been one of the most contentious dimensions of biofuel programmes. There is a general consensus that biofuels production may compete with food crops and cause negative impacts on food security.

The possible competition between biofuel and food production is a valid and real concern. Food crops are recognizably some of the major sources of biofuels and the potential risk is that these crops are usually staples in most communities. Diverting such crops from food to biofuel markets will result in higher food prices. In South Africa, corn was seen as a potential source of ethanol production due to the fact that for a number of years there was a large surplus of corn production. However, this situation changed suddenly a few years ago due to a number of reasons, including drought. Had a decision been taken in South Africa to go into ethanol production using the surplus corn, the country would now be facing a food versus fuel challenge.

In Nigeria, the government launched in its bid to drive an ethanol programme commissioning the use of cassava as the primary raw material. Nigeria is Africa's largest producer of cassava, but nevertheless the country has continued to witness an increase in the cost of cassava derivatives over time which are staple foods in most parts of the country (IFAD and FAO 2004).

Another threat to food security that is often overlooked is inherent in the labour required to produce energy crops. Such labour is likely to be at the cost of labour required for food production.

26.8 Conclusion

There is no single answer to how social impacts of biofuel programmes in Africa can be addressed. However, the most common approach is to have appropriate policies and strategies that provide a roadmap for biofuel programme implementation consistent with other national development needs. It is critically important for African countries to implement biofuel programmes that take into consideration existing socio-economic issues, especially those pertinent to land ownership, economic empowerment and food security. Ensuring sustainability would also require that there are no conflicts between biofuel programmes and objectives of the Millennium Development Goals.

The creation of a robust regulatory framework and responsible decision-making process is needed in order to manage risks and challenges in a transparent way, and to promote and enhance potential benefits of biofuel sector development. This would entail putting in place clear frameworks to create a more stable investment climate, and ultimately to attract more and sustainable investments.

Socio-economic impacts of biofuel development in African countries need to be addressed on community, national as well as on international level. A number of countries have completed or are developing national biofuel strategies, and this is a necessary prerequisite before embarking on a biofuel programme. Social impacts of biofuels must be weighted carefully and tradeoffs must be analyzed in the context of sustainable development.

References

Competence Platform on Energy Crop and Agro-forestry Systems for Arid and Semi-arid Ecosystems- Africa (COMPETE) (2007) First high-level biofuels seminar in Africa. Seminar report, Addis Ababa, 30 July–1 Aug 2007

Cotula L, Dyer N, Vermeulen S (2008) Fuelling exclusion? The biofuels boom and poor people's access to land. FAO and IIED, London

DME (Department of Minerals and Energy) (2006) Draft biofuels industrial strategy of the Republic of South Africa. DME, Pretoria

ENDA (2008) Biofuel development in Africa: illusion or sustainable alternative? http://www.competebioafrica.net/publications/pubs/ENDA%20Biofuels%20Africa.pdf. Accessed 26 Sep 2010

Food and Agricultural Organization of the United Nations (FAO) (2010) FAO's views on Bioenergy. http://www.fao.org/bioenergy/47280/en/. Accessed 18 Aug 2010

Greenpower (2010) International lessons for SA's fledgling biofuels industry. http://www.greenpowerconferences.com/biofuelsmarkets/documents/ArticleInternationallessonsforSA.doc. Accessed 12 Aug 2010

Guardian (2007) Uganda 'averts tragedy' with reversal of decision to clear virgin forest for biofuel. http://www.guardian.co.uk/world/2007/oct/29/uganda.international. Accessed Aug 2010

International Fund for Agricultural Development (IFAD), Agriculutral Organization of the United Nations (FAO) (2004) Cassava industrial revolution in Nigeria; the potential for a new industrial crop. Eds: Phillips T, Taylor D, Sanni L, Akoroda M. Rome, Italy

IPP Media (2010) Tanzania, Kenya coastal communities opposed to cultivation of jatropha. http://www.ippmedia.com/. Accessed 25 July 2010

Malony T, Smith J (2010) Biofuels in Africa: growing complexities. Africa energy yearbook. EnergyNet Limited, United Kingdom

Mingorance F (2006) The flow of palm oil Colombia-Belgium/Europe. A study from a human rights perspective. Coordination Belge pour la Colombie. http://www.cbc.collectifs.net/doc/informe_en_v3-1.pdf. Accessed 13 22 July 2010

Newscientist (2007) Biofuel plantations fuel strife in Uganda. http://www.newscientist.com/article/dn11671-biofuel-plantations-fuel-strife-in-uganda.html. Accessed 28 July 2010

Nicholls T (2008) How the energy industry works; an insiders' guide. Silverstone Communication Ltd, Towcester

Rossi A, Lambrou Y (2008) Gender and equity issues in liquid biofuel production; minimizing the risks to maximize the opportunities. FAO, Rome

Singh M (2006) Economics of biofuels for the transport sector in South Africa. Energy Sustain Dev X(2):40–47

Southern Africa Development Community (SADC) (report authors Takavarasha T, Uppal J, Hongo H) (2005) Feasibility study for the production and use of biofuel in the SADC region. SADC August 2005, Gabarone

Sugrue A (2007) Biofuel production and the threat to South Africa's food security. In: Wahenga Brief, no 11, Apr 2007

Sulle E, Nelson F (2009) Biofuels, land access and rural livelihoods in Tanzania. IIED, London

UN-Energy (2007) Sustainable bioenergy: a framework for decision makers. UN-Energy, New York

Vermeulen S, Cotula L (2010) Making the most of agricultural investment: a survey of business models that provide opportunities for smallholders. FAO and IIED, Rome

Chapter 27
Small-Scale Bioenergy Initiatives: Lessons from Case Studies in Asia, Latin America and Africa

Steven Hunt, Thalia Konaris, Raffaella Bellanca, and Grant Ballard-Tremeer

Abstract A study on the impacts that different types of local level bioenergy initiatives can have on rural livelihoods in different contexts in the developing world was carried out through analysis of 15 international case studies between September and November 2008. The cases were selected from 12 countries in six regions of Latin America, Africa and Asia and included a range of bioenergy resources, including natural bioresources: bioresidues from existing agricultural, forestry or industrial activities and dedicated energy crops, both liquid and solid, commonly known as biofuels. The initiatives match these resources to a range of energy needs including cooking, mobility, productive uses and electricity for lighting and communication - thereby highlighting the scope of bioenergy applications. It was found that small-scale bioenergy initiatives can offer natural resource efficiency and can help create virtuous circles of local energy production, consumption and productivity. Important in their development is long-term planning and regulation and in some cases partial insulation from world energy markets was seen to be justified and necessary in enabling initiative start up, along with a degree of collaboration within the market chain in initial stages. It was seen that bioenergy has the potential to offer flexibility and income diversity, reducing risk to rural producers, and that long local market chains can spread out benefits thus offering new livelihood choices within rural communities. It was also found that moving bioenergy resources up the energy ladder increases their financial value, whilst any new activity raising demand does also raise prices, even those for bioresidues previously considered as waste. Finally, the small-scale bioenergy

S. Hunt (✉)
The Schumacher Centre for Technology & Development, Practical Action Consulting, Bourton on Dunsmore, Rugby CV23 9QZ, UK
e-mail: steven.hunt@practicalaction.org.uk

T. Konaris • R. Bellanca • G. Ballard-Tremeer
Eco Ltd, PO Box 900, London, Bromley BR1 9FF, UK
e-mail: thalia@hedon.info; raffaella@hedon.info; grant@hedon.info

R. Janssen and D. Rutz (eds.), *Bioenergy for Sustainable Development in Africa*, DOI 10.1007/978-94-007-2181-4_27, © Springer Science+Business Media B.V. 2012

initiatives studied appeared to show that local staple food security was not affected with interconnection between food and bioresources/bioresidues initiatives not present, and with biofuels positioned as an additional, rather than as an alternative, income source for rural producers.

Keywords Livelihoods • Small-scale bioenergy • Case studies • Latin America • Asia • Africa • Bioresources

27.1 Introduction

A series of 15 international case studies were conducted between September and November 2008 under a joint initiative of the Food and Agriculture Organization of the United Nations (FAO) and the PISCES Energy Research Programme Consortium funded by the UK Department for International Development (DFID). This chapter summarises the findings of the overall study, as outlined in the final report (Practical Action Consulting 2009). The case studies focused on developing an improved understanding of the linkages between livelihoods and small-scale bioenergy initiatives. The study was developed in consultation with the PISCES Consortium Advisory Group (CAG), made up of leading international participants in the field of energy and development, including members from the IEA, UNEP, ENERGIA, DFID and FAO, as well as policymakers and research organisations in the PISCES target countries of India, Kenya, Sri Lanka and Tanzania.

The focus of the study is on the impacts that different types of local level bioenergy initiatives can have on *rural livelihoods* in different contexts in the developing world. Livelihoods are understood as the enhancement of the full range of *natural, financial, human, social and physical capitals* on a sustainable ongoing basis.

The cases were selected from 12 countries in six regions of Latin America, Africa and Asia. They were selected to highlight the use of a range of bioenergy resources, including natural *bioresources: bioresidues* from existing agricultural, forestry or industrial activities and dedicated energy crops, both liquid and solid, commonly known as *biofuels*. The initiatives match these resources to a range of energy needs including cooking, mobility, productive uses and electricity for lighting and communication - thereby highlighting the scope of bioenergy applications. The approach taken also considers non-energy by-products of production processes where these may form a significant added benefit in terms of livelihoods, revenues and efficiency.

27.2 Methodology

PISCES conceptualises bioenergy systems as energy pathways, which may be illustrated as in Fig. 27.1.

This diagram shows the various bioenergy resources and how they are converted ultimately into energy access and livelihoods outcomes. However, not only does the

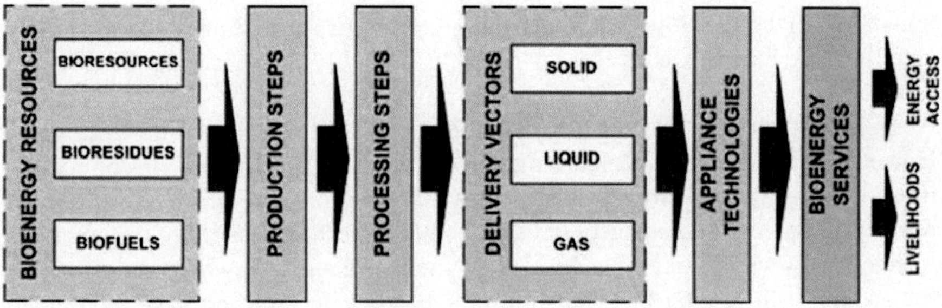

Fig. 27.1 PISCES bioenergy pathways diagram

use of the energy result in livelihoods opportunities via energy access and productive uses in enterprises, but each step and sub-step in the system (as well as wastes, co-products and supporting services) represents a separate livelihood opportunity and has its own interlinked characteristics in terms of possible technologies, capacities required, financial implications, governance issues, access rights, risk characteristics, environmental impacts, etc.

The study has at its heart a *market systems* approach, and in particular the use of market mapping. This approach enables the identification and illustration of the main *market actors* as well as the crucial *supporting services* and *enabling environment,* which contribute to the success or failure of initiatives. Taking the market map for each initiative as the basis, the project then applies the "4Rs" Framework of *Relationships, Rights, Responsibilities and Revenues* to the actors in the system. This method aims to better understand the power dynamics of each case in terms of key issues such as risk, vulnerability, governance and equity. The resulting analysis allows the *impacts* of each initiative on the livelihoods assets of the actors in the chain to be assessed and preliminary conclusions to be drawn. This section elaborates further on this methodology.

27.2.1 Case Study Selection

The 15 cases were selected in consultation between PISCES and FAO on the basis of information gathered via networks, secondary literature, awards programmes, previous research and consultations. These were chosen to cover a cross section of bioenergy types (bioresidues, bioresources and biofuels), with emphases on the emerging biofuels sector. They were also chosen to cover a range of countries, including both least-developed and emerging economies, a range of end-uses with an emphasis on local energy services, a range of ownership, management and business models, and finally all were relatively small-scale initiatives with local participation, leadership and evident focus.

As a result, three case studies involving bioresources, five involving bioresidues and seven involving biofuels were selected. Nine case studies involved initiatives aimed at serving household energy needs, with the remainder split between use in enterprises as a means of production and in public buildings or services (incl.

transport, street lighting, water pumping etc.). The case studies were roughly equally split between end uses of cooking, mobility, electrical appliances and production.

The countries were selected to represent a diversity of situations at the local scale. However, at the national scale the situations are also very diverse in terms of populations, existing energy provision, bioenergy resources, agricultural production and poverty, including food poverty, and many other indicators. Generally, the countries selected show substantial reliance on solid fuels (primarily firewood, but also coal) for cooking, illustrating the vital role of bioenergy in fulfilling basic household energy needs and that its availability and low cost make it indispensable to the poor.

The following case studies were selected. These were renamed for consistent reference by a three-word title containing the country, form of bioenergy and use/relevance: (1) Mali Jatropha Electrification, (2) Senegal Chardust Briquettes, (3) Senegal Typha Charcoal, (4) Tanzania Sisal Biogas, (5) Tanzania Palm Oil, (6) Kenya Afforestation Charcoal, (7) Ethiopia Ethanol Stoves, (8) India Jatropha Electrification, (9) Biodiesel based Water pumping programme in rural Tribal villages of Orissa, (10) Sri Lanka Spice Drying; (11) Brazil Ethanol Micro-distilleries; (12) Guatemala Jatropha Biodiesel, (13) Peru Veg-Oil Recycling, (14) Thailand Jatropha Co-operative, (15) Vietnam Farm Biogas.

27.2.2 Mapping the Market/Value Chain

The market mapping method employed drew heavily from that developed by Practical Action (Albu and Griffith 2005) and was developed using a combination of participatory, interview and research methods. As shown in Fig. 27.2, a market map not only highlights market actors but ensures that enabling environment issues and supporting services to the market chain are captured since these also contribute to livelihoods outcomes and provide broader information about the context within which initiatives are operating.

27.2.3 Relationships, Rights, Responsibilities and Revenues (4R's) Balance Between Market Actors

Based on the full range of actors identified through the market mapping the researchers were asked to consider the *Relationships, Rights, Responsibilities and Revenues* of the key actors in the market system drawing on the 4 R's approach developed by IIED (Mayers 2005). This approach provides a structure for analysing power dynamics between actors as well as ensuring that all of these key aspects are covered for each actor. In this way, important features relating to vulnerability and risk in particular can be addressed (Table 27.1) Balance of Rights, Responsibilities and Revenues of market actors (From the Senegal Chardust Briquettes case study).

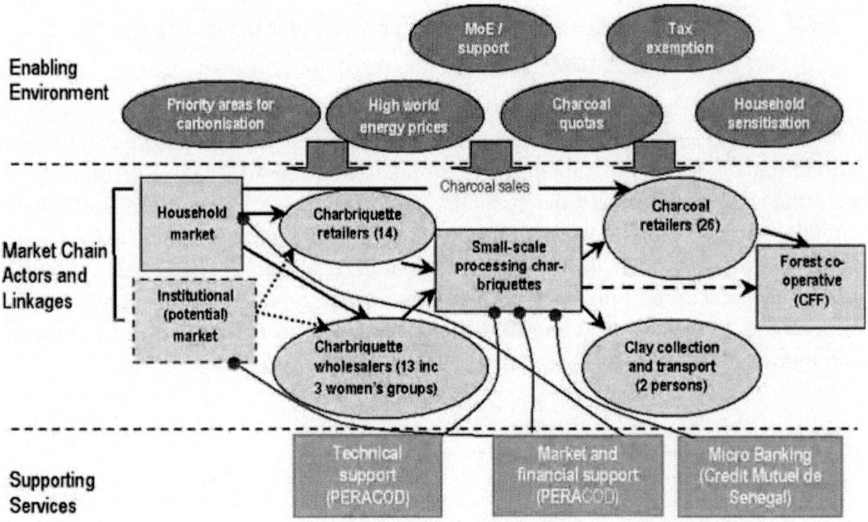

Fig. 27.2 Typical market map (From the Senegal chardust briquett case study)

Table 27.1 Balance of Rights, Responsibilities and Revenues of market actors

	Rights	Responsibilities	Revenues
Ministry of Energy	Regulation framework & fiscal policy (tax exemption in this case), charcoal quotas allocation. Quotas allocation for charcoal making	Ensure that exemption is efficient and properly used	State budget taxation on biomass supply chain
BRADES	Collection, and processing of charcoal residues; contract loans	Char briquette production and commercialisation, loan reimbursement	Income from char briquette sales
PERACOD	Initiation of charcoal residues project	Follow-up of BRADES charcoal residues project	Grant from Germany and Senegal
Mutual credit of Senegal (CMS)	Credit provider	Monitor credit	Profit margin on credit
Forest river cooperative (CFF)	Wood exploitation and charcoal making	Market supply at reasonable price	Income from charcoal and potential for charcoal residues
Women groupings/ retailers (GIE)	Buy char-briquettes	Sale of Charcoal Briquettes	Income for char briquettes sales
Charcoal retailers	Business with charcoal and residues	Buy and sell charcoal and residues	Income for charcoal and chardust sales

From the Senegal Chardust Briquettes case study

27.2.4 *Impacts of the Initiative on the Livelihoods Assets of Market Actors*

After an initial assessment of the vulnerability context for communities involved with initiatives and in the light of the full market map and 4R's analysis, researchers were then asked to identify the contributions of the project to the five livelihoods capitals of participants namely: *Human, Social, Physical, Financial and Natural Capital.* This approach utilised the livelihoods framework supported by DFID (1999) and placed an emphasis on assessing where possible the sustainability of these impacts, thereby enabling to draw conclusions on the impact of the overall initiative on rural livelihoods.

27.3 Comparison and Analysis

27.3.1 *Market Maps*

To compare the approaches of different case studies and their contribution to successful start-up and development of small-scale bioenergy initiatives, as well as their actual or likely impact on rural livelihoods, the following features were considered with the following lessons learnt.

Market system initiation: In tracing back the initiation of the market systems for the cases, an interesting range of initiators and initiation strategies have been discovered. In general, although there is often an individual or institutional driving force, in all cases a coalition of interested parties was established in order to initiate projects, which have in turn led to the establishment of new market systems. These coalitions have been required to overcome barriers to the establishment of initiatives which in general cannot be solved by one institution alone. Some of the key activities involved with initiation seen in the case studies covered include *co-ordination, training and capacity building, technology transfer, marketing/outreach, feasibility study, distribution of initial inputs (e.g. seedlings), and soft loans.*

Market system development: Many of the biofuels cases covered were in relatively early stages and so longer term sustainability issues are both yet to be seen and have strong interactions with larger global issues such as oil prices, evolution of EU subsidies, climate change policy and financing conditions. However, as initiatives grow in size, especially likely in cases linked to some extent with global markets, new opportunities and threats are presented in terms of potential revenues as well as risks and price pressures on producers. Co-operative set-ups and local production and consumption chains are less exposed to both these risks and revenue opportunities, but still have significant growth potential in terms of coverage within rural areas and spread of co-operative systems.

Bioenergy as a component of wider rural market chains: Bioenergy is found to play a range of roles in the market systems, contributing either directly to rural energy access and quality of life, or indirectly as an input into other chains, creating further jobs and income. This categorization revealed different dependencies and vulnerabilities of the bioenergy market for each initiative.

- *Bioenergy as the main output of the chain* (including all biofuels initiatives or bioresources initiatives serving for example household cooking)
- *Bioenergy as a productive input to another chain* helping enhance the efficiency and competitiveness of another market
- *Bioenergy as a by-product of another chain* (incl. all bioresidues initiatives)

Enabling environments: The prevalent enabling environment for bioenergy varies dramatically between countries and indeed between types of bioenergy within the same country. Some important components of enabling environments, cited consistently by initiatives studied, included *government regulation/incentives, standards, and world oil price.*

Supporting services: Supporting services which enable the main market chain to function include *factors of production such as fertiliser and machinery, transportation, legal and contract assistance, technology R&D, bargaining support, training and capacity building, market information provision* and others with *marketing and financing/loans* playing key roles in the success of initiatives. These services are drawn upon by various actors in the chains and some, such as training, are most important at the beginning of the development of a market chain whilst some, such as transportation, are required on an ongoing basis.

27.3.2 Relationships

Given the importance identified of coalitions or partnerships in starting and developing bioenergy initiatives, relationships were clearly a crucial feature of initiatives. Key factors linked to relationships within initiatives and market systems highlighted by the case studies include *leadership and participation* and *levels of formality.* All initiatives had what could be described as leaders. However, no initiative was wholly dependent on one organisation and all have clearly spent time developing participation and support within other linked stakeholder groups and between other actors in the market chains. Also, a range of levels of formality had been noted in the case study approaches, which appeared to be very dependent on the contract enforcement regime in the country.

27.3.3 Balance of Rights, Responsibilities and Revenues

Analysis of the balance of rights, responsibilities and revenues of the actors in the initiative market systems provides a window into the power dynamics of a market system which in turn offers a perspective on where vulnerabilities lie. The reduction of vulnerability amongst rural populations and producers is a key element of the livelihoods impacts which the study seeks to assess.

Distribution of risk: A crucial measure of vulnerability is the extent to which livelihoods are at risk. Risk is a function of the likelihood of an event transpiring and the seriousness of that event. In this respect keen attention is required in a market chain as to who carries the risk of failure. In the case where this is small-scale rural producers without alternative livelihood options, this can be seen as a direct threat to security of livelihoods. All the initiatives covered in this study claim to have addressed these issues in spreading risk between larger players and smaller actors through a number of mechanisms including *diversification, intercropping between energy and food crops, additional growing of short rotation food crops, formation of co-operatives and producer groups for joint bargaining and risk sharing; guaranteed pricing on agreed timeframes.*

Business and management models: Rights, responsibilities and revenues are reflected in the business and management models applied by initiatives, whether they apply to one main actor within the chain or to an umbrella organisation such as a co-operative within which most functions of a market chain occur. Again, several model types were in evidence in the case studies, broadly falling into four categories: *fully commercial – targeting wider markets, fully commercial – targeting local markets, semi-commercial, and volunteer models.*

Land and resource rights: Land and resource rights are a crucial concern, particularly in bioenergy projects involving cultivation of energy crops or access to natural bioresources. In the situation of bioresidues, rights are usually clear and lie with the previous processor of the residue, be it from forestry, agriculture or industry. Different situations were again notable in the cases covered, varying primarily by country, the land laws and allocation systems within each, and sometimes between each actor in the chain. From the perspective of the security of rural producers, the cases fall into the following categories: *land ownership of small farmers is secure, lease or usufructuary rights are available to small farmers, and unclear or no land tenure.*

27.3.4 Livelihoods Outcomes

Although initiatives are at a variety of stages, initiatives were found to contribute to the increase in livelihoods assets of market actors in a variety of ways.

Human capital: The main contribution of initiatives cited was through training and capacity building provided to rural producers, processors and consumers.

Social capital: Participation in the small-scale bioenergy initiatives, whether in co-operatives, outgrower societies or as independent participants in a collective initiative, was consistently shown to build social capital within rural communities. This seems to be a very important component in many rural schemes either during initiation or subsequently when the initiatives take off.

Physical capital: The main increase noted was in the acquisition of processing equipment enabling the conversion of bioresources, bioresidues and biofuels into improved bioenergy services. These include for example improved efficiency kilns, oil seed expellers, transesterification equipment, micro-distilleries and briquetting presses.

Financial capital: In bioresources projects, financial capital is built primarily through the creation of new income generating activities based on existing, previously under- or non-utilised natural capitals. In bioresidues cases financial capital grows through an increase in revenues to the original processor who now receives additional income for a previous waste resource. In biofuels cases there are even more mechanisms and opportunities for increased financial capital gain since not only are processing and retail functions available for wealth creation, but also production of the material itself which is not counted in bioresidue or bioresource projects. Additional financial capital opportunities cited by the initiatives include *carbon financing; government support, utilisation of bioenergy by-products, creation of premium or associated products, and reduced running cost opportunities.*

Natural capital: The cases covered by this study also demonstrate a number of contributions to natural capital, *including reduction in forest depletion, substitution of fossil fuels, and utilisation of waste by-products previously contributing to pollution.* In the **bioresources** cases it appears clear in studied cases, that locally abundant natural capital is harvested in a manner that does not exceed the carrying, and regrowth capacity of that resource and in fact acts to manage that resource in some cases within reasonable limits, for example in the Senegal Typha Charcoal case. In **biofuel** projects the potential for reduction in natural capital is greater as is the case in any agricultural activity, however in the small-scale bioenergy cases covered here there was no indication that this is taking place given the projects scale and orientation.

27.4 Lessons Learnt and Conclusions

The conclusions of the study showed that small-scale bioenergy initiatives can lead to better managed and more efficient use of natural resources, especially what was previously regarded as waste. Additionally, as they grow, such initiatives can help create virtuous circles of local energy production and consumption. Long-term planning, however, and regulation are crucial for their initiation and development, as is leadership and collaboration with other market actors at the initial stages. In some cases partial insulation from world energy markets was seen to be justified and necessary for enabling initiative start-up and initial growth and strengthening.

It was seen that bioenergy has the potential to add to rural producer flexibility and income diversity, reducing risk by acting as additional option rather than as alternative to existing livelihoods options. It was also found that long local market chains can spread out the benefits of an initiative to further market actors thereby widening livelihoods opportunities within rural communities.

It was observed that moving bioenergy resources up the energy ladder can increase their financial value in the market, whilst any new activity raising demand will raise prices, even in the case of bioresidues previously considered as waste.

Finally, the small-scale bioenergy initiatives studied appeared to not affect local staple food security either due to their size or orientation. Bioresources and bioresidues initiatives did not show any interconnection with food whilst biofuels initiatives were positioned as additional option, rather than as a replacement, for rural producers. This may change, however, if initiatives grow in scale and the question should be revisited in future studies.

27.5 Recommendations for Further Work

In summary, there is need for further work in investigating the challenges and opportunities of bioenergy initiatives at the local level and in more detail. Some of the areas recommended for further work to PISCES, FAO and other actors in the sector include the development of specific sustainability criteria for small-scale bioenergy initiatives, further assessment of natural resource efficiency and energy balances, and a more detailed economic analysis for a selection of the cases. It is further recommended that the case studies be examined in more detail from an equity and gender perspective, and that work is carried out to promote the incentives and reduce the constraints faced by market actors in adopting bioenergy technologies and practices. Finally, it is recommended that the case studies approach is replicated and tested in other applicable contexts.

References

Albu M, Griffith A (2005) Mapping the market: a framework for rural enterprise development policy and practice. http://practicalaction.org/docs/ia2/mapping_the_market.pdf. Accessed 31 May 2010

DFID (1999) Sustainable livelihoods guidance sheets. http://www.nssd.net/pdf/sectiont.pdf. Accessed 31 May 2010

Mayers J (2005) The four Rs. international institute for environment and development http://www.policy-powertools.org/Tools/Understanding/docs/four_Rs_tool_english.pdf. Accessed 31 May 2010

Practical Action Consulting (2009) Small-scale bioenergy initiatives: brief description and preliminary lessons on livelihood impacts from case studies in Asia, Latin America and Africa. Prepared for PISCES and FAO by Practical Action Consulting

Chapter 28
Gender Issues of Biomass Production and Use in Africa

Francesca Farioli and Touria Dafrallah

Abstract Energy is a basic necessity for survival and a key input to economic and social development. In Sub-Saharan Africa access to modern energy remains very low and the energy situation is still heavily dependent on traditional biomass that accounts for 80–90% of the countries energy balances. Lack of energy services is correlated with many elements of poverty, such a low education levels, inadequate health care, and limited employment and income generation possibilities. The energy-poverty nexus has distinct gender characteristics. Of the approximately 1–3 billion people living in poverty, it is estimated that 70% are women, many of who live in female–headed households in rural areas. In Sub-Saharan Africa, women have challenging roles on the energy scene as they are in charge of supplying their households with energy amongst other subsistence activities. This chapter looks into the impacts of biomass production and use on women health and livelihood. Literature and research studies by institutions involved in bioenergy and indoor air pollution are considered (World Health Organization, Partnership for Clean Indoor Air, Energia Network, COMPETE, etc.). Current energy policies in Africa seem to ignore the gender dimension of energy, although providing rural women with an affordable, reliable and clean energy source is a priority to effectively alleviate poverty. For any energy policy aiming at poverty reduction it is absolutely crucial not to neglect the fact that men and women have different energy needs due to their traditionally different roles and responsibilities within the households, and due to the unbalanced access to resources and decision-making. Nevertheless, the gender dimension of energy often remains invisible to most policy-makers. In many

F. Farioli (✉)
Interuniversity Research Centre for Sustainable Development –(CIRPS),
Sapienza University of Rome, Piazza San Pietro in Vincoli 10, Rome 00185, Italy
e-mail: Francesca.farioli@uniroma1.it

T. Dafrallah
Environnement et Développement du Tiers Monde (ENDA TM), 54, rue Carnot,
Dakar BP 3370, Senegal
e-mail: touria.dafrallah@hotmail.com

R. Janssen and D. Rutz (eds.), *Bioenergy for Sustainable Development in Africa*,
DOI 10.1007/978-94-007-2181-4_28, © Springer Science+Business Media B.V. 2012

African countries biofuels production has recently gained significant interest. Private companies are investing in biofuels opportunities, as Africa seems to offer a good environment (available land, cheap labour and favorable climate). Unfortunately, policy and regulatory frameworks are not established to monitor the emerging private initiatives on biofuels that seem to focus on exports. This might worsen gender issues as women are economically and socially vulnerable and might be the main group to get marginalized. This chapter identifies relevant policy options related to social aspects of biomass production and use, as well as a set of recommendations how to engender biofuels policies.

Keywords Energy poverty • MDGs • Bioenergy • Health • Livelihood • Gender mainstreaming • Engendering energy policies • Land access • Food security • Income generation • Policy recommendations

28.1 Linking Poverty, Energy and Gender

28.1.1 Energy, Poverty and Development

The linkage between energy and poverty[1] has been a major topic of many investigations and debates at national, regional and international levels. Stakeholders, including development agencies, have come to a consensus that there is an unquestionable connection between access to modern energy services and social, economic and human development. They state that energy can play a crucial role in supporting efforts to achieve the Millennium Development Goals (MDGs) and improving the lives of the poor people across the world.

The eight MDGs consist of eradicating extreme poverty and hunger, achieving universal primary education, promoting gender equality and empowering women, reducing child mortality, improving maternal health, combating HIV/AIDS, malaria and other diseases, ensuring environmental sustainability, and developing a global partnership for development.

The details of the eight MDGs target 18 important areas. Better access to energy is not a specific goal within the MDGs, but energy is an important aspect that runs through many of the targets in the MDGs.

Table 28.1 shows details of the MDGs and associated targets and includes indications on the linkages between energy services and the MDGs. In general terms, energy is crucial for reducing poverty as it provides benefits in terms of improving health conditions, reducing hunger, increasing literacy and education, providing opportunities for income generation, etc.

[1]Poverty can be perceived in a number of ways, for example, in economic terms, poverty indicates an income of less than 1 US$ per day in social terms it means lack of access to adequate levels of food, water, clothing, shelter, sanitation, healthcare and education (World Bank 1996).

Table 28.1 Energy, MDGs and gender linkages

Goal	Targets	Role of energy services to achieve the goals	Gender linkages
Goal 1: Eradicate extreme poverty and hunger	Target 1:Halve, between 1990 and 2015, the proportion of people whose income is less than 1 US$a day Target 2: Halve, between 1990 and 2015, the proportion of people who suffer from hunger	Access to energy services facilitates economic development, livelihood activities beyond daylight hours, locally owned businesses, which will create employment, and income generating opportunities Energy services can improve access to pumped drinking water and energy access for cooking (95% of staple foods need to be cooked) Energy can be used to power labour-saving machinery and increase productivity of enterprises	Women can take benefits of economic development through creation of micro enterprises led by women (job and income generation) Women can save time and burden associated to water collection and gathering fuel for cooking, lighting and heating Women can reallocate their time towards other productive activities (agricultural tasks and micro-enterprises), therefore increasing assets and well-being of the family Women are generally responsible for cooking and feeding their families and often for subsistence agriculture and food processing
Goal 2: Achieve universal primary education	Target 3: Ensure that, by 2015, children everywhere, boys and girls alike, will be able to complete a full course of primary schooling	Energy services reduce the time spent by women and children (especially girls) on basic survival activities (gathering firewood, fetching water, cooking, etc.). Lighting permits home study, increases security and enables the use of educational media and communications in schools, including information and communication technologies (ICTs) Modern energy means better quality of life for teachers	Girls are more likely to be taken out of school to help with domestic and agricultural tasks than boys Spending on schooling, especially for girls, increases with higher incomes for women Girls are more likely than boys to be affected by lack of access to clean water and sanitation facilities, reducing school attendance

(continued)

Table 28.1 (continued)

Goal	Targets	Role of energy services to achieve the goals	Gender linkages
Goal 3: Promote gender equality and empower women	Target 4: Eliminate gender disparity in primary and secondary education preferably by 2005 and in all levels of education no later than 2015	Collection of fuel-wood takes up a lot of time, especially for women Street lighting improves the safety of women and girls at night, allowing them to attend night schools and participate in community activities	Women are less likely than men to have access to education and information and to be included in political and community life
Goal 4: Reduce child mortality	Target 5: Reduce by two thirds, between 1990 and 2015, the under-five mortality rate	Energy is a key component of a functioning health system (lighting, refrigeration of vaccines and other medicines, sterilization of equipment and transport to health clinics) Cleaner fuels and technologies help reduce indoor air pollution, which contributes to respiratory infections that affect children and women	Women and young children spend the most time indoors Women and girls are generally responsible for cooking, often with unventilated open fires
Goal 5: Improve maternal health	Target 6: Reduce by three-quarters, between 1990 and 2015, the maternal mortality ratio	Energy services are an absolute priority for maternity services (medicine refrigeration, sterilization equipment, clinics) Access to communication facilitated by energy services brings information on sex education and contraceptives	Excessive workload and heavy manual labour (heavy loads of fuel-wood and water; arduous and repetitive agricultural and food processing tasks) may affect pregnant women's health and well-being
Goal 6: Combat HIV/AIDS, malaria, and other diseases	Target 7: Have halted by 2015 and begun to reverse the spread of HIV/AIDS	Electricity is needed for pumped and boiled water contributing to safe water supply	Women are more subject to diseases due to more vulnerable living conditions
	Target 8: Have halted by 2015 and begun to reverse the incidence of malaria and other major diseases	Modern energy services lead to reduced air pollution and fecilitate refrigeration for vaccines and food preservation	

Goal 7: Ensure environmental sustainability	Target 9: Integrate the principles of sustainable development into country policies and programme and reverse the loss of environmental resources	Improved energy efficiency and use of cleaner alternatives can help to achieve sustainable use of natural resources, and to reduce emissions, which protects the local and global environment	Women and girls are generally responsible for gathering fuel-wood and collecting water
	Target 10: Halve, by 2015, the proportion of people without sustainable access to safe drinking water and basic sanitation	Local domestic environment is improved with modern energy services	Environmental degradation and deforestation impacts most women and girls who undertaken longer and unsafe travels to reach places to collect water and fuel-wood
		Land degradation is reduced with less biomass used for cooking	
		Local pollution is reduced (especially in built-up areas)	
	Target 11: Have achieved, by 2020, a significant improvement in the lives of at least 100 million slum dwellers	Renewable energy and energy efficiency help alleviate global warming and climate change	

Sources: Adapted from Ines Havet, "Linking Women and Energy at the Local Level to Global Goals and Targets," Energy for Sustainable Development VII (September 2003)

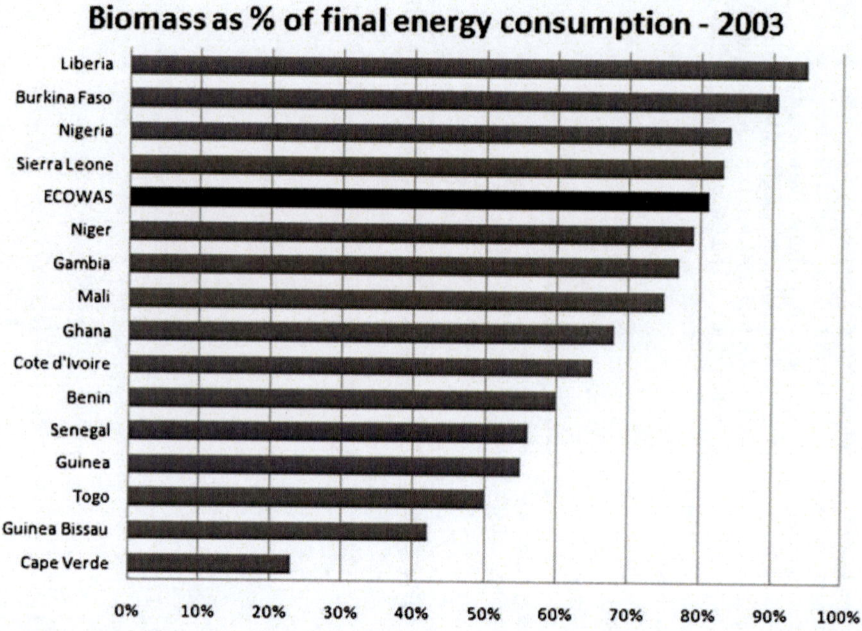

Fig. 28.1 Share of biomass in the ECOWAS countries energy balance (Source: ECOWAS/UEMOA 2006)

Energy poverty indicates lack of access to adequate, affordable, reliable, high quality, safe and environmentally harmless energy services to support economic and human development, and it is interrelated with all other poverty aspects (Kovacevic 2007). Without access to modern forms of energy for lighting, cooking, heating and cooling, refrigeration, pumping, transport, communication and productive purposes, people must spend much of their time and physical energy on basic subsistence activities. At local and national levels, a reliable energy supply is essential for economic stability and growth, jobs, and improved living standards. Limitations on the availability of energy services create barriers to socio-economic development (Muchiri 2008).

The challenge of modern energy access in Africa is important, particularly in the Sub-Saharan part, in view of the fact that only 24% of the Sub-Saharan African population has access to electricity compared to 40% in other low-income countries. Furthermore, the region's expansion of electrification is proceeding slower than in other low-income countries (IMF 2008). Biomass accounts for an average of 80% in the energy balances in the ECOWAS counties (ECOWAS 2006). This share reaches more than 90% in some countries and even in Nigeria, an oil producing country, biomass account for more than 80% in the energy balance (Fig. 28.1).

It is worth highlighting that the first energy priority for the poor is domestic energy for cooking and heating. To fulfil these needs, they currently heavily rely on traditional biomass (wood, charcoal, animal residues, etc.), whilst they lack modern energy in terms of availability, affordability, accessibility, and acceptability (WEC 2000).[2] Furthermore, up to a third of the incomes of the poor is spent on energy, mostly used to cook food.

28.1.2 Energy Poverty and Gender

Energy poverty is a problem that has a disproportionate effect on women and girls, especially in rural areas. In fact, the energy-poverty nexus has distinct *gender characteristics*.[3] Of the approximately 1.3 billion people living in poverty, it is estimated that 70% are women, many of whom live in female-headed households in rural areas (Clancy et al. 2003). For any energy policy aiming at poverty reduction it is absolutely crucial not to neglect the fact that men and women have different energy needs due to their traditionally different roles and responsibilities within the households, and due to the unbalanced access to resources and decision-making.

In 1995, the UN's Fourth World Conference on Women, held in Beijing, concluded that throughout the world women continue to have fewer options and opportunities than men. Unequal treatment of men and women, and their differentiated social and economic roles, has also led to higher levels of poverty for women than for men in many countries (Muchiri 2008).

The International Network on Gender and Sustainable Energy (ENERGIA) established in preparation of this Conference (Yianna 2006), supports that gender equity in energy systems should be guaranteed. It is recognized that (Muchiri 2008): (1) Women and men have different roles in the energy system: women bear the main burden of providing and using fuels (dung, raw biomass) for cooking. A situation made worse by fuel scarcity and negative health and safety impacts (such as indoor air pollution, which kills an estimated 1.5 million women and children in the developing world each year); (2) Women bear the invisible burden of the human energy crisis (i.e. their time and effort in water pumping, agricultural processing and transport). They need modern and more efficient energy sources to improve their work and quality of life both within and outside the home; (3) Women have less access than men to credits, extension services, land and training, necessary for improving energy access to support their livelihoods and income generation from micro enterprises; (4) Women and men have different kinds of knowledge and experience of energy, either through their traditional roles, their new roles or increasingly as professionals in the energy sector; (5) Since women experience poverty differently to men, they may need different energy policies to help them escape energy

[2] WEC identifies what it calls 'three energy goals: accessibility, availability, acceptability' in WEC (2000).

[3] Gender refers to socially constructed roles and relations between men and women. This includes the different responsibilities of women and men in a given culture or location. These roles vary within and between cultures, ethnicity and class and change over time (Muchiri 2008).

poverty: New energy technologies can even have unintended negative consequences for women, as has happened in the past with other technologies.

In Africa, men and women play a different role in the energy system. Women have the responsibility of ensuring the supply of energy to their households by collecting, managing and using biomass for cooking and heating. This situation is worsening considering the reduced availability of biomass resources, insecurity and most of all, the negative effects (of the use of traditional biomass in households) on health. The following section provides significant evidences with respect to the linkage of health and energy.

28.2 Impacts of Bioenergy Production and Use on Women Health and Livelihoods

In many developing countries, women are responsible for both securing household energy and producing crops. Consequently, developments in bioenergy and related biofuels markets have the potential to benefit women if well planned. Yet, if gender and poverty considerations are not incorporated into bioenergy policies and practices, the livelihoods of women and their families could be threatened.

Men and women within the same household as well as male- and female-headed households, could face different risks, particularly with regard to their access to and control of land and other productive assets, their level of participation in decision-making and socio-economic activities, employment opportunities and conditions, and food security. This reflects different roles and responsibilities of men and women within rural economies, as well as pre-existing socio-economic inequalities (Rossi and Lambrou 2008).

28.2.1 Health Issues for Women

The traditional uses of bioenergy for cooking and heating through direct burning of wood and other biomass, affect the health of women more severely than men and contribute to the relative disempowerment of women (UN-Energy 2007). Smoke inhalation from cooking with traditional biomass cook-stoves is one of the leading causes of disease and death in developing countries. Indoor smoke is responsible for an estimated 3.7% of the overall disease burden (determining pneumonia, chronic respiratory disease and lung cancer) making it the most lethal killer after malnutrition, unsafe sex and lack of safe water and sanitation (WHO 2005).

Furthermore, women spend three times as much time transporting fuel and water than men, and regularly carry four times as much as men in volume (IUCN 2007). The shift to modern bioenergy services may benefit especially women, by reducing the time spent collecting firewood and charcoal and by limiting exposure to indoor pollution and the associated health risks (FAO 2009). The use of improved cook-stoves and modern bioenergy sources, if made available at an affordable price to consumers, reduces safety and security risks and improves the living conditions in homes due to cleaner air (UNDP 2004).

If the shift to modern bioenergy services cannot be realized, the potential reduction in the availability of firewood and clean water due to biofuel expansion, would place an additional burden on rural farmers' work and health, forcing women and girls to travel longer distances and allocate more time to these tasks, thus reducing thc time available for other income-generating activities and education (UNDP 2004).

28.2.2 Access to Land

Expanded biofuels feedstock production will lead to increased levels of deforestation. Forests contribute to the livelihoods of many of the 1.2 billion people living in extreme poverty and 70% of these are women. Uncontrolled biofuel production could worsen the living conditions of women living in poverty, in particular those that rely on the forest for their survival (IUCN 2007).

There are significant gender gaps in land ownership. For instance, in Cameroon, women undertake more than 75% of agricultural work, but own less than 10% of the land. In Brazil, the percentage of land owned by women is 11%, in Peru 13%. Similar disparities have been identified in Tanzania, Kenya, Nigeria and other countries in Sub-Saharan Africa. The land access issue involves concerns to secure land rights across the developing world with only 5% of women farmers owning their land (IUCN 2007). In addition, women's land is often registered to male members of the family, and widowed women and single mothers risk being thrown off the land or denied land titles (Cotula et al. 2008). In absence of a secure land right, the growing global demand for liquid biofuels, combined with high land requirements, might put pressure on so-called "marginal" lands, contributing to displacement and a negative gender impact. On these lands, women have traditionally grown crops for household consumption, rituals and medicinal uses. Also, women's ability to participate in land-use decision-making may be reduced as the amount of land they control will decline (Rossi and Lambrou 2008).

In addition, the increase of land price due to the expansion of biofuels production can move people to sell their plots of land. Therefore, it is important to quickly expand small-scale schemes to show farmers that they do not need to sell their land, and that using it to grow crops such as jatropha provides a sustainable way of earning their living.

Generally women lack access to formal credit schemes, thus being limited in their ability to acquire agricultural inputs. In Nigeria, for instance, only 3% of women receive credit from banks, against 15% of men. Moreover, although the average value of the loan obtained by women is only 42% of that of men, the percentage of collateral required is regularly higher for women.

28.2.3 *Working Condition of Women and Men*

There is evidence that, in some cases, working conditions on plantations (large-scale production schemes) including those of biofuels feedstock tend to have a differentiated gender impact. Landowners prefer women workers, as they are able to pay them less than their male counterparts and find them a docile and dependent workforce, therefore being more exploitable (ILO/FAO/IUF 2007). A significant (and growing) number of agricultural workers in developing countries are employed on seasonal and often casual or temporary basis (with limited, if any, social security, including medical assistance). An increasing share of these workers is women. Small-scale production schemes within the rural communities allow gaining benefits for women. If well planned, these schemes can strengthen women's empowerment in several phases of the supply and in the management of bioenergy plants.

28.2.4 *Food Security and Vulnerable Groups*

The rising demand for liquid biofuels could also make the prices of agricultural commodities and food less stable, exposing a significant number of households and individuals to the risk of food insecurity. Sudden increases in food prices (determined by large-scale production) would have negative repercussions in particular for poor households and vulnerable groups. Women-headed households tend to be particularly exposed to chronic and transitory food insecurity due to their limited access to income generating activities (Energy and Transport Branch Division for Sustainable Development United Nations Department of Economic and Social Affairs 2007). This could have impacts on nutrition of poor people, especially women who are already more vulnerable to nutritional problems, (for example, 50% of women and children in developing countries are anaemic) due to physical, social, economic, gender and cultural issues (e.g. pregnancy, lactation, inequitable food distribution within families) (IUCN 2007).

In addition, the establishment of energy crop plantations on "marginal" lands might negatively affect women's ability to meet household obligations, including traditional food provision and food security.

28.2.5 *Energy Pricing*

At the national level, decisions about access to energy supply are male dominated. Income disparities exist between men and women. Most rural women have limited economic control at the household level whilst men control household income. When energy has to be purchased, men enter the energy decision-making process. This means that energy pricing will impact men and women differently (Muchiri 2008).

From an economic point of view, pricing of domestic fuel affects women because they are mainly responsible for the fuel procurement. Pricing should therefore promote efficient energy use and alternative fuels ensuring women's participation in productive end uses.

28.2.6 Income Generation, Education Level and Rural Development

Especially in small-scale production schemes, biofuels production and use can stimulate income generating activities including seed cake for fertilizer and soap production, enhancing rural development. Women can be involved in feedstock production and in the management of plants which help strengthening their empowerment (becoming for example energy entrepreneurs). They can improve the productivity of their activity by the use of mechanized equipments. Access to modern energy services (energy for cooking, heating, pumping, grinding, transport communication, and energy for small businesses development) improves girl's school performance (school attendance and homework) by freeing them from exhausting and time consuming household tasks.

28.3 Impact of Bioenergy on Gender: Two Case Studies

This section focuses on two case studies Africa (Mali) and Asia (Cambodia) that give evidence of successful gender sensitive bioenergy projects.

28.3.1 Multifunctional Platform Project in Mali

This project was initiated by UNDP and UNIDO in collaboration with the Government of Mali in 1996. The platform was designed to take into account multiple end uses for energy in rural economies, and to provide a substitute for human energy. The platform consists of a small diesel engine mounted on a chassis, to which a variety of end use equipment can be attached, including grinding mills, battery chargers, vegetable or nut oil presses, welding machines and carpentry tools. It can also support a mini grid for lighting and electric pumps for a small water distribution network or irrigation system. With its many functions, it can be used for a variety of services that can generate incomes for the group operating the platform. Due to its extreme flexibility the platform can easily be modified to suit the specific needs of each village and run on straight vegetable oil produced from the locally available jatropha plants.

According to the UNDP report on "Access to energy and poverty reduction" (UNDP 2004), the multifunctional platform fueled with jatropha oil provides the following positive aspects: (1) added value to local resources from processing of the jatropha fruits (2) possibility to grow jatropha on low-fertility soils, (3) planting jatropha creates a fence against wind and soil erosion (i.e. living hedges), (4) reduction of cash outflow from the villages (fuel and soap from outside can be substituted with jatropha oil and soap), (5) creation of opportunities for income-generating activities such as soap making with sediment and oil, (6) by-product (cake) provides an organic fertilizer with more nitrogen and dry matter than conventional mineral fertilizers, and (7) jatropha biodiesel offers GHG benefits with respect to fossil diesel.

Reports on poverty reduction impacts of MFPs show positive effects on women's social conditions, entrepreneurship, education, nutrition, and health. Some of the evidences on women's living condition improvement after the installation of multifunctional platforms are shown below.

Increased productivity: MFPs enable increased productivity of crop processing as cereal, rice husking or activities such as oil extraction, production of food pastes and shea nut butter, allowing women, not only to produce more efficiently for their household consumption, but also to get engaged in the sales of their products at local markets. Some surveys indicate that, in rice processing, the average annual production per woman in a normal year of rainfall increased from 275 to 600 kg of paddy rice after the installation of platforms. Half of this production is used for household consumption. The remaining half of paddy rice is directly sold at local markets generating a monetary income of 50,000 CFA per season per woman (Diagana 2001).

Freeing-up women's time: The husking of 28 kg of paddy rice is accomplished in about 1 h using the MPF, instead of 48 h by hand (Diagana 2001). The surveys show that the accumulated time saved in a week for cereal processing (millet, sorghum and maize) by women is equivalent to an 8 h work day (Havet 2003).

Income generation opportunities: Increased income generating activities include oil extraction and food pastes, shea butter and soap production. Oil from jatropha seeds can be used as fuel to run the engines, and is expected to be used in at least 15% of the installed platforms. Additional potentials include increasing and diversifying energy sales and enhancing the energy clients' ability to pay for energy services by enhancing the use of the platform's end use equipment. The platform's services help release time to generate income which is needed to pay for the energy service itself. In some villages, the economic opportunities generated by the platform include the ability of women to have access to small individual plots that they now have time to tend and control for their own use (Burn 2001).

Increased education and literacy levels of village women: Clear signs of improvement have been noticed on two different levels: firstly, every implementation of a platform is anticipated by literacy training of women who are its operators and therefore need to read and write. Secondly, the platform indirectly contributes to the improvement of girl's school performance, by freeing them from exhausting and time consuming household tasks, leaving more time for school attendance and studying.

Better health and safety: Platforms can contribute to improve health and safety in many ways. Night lighting offers better safety conditions for women that must work during the night. Refrigeration reduces risks from food-borne diseases and helps conserving vaccines and other medicines. The use of a mechanical press increases productivity of jatropha by-products for medicinal purposes. Oil produced by pressing jatropha seeds may be used as cooking and heating fuel, thus reducing risks from indoor-air pollution.

Food security: The increased productivity by the platforms enables women to store more crops to be used in case of severe drought and food shortage. The cultivation of jatropha for fuel does not affect food security due to the fact that the plant is not edible. On the contrary, if planted as a hedge, it can prevent animals from damaging crops. Moreover, if used as organic fertilizer, the seed cake can contribute to further improvement of agricultural productivity.

Empowerment of women: MFP projects strengthen women's capacity to manage the process of development and change. This is expressed by organizational capabilities of women's group that can choose a Women Management Committee composed by members on the basis of their demonstrated abilities and aptitude for management and adaptation to change. Men must get aware of the benefits they could gain from the platform, and the beneficiary community must be informed about existing financial and management support facilities as well as the ways and means to access banks loans or donor support.

28.3.2 Jatropha Oil for Decentralized Rural Electrification in Cambodia

In Cambodia, an entrepreneurial farmer is growing jatropha and extracting oil from the seeds to run a diesel generator that supplies electrical power for a mini-grid serving over 80 homes. This project was organized by a Cambodian NGO called 'Solidarity and Community Development' (SODECO), which works on development issues in rural communities, mainly by training women at the village level. The diesel generator has been adapted to operate efficiently on pure plant oil.

Although it was a woman parliamentarian who encouraged the male farmer to switch from diesel to jatropha oil, originally there was no special emphasis on gender equity with regard to the benefits of the biofuel-generated electricity. However, women did share in the benefits of lower-cost electricity, and in new opportunities for income generation related to growing and processing jatropha trees and selling them for oil production. In addition, a replication of the project, at 50 km distance from the initial project, is targeted to support women's income generating activities such as silk production and sewing activities.

In September 2009, women started to receive electricity in exchange for jatropha seeds. This is expected to greatly increase the productivity of their silk-related and sewing activities. An important challenge is the lack of funding for fast and wide replication of the project to other remote rural areas deprived of access to energy (Karlsson and Banda 2009).

28.4 Gender in Energy Policies and Biofuels Programmes

A review of the energy policy documents[4] shows that in general, policy documents are more focused on increasing investment and infrastructures, securing supply/diversification of energy sources, promoting renewable energy and energy efficiency, managing the environment, improving access to energy, and promoting the institutional framework. Poverty alleviation and promotion of rural development through energy are also mentioned in the documents. Other energy policies focus on enhancing research and development, promoting employment, and developing subregional, regional and international cooperation as well as capacity building. Efficient use of biomass is considered to reduce the pressure on forests resources used for cooking and heating. On the supply side, bioenergy is addressed in energy policies in relation to forest management and environmental protection.

Gender has not been a particular/effective dimension in most energy policy formulation processes. This general conclusion was drawn by numerous organizations researching on gender and energy.

"*Gender mainstreaming*" was established at the Fourth UN International Conference on Women held in Beijing as the internationally agreed strategy for governments and development organizations to promote gender equality. It is a process to ensure that the concerns and needs of both women and men are considered in all planning and policy making and that all policy makers are aware of the needs of women and men and their roles and responsibilities (Feenstra 2002).

In order to identify gender gaps in national energy-related policies and programmes, energy policy gender audits were carried out in some African countries by the members of the network Energia (Energia 2007). The main findings of the gender audits in the specific cases of Senegal and Kya were: (1) Engendering energy policy is constrained by the limited availability of gender-disaggregated data on energy development and utilization; (2) Even though awareness of gender at the policy level has increased, there is still a need for policy makers to fully mainstream gender perspectives into programmes and activities; and (3) Gender

[4]The policy documents assessed were those of Burkina Faso, Central Africa Republic, Ghana, Mali, Niger, Senegal, Sierra Leone, and Togo (2007).

mainstreaming in policies is supported by various institutions, including those in the energy sector. This represents an opportunity to develop concrete activities geared towards achieving this objective (Energia 2007).

In the bioenergy sector, biofuels development is becoming an increasingly important area in national and international economic development. However, biofuels issues are not well discussed in most national policy frameworks/programmes in Africa.

28.5 Key Policy Recommendations

Based on the review of energy policy documents and good practice surveys conducted in some countries it is recommended to consider the following elements to engender biofuels policies.

Definition of a gender analysis framework on energy (at national and local level): Women make decisions relating to domestic energy uses. The burden of energy shortages, price increases and cost-recovery plans tend to fall disproportionately on women. Therefore, there is a need to address the current gender differences prevailing. The suggested framework is useful to explore the current situation in terms of (1) gender differences, (2) division of labour and responsibilities, (3) time and budgets, (4) access to productive resources and assets, (5) income earning opportunities, (6) decision making and empowerment.

Land access and ownership:
Land access issues are likely to be far more acute in countries where much of the population depends on land and natural resources and where poverty has a significant rural dimension. Unclear ownership rights represent a major obstacle for biofuels development due to the fact that increased production rising considerable income might result in land related conflicts. Specifically women involved in feedstock seed collection and processing are at risk due to the fact that they are not allowed to own or inherit the land they work on. Often, there is still a huge gap between the existing regulations and their application in practice. Most people are not aware of their legal rights and keep living under local traditional law. Current tendencies focus on transferring decision making power regarding land and natural resources management to local authorities which are closer to the people and more familiar with the local customary law.

To improve access to land and ownership, policies should: (1) strengthen the protection of land tenure in the agricultural and land use sectors and not only in the specific biofuels sector, focusing on gender disparity, (2) clarify definitions of idle, under-utilized, barren, unproductive, degraded, abandoned and marginal lands in the country contexts to avoid allocation/dis-allocation of lands on which local user groups depend for livelihoods, and (3) support the small-scale inclusion in the biofuels industry with financial incentives to favor the agro-system adoption (food and fuel crop cultivation that reduces conflicts on land) and to boost rural development through local livelihoods improvement.

Access to finance, credit and markets:
It is recommended that energy policies should improve access to financing and credit facilities to promote energy-based business opportunities (micro-credit, collateral and legal issues, targeted lending institutions). Furthermore, information on markets and consumer demand for energy products need to be enhanced to assist women in becoming energy entrepreneurs (marketing and business training, consumer and trade organizations, information services).

Participation in bioenergy policies:
Planning, implementation of bioenergy policies should involve coordination between sectoral policies based on the assessment of global energy needs of women and men within communities. In much of Sub-Saharan Africa there is a lack of strategies to support small-scale biofuels development as existing biofuels policies focus mainly on subsidies for large productions. At the local level potentials for biofuels development to meet local energy needs has not yet been widely recognized. Special efforts should ensure that local population, especially women, capture the benefits of energy services from biofuels development, as well as associated income and job opportunities. The main challenge of this bottom-up approach is to identify and work with appropriate legal and institutional mechanisms as well as to coordinate and facilitate the collaboration between various local and national decision-making bodies.

Capacity building of women:
This includes (1) promoting the use of efficient bioenergy stoves and their availability, (2) providing special trainings and special fellowships for women in various institutions/universities to set frameworks of bioenergy professionals, (3) building capacity and assistance to manage energy programmes, policy, and projects with a gender perspective, (4) enhancing rural electrification, including decentralized programmes, (5) launching large awareness programmes in the sectors of energy and water.

References

Burn N, Coche L (2001) Multifunctional platform for village power: Mali. In: Karlsson GV (ed) Generating opportunities: case studies on energy and women. United Nations Development Programme, New York

Clancy J, Scutsch M, Batchelor S (2003) The gender – energy – poverty Nexus. Finding the energy to address gender concerns in development, DFID Project CNTR998521

Cotula L, Dyer N, Vermeulen S (2008) Fuelling exclusion? The biofuels boom and poor people's access to land. FAO and IIED, London

Diagana M (2001) Impact Study of the Multifunctional Platform on the Living Condition of Women, Mali, Multifunctional Platform for Poverty Alleviation, UNDP Project MLI/99/001

ECOWAS/UEMOA (2006) White paper for a regional policy on energy access. Geared towards increasing access to energy services. UNDP

Energia News (2007) The African Agenda. Gender and energy access, 10(1) ENERGIA Secretariat Leusden, The Netherlands

Energy and Transport Branch Division for Sustainable Development United Nations Department of Economic and Social Affairs (2007) Small-scale production and use of liquid biofuels in Sub-Saharan Africa: perspectives for sustainable development, Commission on Sustainable Development, New York, 30 April – 11 May 2007

Feenstra M (2002) Towards a gender-aware energy policy: a case study from South Africa and Uganda. Department of Public Administration and Political Science, University of Twente, Enschede

Havet I (2003) Linking women and energy at the local level to global goals and targets. Energ Sustain Dev VII(3):75–79, Available on line at http://www.ieiglobal.org/esd.html

ILO, FAO, IUF (2007) Agricultural workers and their contribution to sustainable agriculture and rural development. Geneva

International Monetary Fund Survey Magazine (May 22, 2008) Africa's power supply crisis: unraveling the paradoxes

IUCN (2007) Gender and bioenergy, international union for the conservation of nature. Available at http://www.genderandenvironment.org/admin/admin_biblioteca/documentos/Factsheet%20BioEnergy.pdf

Karlsson G, Banda K (2009) Biofuels for sustainable rural development and empowerment of women. Cases studies from Africa and Asia ENERGIA, Leusden, the Netherlands

Kovacevic J (2007) Biofuel development: implication, impact and gender-sensitive policy making. Master thesis, Master on International Cooperation, Human Rights and EU Policies University of Rome 3

Muchiri L (2008) Gender and equity in bioenergy access and delivery in Kenya. Study for the PISCES RPC, Practical Action Consulting East Africa, Apr 2008

Rossi A, Lambrou Y (2008) Gender and equity issues in liquid biofuels production. Minimizing the risks to maximize the opportunities. FAO, Rome

UN-Energy (2007) Sustainable bioenergy: a framework for decision makers. http://esa.un.org/unenergy/pdf/susdev.Biofuels.FAO.pdf

UNDP (2004) Reducing rural poverty through increased access to energy services. UNDP, Bamako

WHO (2005) Indoor air pollution and health. Fact sheet N°292 June 2005. Available at http://www.who.int/mediacentre/factsheets/fs292/en/

World Bank (1996) Rural energy and development for two billion people. Washington DC, USA

World Energy Council (WEC) (2000) Energy for tomorrow's world: acting now. World Energy Council (WEC), London

Yianna L, Grazia P (2006) Energy and gender issues in rural sustainable development. FAO, Rome

Chapter 29
Financing of Biofuel Projects in Africa

Michael Hofmann, Glynn Morris, Grant Ballard-Tremeer, and Kaysara Khatun

Abstract The development of biofuel ventures offers significant opportunities for Africa to improve the low yield subsistence farming-dominated agricultural sector, mostly with regards to energy, environmental and socio-economic aspects. Taking into account a number of specific characteristics (i.e. reliance on feedstock, food versus fuel debate, strong dependence on related policy frameworks and high risk profiles), barriers related to policies, financing, trade, and other non-technical issues are particularly relevant to the development of biofuel ventures in Africa. Strategies to overcome these barriers include measures such as the removal of regulatory risk through government policy support, creation of financial incentives for biofuels, access to financing, improved trading regulations and infrastructure and the development of related markets. Financing options for biofuel ventures are influenced by the mode of service of the venture. Project-based fuel supply businesses generally require project-based finance options whereas energy service company (ESCO) approaches can attract a wider range of finance options and access carbon finance or green certificate revenue by virtue of scale opportunities.

M. Hofmann (✉)
Camco, 172 Tottenham Court Road, London W1T 7NS, UK
e-mail: michael.hofmann@camcoglobal.com

G. Morris
AGAMA Energy, The Green Building, 9b Bell Crescent Close,
Westlake Business Park, Westlake 7945, South Africa
e-mail: glynn.morris@agama.co.za

G. Ballard-Tremeer
Eco Ltd, PO Box 900, Bromley, London BR1 9FF, UK
e-mail: grant@ecoharmony.com

K. Khatun
Basque Centre for Climate Change (BC3), Alameda Urquijo 4, 4° – 1a,
Bilbao 48008, Spain
e-mail: kaysara.khatun@bc3research.org

R. Janssen and D. Rutz (eds.), *Bioenergy for Sustainable Development in Africa*,
DOI 10.1007/978-94-007-2181-4_29, © Springer Science+Business Media B.V. 2012

Keywords First generation biofuels • Second generation biofuels • Biofuel policy • Biofuel trade • Carbon finance • Carbon markets • Energy security • Energy service company (ESCO) • Food versus fuel • Green certificates

29.1 Introduction

The development of biofuel ventures offers significant opportunities for Africa to improve the low yield subsistence farming-dominated agricultural sector, mostly with regards to energy, environmental and socio-economic aspects. In particular, an increased investment in modern biofuel projects in Africa would result in a number of improvements, including reduced dependence on fossil-fuel based transport fuel imports, increased income generation and employment, infrastructure improvements in agriculture and optimisation of related policy frameworks.

Despite the evident potential of biofuels in Africa, these ventures have been facing complexities across the entire value chain, mostly related to policy development, land competition, supply chain, and above all access to financing.

This chapter presents an overview on financing of biofuel ventures in Africa, followed by approaches on how to overcome financing barriers. The final part of the chapter provides an example on how to improve financing of biofuel ventures in Africa.

29.2 Characteristics and Barriers of Biofuel Ventures in Africa

29.2.1 Characteristics of Biofuel Ventures in Africa

Every commercial venture requires close consideration of the underlying specific characteristics and its risks and opportunities. The specific nature of biofuel ventures is particularly complex, especially compared to other types of renewable energy ventures. This is due to the following four main aspects:

- *Reliance on feedstock*: The viability of biofuel ventures is strongly dependent on the cost effective, long-term supply of feedstock, an issue not applicable to other types of renewable energy ventures (such as wind, solar, hydro, marine, geothermal).
- *Food* versus *fuel debate*: The production of biofuels can have negative implications on the food situation in a country due to the argument that it could divert crops (e.g. corn, sugarcane), farmland and other resources (e.g. water, labour) away from food production, thereby potentially causing food price increases, food shortages, and an increased dependence on food imports.

- *Strong dependence on related policy framework*: Due to the lack of biofuel cost competitiveness and the market power of fossil fuel industries, combined with issues such as land and resource competition as well as food security, there is a strong dependence of the demand for biofuels on an effective policy framework, e.g. in the form of mandatory blending requirements.
- *Perceived risk profile of biofuel ventures*: Given the involved complexity, biofuel ventures are perceived as risky by the financing sector, which results in the application of high risk profiles, translating into stricter requirements with respect to collateral and/or lending terms for the provision of financing for bioenergy projects. This in turn exacerbates the difficulties for project developers to access external funding from the financial sector.

29.2.2 Barriers to the Development of Biofuel Ventures in Africa

The main barriers to the development of biofuel ventures in Africa are related to four main categories: (1) *policy* (i.e. policy frameworks), (2) *financing* (access to/supply of financing, the carbon markets, etc.), (3) *trade* (options affecting trade in the areas of infrastructure, policy and markets), and (4) *other non-technical issues/markets* (i.e. the link between the financing of biofuel ventures and any relevance to the development of related markets in Africa – taking into account demand, supply and carbon market-related aspects).

Barriers specific to the biofuel-related *policy* development in Africa are (1) competition of biofuels with other areas (e.g. security of food supply; land/resource competition; market/lobbying power of fossil fuel industry), (2) issues with a direct impact on investors (e.g. political/regulatory risks; unfavourable regulation regarding free movement of investments and/or profits in and out of countries), and (3) other issues such as the lack of awareness and/or understanding regarding the benefits of biofuels, unsuitable sites and insufficient management capacity.

The barriers with respect to *financing* are related to (1) access to/supply of financing (e.g. strict requirements set by the financing sector in terms of collateral/risk mitigation; lack of in-house expertise within the financing sector regarding biofuel venture evaluation), (2) the carbon markets (e.g. widespread inadequate on-the-ground carbon market-related capacity and expertise; high and even increasing transaction costs of carbon projects; financial/regulatory host country risks; underperformance due to faulty monitoring/delays in issuances), and (3) other areas (e.g. infrastructural issues; venture ownership issues).

Barriers with a negative impact on biofuel *trading* activities in Africa mainly concern the following three areas: (1) infrastructure/logistics-related barriers (e.g. bad land transport infrastructure; port inadequacies; insufficient availability of suitable vessels), (2) regulatory/policy barriers (e.g. strong dependency of international trade prospects of biofuels on European and American trade and agricultural policies; import tariffs in many potential target markets; export tariffs mostly levied on feedstock/raw materials), and (3) other barriers including uncompetitive production costs of biofuels in Africa and insufficient technology transfer and capacity building.

The barriers regarding the development of biofuel ventures in Africa associated to *other non-technical issues* are fourfold: (1) carbon market-related aspects (e.g. general uncertainty regarding the future of the carbon price and of global carbon markets as a whole; widespread scarcity of relevant skillsets and knowledge of carbon financing in Africa; lack of a supporting framework for carbon trading activities), (2) other aspects (e.g. insufficient collaboration between project developers and RD&D (Research, Development & Demonstration) activities; land ownership issues), (3) demand side aspects (e.g. generic difficulties of project developers to get access to financing), and (4) supply side aspects (e.g. issues affecting investor security; strict loan security requirements for project developers).

29.3 Strategies to Overcome Barriers of Biofuel Ventures in Africa

Overcoming barriers of biofuel ventures in Africa not only represents a promising means to improving aspects such as transport fuel-related energy security, income generation or the insufficient infrastructure, but most importantly, it presents opportunities for African countries to attract associated investment and technology flows. This section provides a number of strategies to the barriers identified in the previous section namely policy, financing, trade and development of related markets.

The potential strategies for biofuel-related *policy* developments in Africa cover the following two main aspects: (1) removal of regulatory risk through (direct) government policy support (e.g. establishment of legally binding blending requirements, creation of financial incentives for biofuels, including subsidies and grants; creation of a stable and reliable legal framework supportive of biofuels; support to advance biofuel-related RD&D (Research, Development & Demonstration) activities), **(2) setting up of pilot and demonstration projects, best practice dissemination support and governmental loan guarantee programmes**.

The importance of a well thought-out biofuel policy on the impact on related investment decisions can also be illustrated by its potential to support solutions that address one of the most important issues in the development of biofuels, the food versus fuel issue. A significant means of addressing this issue is the development of second generation biofuels. Second generation biofuels have the advantage of *not* using feedstock that could also enter either the animal or human food chain. Instead, non-food crops (or their residues) are utilised, thereby providing a solution to the food versus fuel debate. However, second generation biofuels are currently at the (policy susceptive) RD&D (Research, Development & Demonstration) stage, so any policies supporting their further development and potential market entry need to take into account the underlying risk profiles of both first and second generation biofuels. These are illustrated by Fig. 29.1 and demonstrate a shift from a current susceptibility to changes in *feedstock*, i.e. commodity prices (for first generation

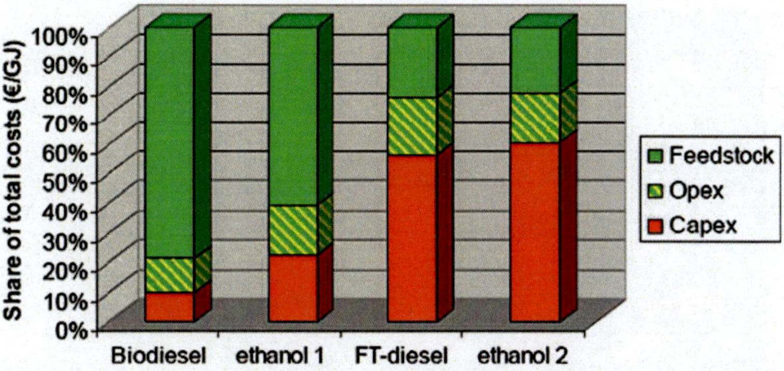

Fig. 29.1 Relative shares of feedstock costs, operational expenditures (Opex) and capital expenditures (Capex) in total biofuel cost price for different biofuels (Bindraban et al. 2009)

biofuels) to a susceptibility to changes in CAPEX, i.e. investment costs (for second generation biofuels). A bioenergy-related policy framework that attracts investment flows into these technologies will therefore also be beneficial for the development of second generation biofuels, thereby contributing to finding solutions for the food versus fuel issue.

The potential strategies which can contribute to facilitating the *financing* of biofuel projects in Africa are mainly aspects related to: (1) access to/supply of financing (e.g. creation of a comprehensive online and offline database of potential financiers/investors that are active in the biofuel markets in Africa, development of/provision of insurance products to mitigate certain types of biofuel-related risks; establishment of special purpose national/regional biofuel funds/investment facilities with financial institutions; provision of training to improve the financing sector's understanding regarding biofuels), **(2) the carbon markets** (e.g. shortening of the Designated National Authorities' (DNA) approval processes and increase of staff (availability) within DNAs; continued efforts to develop risk mitigating mechanisms for carbon financing; a strong supporting framework, including adequately skilled human resources and regime certainty, to promote the further development of carbon credit trading activities in Africa), and **(3) other areas** (e.g. creation of 'one-stop shops' for biofuel project developers and project investors, i.e. nationally/regionally centralised biofuel facilities in the form of both physical, on-the-ground, as well as online portals serving as central hubs of information exchange regarding biofuel aspects, other types of biofuel enterprise development support, including provision of standard documentation for biofuel market players, such as spreadsheet models, legal documentation, risk analysis tools/risk mitigation instruments or basic engineering support).

Potential solutions that can promote biofuel *trading* activities in Africa are mainly related to the following areas: (1) regulation/policy (e.g. subsidies/grants to build the necessary biofuel production infrastructure for biofuel feedstock production and fuel-excise taxes/tax credits/grants awarded for the production, sale and blending of biofuels; minimisation/removal of non-tariff barriers; lowering of unjustified technical standards of biofuels; enforcement of sustainability criteria/certification systems for biofuels; further trade liberalisation activities); **(2) infrastructure** (e.g. improvement

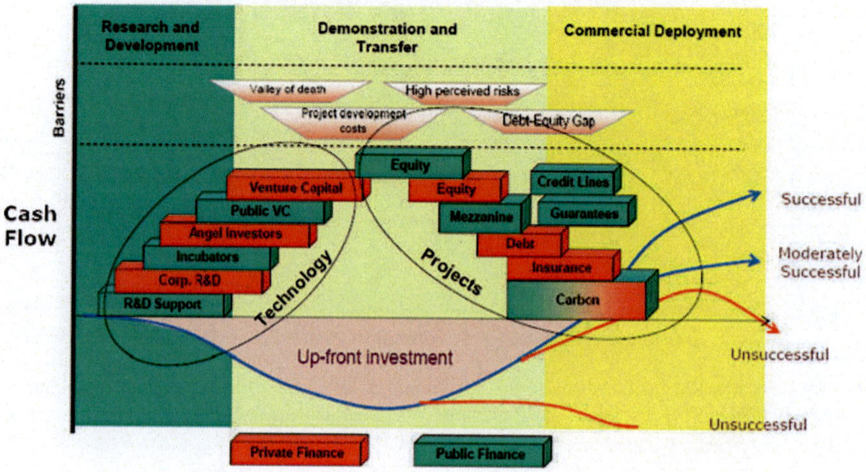

Fig. 29.2 Financing options at various clean energy project development stages (UNEP 2010)

and expansion of distribution and transportation infrastructure; strategic locating of biofuel-related infrastructure; improvement of port handling and maritime shipping capacities) and **(3) economy/markets** (e.g. increase in scale of production/factory size, thereby improving (cost) competitiveness of African biofuels; development of niche markets for biofuels, such as airline biofuel trial projects).

The potential strategies supporting the development of biofuel ventures in Africa related to the development of markets are fourfold: (1) carbon market-related aspects (e.g. carbon market capacity building activities/support facilities benefiting both the financing sector as well as the project development sector; increase of budget and experience of DNAs; support to kick-start carbon trading activities in Africa), **(2) other aspects** such as improved collaboration between biofuel-related research activities and project development, boosting infrastructure investment to enhance market access for biofuels and improvement of land ownership-related regulation, **(3) demand side aspects** (e.g. counter the frequent lack of general venture development skills prevalent across mostly small-scale biofuel entrepreneurs in Africa; ideally, these services would be offered free of charge as part of for instance a separate governmental small enterprise development agency), **(4) supply side aspects** (e.g. development of a market that bundles small biofuel projects, at the same time resulting in diversification and improved risk mitigation, thereby increasing small projects' ability to attract financing; establishment of development insurance on projects (pays out on failure), in turn facilitating project financing of biofuel projects; investor-supportive regulation with respect to reinvesting profits and the movement of investments in and out of countries; improving the suitability of financing instruments to the respective project development cycle in order to (1) maximise the effectiveness of public (grant/subsidy) funding made available for biofuel project development and (2) improve the pre-investment analysis capabilities of the private financial sector in that respect).

Figure 29.2 illustrates a number of financing options best suited to the respective requirements at various development stages of clean energy projects.

29.4 Financing Implications for Different Modes of Service Delivery

The nature or choice of business model for a biofuel enterprise, small or large, influences the funding requirements and the financing implications for the venture. Typical business models can be characterised in terms of an enterprise or venture which supplies *fuels* or provides *energy services*. The key differentiating characteristic of the business models is whether the venture is offering energy to customers in the form of an energy carrier such as biodiesel, gelfuel or bioethanol or an energy service such as cooked meals, heated water or lighting.

The former model focuses on the supply chain up to the sale of the energy carrier (the biofuel itself), whereas in the latter model, an energy service company (ESCO) encapsulates the whole supply chain for the energy service including distribution and energy conversion. Figure 29.3 illustrates the example of different energy service supply chains for the service of 'cooked meals'. The diagram illustrates the conventional point of sale for energy supply ventures, namely for the energy carrier.

The key difference in the financing requirements of these two broad categories of biofuel ventures is that the former addresses individual energy production facilities on the basis of financing specific production equipment and systems, whereas the latter is focused on an overall service business which is not dependent or based on any particular plant or equipment but rather on the overall operations required to provide the energy service.

The ESCO mode of service delivery is more suited to institutional customers (such as hospitals, schools, canteens) whereas the fuel supply mode of biofuel ventures would be applicable to any customer for the fuel. Examples of ESCO ventures for individual households include the off-grid concession-based electricity utilities in South Africa which provide light (and television) by means of utility-owned standalone solar home systems (SHS) (Banks 2007).

29.4.1 Financing Energy Supplier Ventures

The predominant experience with business models for biofuel operations in Sub-Saharan Africa has been with the production and sale of fuels from individual plants. In these businesses the focus for financing has been on project development and project finance of individual centralised biofuel infrastructure investments such as a biodiesel production plant, a bioethanol plant or a biogas from waste water treatment production facility. In the case of small producers the infrastructure investment may be a single multi-functional platform, a small biodiesel plant or even a household biogas digester. The funding and finance has essentially been based on an asset finance approach for fixed or moveable assets. Typical financing mechanisms would include: capital subsidies, subsidised loans, lease agreements or hire-purchase schemes.

In the case of biofuel ventures for less developed economies the key challenge for asset-based finance is the difficulty of securing collateral to manage risk on

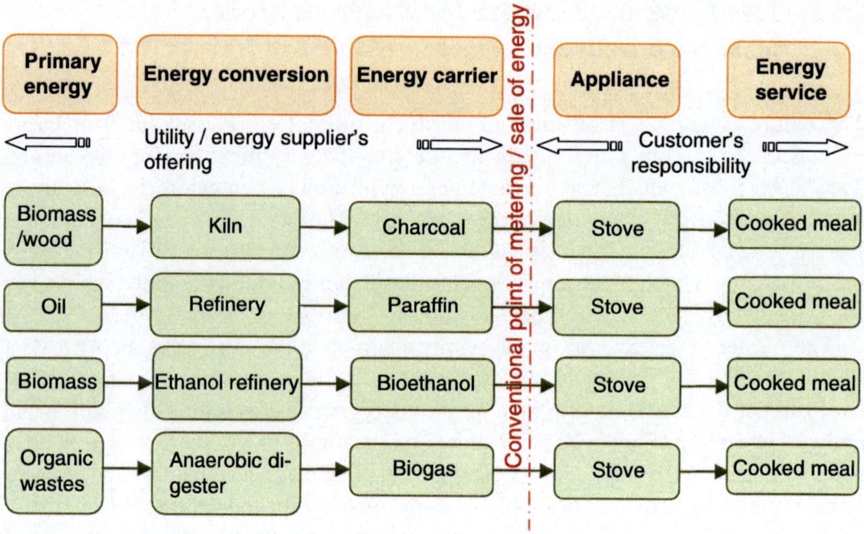

Fig. 29.3 Typical energy service supply chains for cooked meals

individual finance agreements. Sufficient scales are achieved by aggregating several finance agreements.

29.4.2 Financing Energy Service Businesses

In the case of the energy services business model, the service provider contracts with customers for a specific energy service such as litres of hot water, hours of cooking/baking per day or hours of lighting per day. Clearly there could be cases where the service provider could offer a combination of energy services.

In the case of an energy services business model the financing is required within the context of a services business. The benefits of the services approach is that the business model includes more opportunities for capturing efficiencies in the supply chain and retaining the financial benefits of these savings in terms of increased margins. The disadvantage of this approach is that it is a significantly more complex business operation including contracting (service agreements), logistics for distribution and also more complex customer service interactions.

The finance options for energy service businesses require a careful assessment of the overall business model and the projected market for the energy services. Whilst the assessment of the business model is likely to be more complex than in the case of an asset-based finance agreement, the scope of achieving scale and efficiencies is larger, offering opportunities of greater efficiency in the financing, too. Furthermore, ESCO operations are eligible for operational subsidies, in addition to asset finance subsidies, and tax or rates rebates on the basis of the public benefits of the service which is provided. The scale-driven opportunities for accessing carbon-based funding and finance are an additional advantage of this approach. These options provide additional opportunities for optimising the financing of the overall venture.

29.4.3 Financing Distributed Energy Infrastructure

A further challenge in the financing of biofuel ventures arises from the fact that the energy infrastructure may comprise of many small-scale systems which may be located at the point of use distributed over a wide geographical area. A typical example is a biogas digester programme where the energy infrastructure assets (the digesters and cookers) are located close to (and inside) households.

As outlined above, the mode of implementation of the programme influences the financing options for the programme. If the systems are treated as assets to be owned by the associated homeowner, the financing options are the typical asset-based finance options such as micro-lending schemes. The experience to date is that this kind of finance requires specific experience on the part of the financier which is not generally available. If the programme were implemented as an energy services programme with an ESCO owning and financing the entire infrastructure (e.g. digesters and cookers) as part of the services offering, the financing requirements are immediately more attractive in terms of scale and risk mitigation. The range of financiers of this type of venture is also wider and more generally available. In this case, complexity issues are contained and managed within the ESCO and the financing requirements become more interesting to infrastructure investment financiers.

Clearly, the revenue requirement for funding and financing the capital and operational costs of the energy service provision are a function of the scale, risks and experience of the venture. These are assessed on the basis of existing criteria and due diligence procedures which have been developed and applied over many years to this type of venture.

29.4.4 Accessing CDM and Green Certificate Funding

The different modes of implementation of biofuel ventures also influence the opportunities for access to CDM and schemes for Renewable Energy Certificates. The larger opportunity for scale in the ESCO approach increases the likelihood of the venture being financially viable due to the dilution of the overheads for registration.

In the case of registration of a biofuel venture as a carbon project the options include either a CDM project registration under the UNFCCC, most likely in terms of the approved small-scale methodologies,[1] or as a voluntary carbon project. In both cases the project could secure better terms for the carbon emission reductions

[1] http://cdm.unfccc.int/methodologies/SSCmethodologies/approved.html

(CERs or VERs) under the Gold Standard programme for carbon projects with developmental achievements.[2] The process for registration of a CDM project starts with an application to the Designated National Authority (DNA) in the respective country. The costs of registration of the venture as a carbon project will favour larger projects in terms of the overall financial viability.

In the case of green certificates, the venture could qualify in terms of a voluntary market for RECs under a Domain Protocol and Issuing Body in the country for implementation of the venture. In practice and at the time of writing, these have only been established in South Africa.

29.5 Conclusions

In order to realise the significant energy, environmental and socio-economic opportunities that the development of sustainable biofuel ventures offer for Africa, there is a need to take into account the underlying specific characteristics of these types of ventures as well as to overcome the various types of barriers in the African context. The development and implementation of key strategies to improve financing of sustainable biofuel ventures in Africa involves the four interrelated areas of policies, financing, trading and other non-technical issues and is therefore a complex and multi-faceted undertaking. However, the potential rewards include a considerable improvement of a range of issues, including the food versus fuel debate, an insufficient related policy framework and access to financing.

Whilst the strategies in the above-mentioned four main areas cover a broader context of how to improve the financing of biofuel ventures in Africa, there is a need to also consider the implications of specific types of biofuel ventures, i.e. in the context of the supply of fuels or the provision of energy services, regarding the respective financing options. For the supply of fuels, financing requirement include project development and project finance of individual centralised biofuel infrastructure investments. Finance options for the much more complex energy service businesses require a careful assessment of the overall business model and the projected market for the energy services including the operational costs of the energy service provision.

In conclusion, through the implementation of strategies to improve the financing of sustainable biofuel ventures, African countries can attract associated investment and technology flows, thereby not only improving aspects such as energy security, the insufficient infrastructure and fossil-fuel related emission reductions, but most importantly, generating income and employment, thus contributing to poverty eradication.

[2] http://www.cdmgoldstandard.org/

References

Banks DI (2007) Perspectives on providing rural energy services in rural areas. In: Proceedings of hybrid electricity systems powering mini-grids: a Southern African perspective, Desert Research Foundation of Namibia (DRFN), Namibia

Bindraban P et al (2009) Can biofuels be sustainable by 2020? An assessment for an obligatory blending target of 10% in the Netherlands. Available at: http://np-net.pbworks.com/f/Bindraban+et+al+(2009)+Can+biofuels+be+sustainable+by+2020,+Scientific+and+policy+analysis.pdf

UNEP (2010) Scaling up climate financing. Available at: http://www.carbonexpo.de/global/dokumente/carbon_expo/en/conference_programm/270510_Finance_1045_Hodes.pdf

Chapter 30
Economics of Modern and Traditional Bioenergy in African Households: Consumer Choices for Cook Stoves

Francis X. Johnson and Takeshi Takama

Abstract The overwhelming majority of African households use traditional biomass in the form of wood-fuel or charcoal to meet their daily cooking needs. Modern options such as LPG or ethanol can provide considerable benefits for health and environment. The case of ethanol is interesting as a renewable source with lower GHG emissions and also having the potential to be a locally produced resource. The purchase cost of such stoves is considerably higher whilst the fuel costs will generally be lower. Previous research on household adoption of new cook stoves has tended to focus on demographic or socio-economic factors such as education and income in trying to explain consumer choice. Such variables change only slowly and thus generally cannot support rapid introduction of improved stoves. A discrete choice model was developed aimed at focusing more on the characteristics of the cook stoves themselves and the way in which they are used, which are referred to as "product-specific" attributes. The methodology is outlined here followed by a brief summary of the model applications in three countries: Ethiopia, Mozambique and Tanzania. This approach could improve the understanding of the underlying economic issues and thereby contribute to better design of cook stove programmes and help stimulate a market transformation towards cleaner and more efficient cook stoves.

Keywords Cook stoves • Economics • Ethanol • Traditional biomass • Consumer choice • Wood-fuels • Discrete choice • Socio-economics • Energy access • Fuel costs • Purchase price • Charcoal • Africa • Safety • Smoke

F.X. Johnson (✉)
Stockholm Environment Institute, Kräftriket 2B, Stockholm SE-106 91, Sweden
e-mail: francis.johnson@sei.se

T. Takama
Japanese International Development Agency, Jl. Angkasa I No. 2, Kemayoran, Jakarta Pusat 10720, Indonesia
e-mail: takeshi.takama@sei.se

R. Janssen and D. Rutz (eds.), *Bioenergy for Sustainable Development in Africa*,
DOI 10.1007/978-94-007-2181-4_30,

30.1 Introduction

Nearly half of the world's population continues to rely on traditional biomass for some or all of their energy needs, with about half of that group having no access to electricity. In Sub-Saharan Africa the share of people relying on traditional biomass is greater than 80% and nearly as many lack electricity access. The WHO estimates that traditional biomass will continue to lead to over 1.5 million premature deaths per year in 2030, greater than the estimates for premature deaths from malaria, tuberculosis or HIV/AIDS (WHO 2008).

Clean cooking options can include the introduction of clean-burning efficient stoves fired with wood chips, renewable fuels such as bioethanol and non-renewable fuels such as LPG. Whilst the annual fuel costs of advanced stoves are lower than those incurred by traditional biomass users, the purchase costs can be considerably higher, and thus most households do not have sufficient disposable income nor access to credit that would allow them to pay the additional costs upfront even with short payback time of 1 or 2 years (Schlag and Zuzarte 2008).

In this chapter the characteristics of consumer choice for cook stoves and the associated fuels in African households are described. In particular, the economic values associated with the tradeoffs that households make when they purchase a cook stove are assessed. The combination of a household survey approach with an econometric modelling technique makes it possible to conduct an experiment on the willingness of households to pay more for improved stoves and/or fuels. A methodology for conducting this type of household energy economic analysis (HHEA) was designed and applied in three case studies in Ethiopia, Mozambique and Tanzania (Takama et al. 2011). A brief summary of this study is given here along with the relevant thematic and policy context.

30.2 Household Energy in Africa and its Impacts

Improving the quality of energy supply in Africa is central to sustainable development and poverty reduction efforts. As of 2009, an estimated 653 million people in Sub-Saharan Africa depended on traditional biomass to meet their daily cooking needs and this number is expected to increase to 918 million by 2030 under current trends (IEA-WEO 2010). Although this represents a decrease in percentage terms from 80% to 70% due to high population growth, the movement of 265 million people into a condition of energy poverty presents serious economic, environmental, health and ethical concerns. Access to modern energy sources will not only improve quality of life and create income-generating opportunities, but reduce the serious health impacts and environmental degradation often associated with the inefficient use of biomass (Smith et al. 2004).

The transition away from biomass to modern fuels has often been termed as the "energy transition" or is associated with climbing the "energy ladder" towards modern energy forms (Leach 1992; Chapter 1 - Smeets et al. this book). The Stockholm Environment Institute (SEI) and its predecessor, the Beijer Institute, conducted many studies on stove use, household energy, health and environment in various African countries (Ellegård 1997; Kaoma et al. 1994; Lopes and Ellegård 1990; Chiwele et al. 1994). More recent studies on household energy have looked in detail at household interventions to identify the key characteristics associated with health improvements (Bailis et al. 2007).

In rural areas, 85–95% of the population depend on firewood as the primary source of energy whereas in urban and peri-urban areas charcoal is more likely to be the fuel of choice (UNDP 2009). The traditional three-stone stove is widely used in SSA, and has a relatively low efficiency of about 10–14%. Because of its advantages over firewood, there have been a number of efforts to promote charcoal in household cooking. Charcoal is often used in traditional metal stoves, which have slightly higher end-use efficiencies of 15–18% (Malimbwi and Zahabu 2007). However, due to the loss of energy during charcoal production about 5 times the amount of wood is needed when cooking on charcoal as compared to firewood when assuming a charcoal kiln efficiency of 15% (Smeets et al. this book).

Higher efficiency stoves that use wood or charcoal are useful to consumers in that they will generally result in cost savings when fuel is purchased, as well as health and environmental benefits due to less smoke and less demand for wood from forests. However, the benefits for climate may be minor in such cases, since these stoves will still emit considerable amounts of black carbon, which is now recognised as a significant contributor to climate forcing. This is particularly true in regions near mountainous areas where they deposit on snow and reduce reflectivity. Furthermore, efficient stoves have additional climate forcing impacts through the decrease of white particles (due to lower smouldering) that have cooling effects (Ramanathan and Carmichael 2008). Since electricity is rarely available to rural households in Africa, fuel-switching can offer options that address climate impacts as well as improving health and environmental effects.

30.3 Economics of Cook Stoves and Fuel Choice

The economic implications of household energy from a consumer perspective are perhaps less understood than the health and environmental impacts, since detailed assessment at household level is required, which can be time-consuming, complicated, and beset with uncertainties. Knowledge of the determinants of fuel choice at household level therefore remains somewhat elusive (Heltberg 2004; Kohlin and Gupta 2006). The so-called "energy ladder" model implicitly assumes that residential consumers use the most sophisticated fuels they can afford. Wood is thus the "energy of the poor" and other factors are thought to have little effect on fuel selection. The energy ladder model does not really consider the often important role of consumer choice in fuel and stove selection, including the influence of structural factors such as initial purchase price or stove performance (Van der Horst and Hovorka 2008).

Research on the determinants of stove choice has tended to focus on socio-economic factors such as income, age, gender and education, whereas product-specific factors such as safety, indoor smoke, usage cost and stove price have largely been ignored (Takama et al. 2008). Previous research has neglected to include quantitative estimates of such product-specific factors due to the lack of a reliable and consistent methodology. The research described here involved the design and application of a methodology that included both socio-economic and product-specific determinants of fuel choice. The quantitative analysis is complemented with qualitative assessments based on focus groups, interviews and other social science methods (Takama et al. 2011).

The primary objective of the research was therefore to evaluate the determinants of household fuel and stove choice in selected case study regions in Sub-Saharan Africa in order to improve the understanding of the household market for improved cooking fuels and stoves. Such improved scientific understanding can then be used to better inform policy makers, programme developers and product designers in supporting more efficient, modern and clean cooking stove initiatives. In each case study, the research methodology includes a stated preference based survey of 100–300 households and an econometric modelling analysis based on discrete (qualitative) choice theory (McFadden 1974).

30.4 Model Design

Choice modelling offers a quantitative/statistical approach to understanding the behaviour of consumers based on microeconomic theory in combination with empirical data about the attributes of products (in this case, the combination of stoves and fuels) and the various alternatives. In theory there may be many alternatives when buying a cook stove, although in practice only a few options may be easily available. The approach is flexible enough to consider many options or only a few, depending on the nature of the problem at hand.

In the simplest case, there are only two alternatives, for example a specific type of firewood stove and a specific type of ethanol stove, and two attributes, such as the price of stoves and level of smoke. Those concerned with both price and smokiness (due to health impacts) face trade-off between attitudes (i.e. if an ethanol stove is very clean, high price may be accepted and *vice versa*). The preference between the two attributes for a stove might be depicted as in Fig. 30.1. It depicts so-called indifference curves, which describe sets of stove characteristics such that the consumer has no preference between them (i.e. is indifferent amongst the options along that curve). Furthermore, the consumer does not care about the fuel itself, but about the *attributes*, which in this case are smokiness and price. It is also important to note that for technical reasons, fuels must analysed in combination with the stoves, since some stoves cannot accommodate different fuels and furthermore the performance and hence the attributes would change.

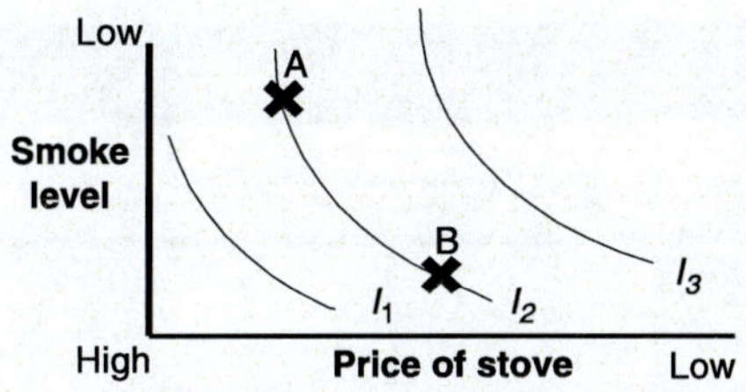

Fig. 30.1 Indifference curves of cooking stove choice. *Stove A* and *Stove B* on the I_2 curve give the equal satisfaction, i.e. the consumer is indifferent between the two options. Moreover, any stove on the I_3 is better than any stove on I_2 for the consumer because those stoves have less smoke AND lower price (Source: Takama et al. 2011)

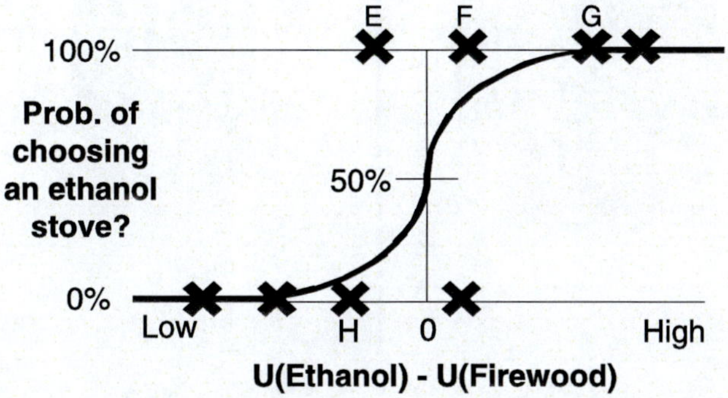

Fig. 30.2 Stochastic (discrete choice) model for ethanol vs. firewood. The model, as depicted by the curve, facilitates the inclusion of a random component so as to describe the actual choices (x's) in an internally consistent manner in accordance with the theory (Source: Takama et al. 2011)

Conventional consumer choice theory states that if the utility of a good is greater than that of another good, a consumer will choose the former good. Thus, if buying an ethanol stove brings higher utility than buying a firewood stove (i.e. U(Ethanol) > U(Firewood) > 0), a consumer will buy an ethanol stove. *Discrete Choice Analysis* (DCA) combines the multi-attribute utility of microeconomic consumer choice theory with the statistical random utility maximisation model. The random component means that there will be some probability distribution associated with the choice between the two stoves, as depicted in Fig. 30.2. Policy makers and stove developers will be more interested in the macro level outcomes such as the market share and price elasticity of a stove than probability of buying a particular stove. The DCA will answer these questions as the probability of a stove choice is the same as the market share of the stove in *probability theory* and the slope of the "S" curve is its price elasticity.

A Stated Preference (SP) survey technique was used to provide the data for the DCA model. In an SP survey, an analyst asks people to choose alternatives with

Fig. 30.3 Example of stated preference survey card used in Ethiopia case study. The choice experiment in the stated preference survey is such that each respondent views the card and chooses between the two options based on the specified attributes. Each choice requires a separate card so that pair wise comparisons between the stove/fuel options are made (Source: Takama et al. 2011)

stated attributes in a hypothetical situation using questionnaires, cards, telephone interviews, web, etc. For example, in Ethiopia, there were options for ethanol, kerosene and firewood stoves, and the respondents are asked to choose between such options. Figure 30.3 shows an example pair wise comparison between a firewood

Table 30.1 Summary of product specific and socio-economic attributes

	Product specific attributes	Socio-economic attributes
Specific to	Product	Person
Characteristics	Universal in nature	Specific to context
Variation in choice	Within individuals	Between individuals or groups
Change in short-term	Relatively easy	Difficult
Useful for	Product design, demand forecast, policy formulation	Market segmentation/profiling and policy formulation

Source: Takama et al. (2011)

stove and a kerosene stove. If the consumer is willing to pay a greater upfront cost (purchase price) or is very concerned about the monthly cost for fuel, he/she might choose the kerosene stove. Otherwise, it is better to choose a firewood stove.

It was decided to focus on the product-specific aspects in the choice experiment. As summarised in Table 30.1, the product-specific and socio-economic attributes have different characteristics and roles in the promotion of modern energy efficient cooking stoves. Therefore, an innovative approach and method is required to categorise and study each. Furthermore, the quantification of these attributes is necessary to provide guidance for practical project implementation.

30.5 Application of the Model

In this section, the basic context and application of the model for the three case studies are summarised. A longer discussion is found in the detailed research report (Takama et al. 2011). The basic structure of the model was the same across the three case studies, although the parameters and choices differed slightly (for example, kerosene was not an option in Mozambique). The general structure is given and discussed in the Annex to this chapter.

30.5.1 Ethiopia

Alternatives for the cooking stove choice experiment were evaluated based on four criteria: (1) number of users; (2) distribution across different income groups; (3) usage level; and (4) relevance to the study objectives. In Ethiopia, essentially every household uses charcoal for the traditional ‘coffee ceremony’ and therefore when assessing fuel-switching, charcoal is less appropriate since no one would actually switch away from it. Thus, on the basis of the above criteria and discussion, kerosene and wood were selected as the relevant cooking fuel alternatives along with ethanol. One reason for including ethanol in such an experiment is that it is a relatively new fuel and therefore experimental techniques are useful to understand market behaviour.

Some special characterisation was needed for ethanol and kerosene in Ethiopia in terms of risk, because of recent negative experience. In 2005, a local company produced kerosene and ethanol blend as cooking fuel and termed it as K-50. It developed a reputation for being explosively volatile due to a poor blend and low quality stoves. Now, many respondents associate ethanol with explosions and consider it unsafe (Kassa 2009). Both ethanol and kerosene stoves were consequently viewed as risky. The perceptions created by the experience clearly were having strong influence on consumer preferences, thus illustrating some of the intangible factors related to the specific context and circumstances beyond just the attributes themselves and the associated tradeoffs.

30.5.2 *Tanzania*

In Dar es Salaam, charcoal is by far the most commonly used fuel for cooking, with approximately 75% of population using it as the primary cooking fuel. In a discrete choice model, there will be a heavy bias when one choice is so dominant. Consequently, charcoal was not included as an alternative and furthermore, its exclusion made the case study consistent with the Ethiopian case. Having excluded charcoal, firewood was selected instead. Firewood is becoming a less common cooking fuel in Dar es Salaam and other urban areas, but is still widely used in rural locations. Thus, it was chosen to represent a traditional fuel with many similar properties to charcoal. In addition to firewood, kerosene was selected as an important fuel (second after charcoal in Dar es Salaam), and ethanol, as a modern but renewable fuel.

Ranges of prices for stoves and monthly costs were obtained from internet research and confirmed by a field survey in Dar es Salaam. In comparison to the Ethiopian study, stove prices included four levels instead of three, and indoor smoke and safety risks also featured four levels, both uniformly on a 0–3 scale.

30.5.3 *Mozambique*

The stated preference-based survey was administered in the municipal district of Catembe, an extension to the city of Maputo, with a distinct blend of urban, peri-urban and rural characteristics. One difference was the choice of alternatives, as charcoal replaced kerosene. A focus group discussion amongst household cooks in Maputo served to confirm the choice of alternatives and levels of attributes chosen (Takama et al. 2011).

30.6 Results

The model was evaluated using the BIOGEME software, which provides parameter estimates and full statistical analysis. The estimates and the *P*-values for statistical significance are given in Table 30.2. The table shows that two constants and the five parameters are significant at 5% levels in the three cases. A similar analysis was

Table 30.2 General parameters derived from the BIOGEME model

		Ethiopia		Tanzania		Mozambique	
Parameter	Description	Coeff.	P-value	Coeff.	P-value	Coeff.	P-value
α^{E}	ASC for Ethanol	.415	.05	.666	<.01	.427	<.01
$\alpha^{K/C}$	ASC for Kerosene or Charcoal	–.392	.04	–	–	.558	<.01
α^{W}	ASC for Firewood	Fixed	–	Fixed	–	Fixed	–
β^{cost}	Usage Cost	–.029	<.01	–.024	<.01	–.009	<.01
β^{price}	Stove Price	–.024	<.01	–.022	<.01	–.008	<.01
β^{risk}	Explosion risk	–.272	<.01	–.296	<.01	–.357	<.01
β^{risk_b}	Burn risk	–.256	<.01	–.298	<.01	–.118	<.01
β^{smoke}	Indoor Smoke	–.101	<.01	–.214	<.01	–	–

Source: Takama et al. (2011)

Table 30.3 Trade-off between attributes (MWTP) based on stove price unit

		β^{cost}	β^{risk}	β^{risk_b}	β^{smoke}
Ethiopia	General	1.208	11.333	10.667	4.208
	Low	0.956	5.244	6.622	[−1.511]
	Middle	1.167	11.333	8.433	5.167
	High	[1.875]	[42.000]	[38.750]	[40.250]
Tanzania	General	1.091	13.455	13.545	9.727
	Low	1.160	9.760	7.640	6.400
	Middle	0.920	13.800	14.840	10.960
	High	1.385	23.308	28.385	14.615
Mozambique	General	1.125	44.625	14.750	
	Low	1.143	48.000	12.400	
	Middle	1.000	27.222	14.667	
	High	1.250	64.375	19.625	

Source: Takama et al. (2011)

done for each of three income strata: low, medium and high (see full report for details: Takama et al. 2011). The constants in some of the stratified cases were insignificant, which suggests that they do not explain well the residual preferences within those income strata.

The trade-off between attributes can be normalised as the ratio of an individual attribute coefficient and the stove price coefficient (Table 30.3). In this way, it is possible to see how consumers relate the other attributes to the stove price, which is often a crucial factor in adoption/purchase. The trade-off can be interpreted as the Marginal Willingness to Pay (MWTP) for the other attributes, i.e. for one more unit of the other attribute relative to the stove price. The amounts were normalised to US$ for comparison purposes, although it should be noted that differences in purchasing power across countries and regions makes comparisons across countries in this manner quite difficult. The values in square brackets are insignificant values, as either a subject coefficient or a stove price coefficient was determined to be not statistically significant.

There is a general trend that is not especially surprising, that the willingness to pay for non-monetary attributes such as safety seems to increase with income. In Ethiopia,

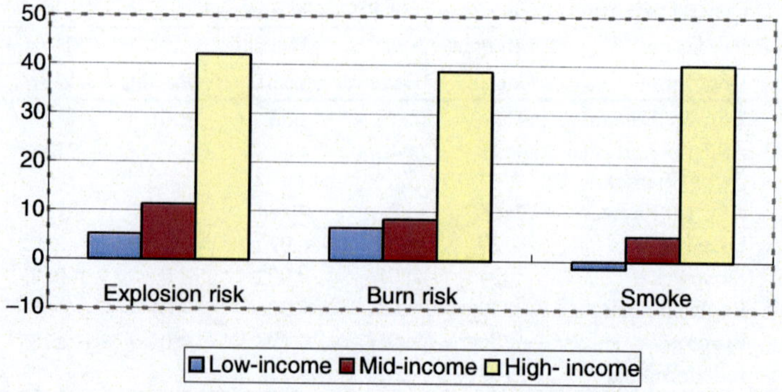

Fig. 30.4 MWTP of soft attributes between income groups in Ethiopia (Source: Takama et al. 2011)

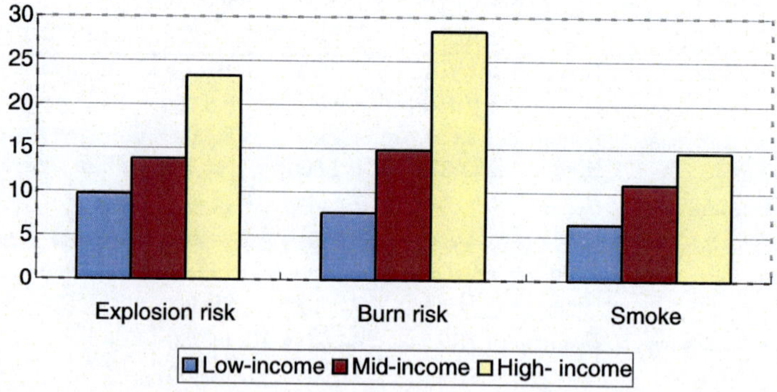

Fig. 30.5 MWTP of soft attributes between income groups in Tanzania (Source: Takama et al. 2011)

low-income households want to pay only 0.956 US$, compared to 1.167 US$ and 1.875 US$ for the middle and high-income groups, respectively. It reflects the high discount rate (preference for near-term benefits rather than longer-term) and the high capital cost constraint (i.e. difficulty to pay upfront) for lower-income consumers.

The difference in the trade-offs for the non-monetary attributes across the socio-economic classes is quite striking in some cases (Figs. 30.4 and 30.5). In Ethiopia, the high-income group was willing to pay nearly ten times as much for reducing risks (or smoke) compared to low or medium-income consumers. The negative value for smoke for the low-income group technically means that they are willing to pay for more smoke! In fact, in some cases smoke is seen as a way to keep away mosquitoes. However, the very small value and the counter-intuitive nature of such interpretation means simply that low-income consumers are either not aware or simply cannot afford to be concerned with the effects of smoke. In Tanzania, the low-income group is willing to pay only 6.4 US$ for a unit reduction in the smoke level, whilst the wealthier group is willing to pay more than two times that amount (14.615 US$).

30.7 Discussion

Previous research on cook stove and fuel choice has tended to focus on demographic or socio-economic factors, which do not change significantly in the short-term. In some respects, this is because poor consumers are viewed as not really making choices. As a result there has been a tendency in cook stove programmes to provide some stoves for free in some regions and then expect consumers to adopt these advanced stoves due to their advantages. Whilst this might be logical if consumers were perfectly rational and equipped with perfect knowledge about the stoves and fuels, such situations never occur. Two polarised perceptions of households in developing countries emerge: either they will rationally adopt the "best" option when presented with it, or they are incapable of understanding the options and therefore only some type of forced transformation can impact the transition away from traditional biomass.

The approach is aimed at neither of these polarised cases. Rather, the idea is to aim at a better understanding of the poor household as a consumer so as to determine how this market might be transformed in a way that respects their preferences, but at the same time might take advantage of the willingness to pay of some groups or some segments for attributes that have public as well as private benefits. In order to accomplish such analysis, both monetised and non-monetary attributes (e.g. safety, smoke) must be included. The socio-economic and product-specific attributes must be given at least equal weight in the analysis and probably much greater weight to the product-specific attributes given that they have received less attention. The four types of attributes are depicted in Fig. 30.6. Previous research has focused mainly on Quadrants 1 and 3 (Q-1 and Q-3) whereas this study has identified the importance of Q-2 and Q-4 by formally evaluating the trade-offs amongst such attributes.

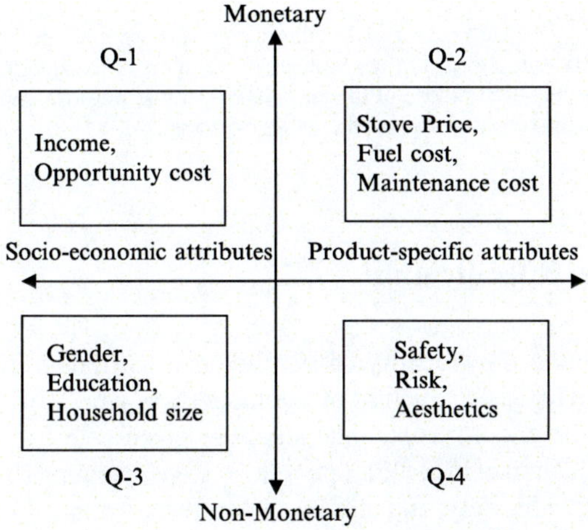

Fig. 30.6 Four categories of clean cook stove determinants with examples (Source: Takama et al. 2011)

There has been a tendency to look for perfect solutions to the problem of improving energy access through better cook stoves. This tendency is partly related to the fact that research has focused on socio-economic variables and has not looked in much detail at the product-specific variables affecting consumer choice. By looking at the household energy options in a more formal way and placing the consumer in the centre, the designers of stove programmes and projects as well as policy makers and donors could be better informed about the opportunities for transforming these markets. In such way, the transformation could produce, at lower costs, the public and private benefits that seem to be clearly available.

30.8 Conclusions

The research summarised here involved the development and application of a household energy economic analysis model that was based on a stated preference survey and a discrete choice analysis. The model applications focused on clean cooking options and household energy demand in Ethiopia, Tanzania, and Mozambique. The model and its application reveal that there is some willingness to pay for health and environmental improvements, mainly at medium and high income levels, but even in some cases at low-income levels. The most significant findings in this research was the importance of distinguishing between socio-economic and product-specific attributes as the determinants of cook stove and fuel choice, with the former responsible for variation in choice between individuals or groups, and the latter for variation within individuals or groups. The model can support the design of cook stove programmes and energy access policies.

Acknowledgments The study presented in this chapter has been made possible through financial support provided by the Swedish International Development Cooperation Agency (SIDA). However, the views are strictly those of the authors. SIDA was not involved in the design of the analysis and does not necessarily support the views expressed.

Annex: Model Specification

The model is developed by identifying key factors or attributes related to the stoves and/or their use. The basic structure of the model was similar in the three applications considered here, but the example structure given below is for the Ethiopian case. The factors identified in the Ethiopian study were operating (fuel) cost, purchase price, smokiness, and safety, and similar factors were chosen in the Tanzania and Mozambique case studies.

Utility in the model is defined by using the non-stochastic variable V to represent the observed Utility, plus an error term (i.e. $U = V + \varepsilon$). The following utility functions were thus specified for the three stove/fuel alternatives:

Ethanol:	$V_i^E = \alpha^E + \beta^{cost}(cost\_e) + \beta^{price}(price\_e) + \beta^{smoke}(smoke\_e) + \beta^{safety}(safety\_e)$
Kerosene:	$V_i^K = \alpha^K + \beta^{cost}(cost\_k) + \beta^{price}(price\_k) + \beta^{smoke}(smoke\_k) + \beta^{safety}(safety\_k)$
Wood:	$V_i^W = \alpha^W + \beta^{cost}(cost\_w) + \beta^{price}(price\_w) + \beta^{smoke}(smoke\_w) + \beta^{safety\_w}(safety\_w)$

Each variable name represents a relevant factor and its suffix represents one of the three alternative fuels that were available to be used for the stoves in this application. A β represents a weighting parameter of a relevant attribute. An α represents a relevant alternative specific constant (ASC), with the ASC of fire wood fixed as zero. The ASC provides a simple way of including other effects that are not captured in the specified variables and parameters. All alternatives use the same parameters except for the safety of firewood, β^{safety_w} because the safety issues are different between liquid and solid fuels (i.e. firewood does not have an explosion risk). The form of the fitted model for choosing the ethanol stove is therefore:

$$P(\text{Ethanol}) = \frac{\exp(U_i^E)}{\exp(U_i^E) + \exp(U_i^K) + \exp(U_i^W)}$$

The other logistic forms for the other two options are similar to the one above. U_i^* is the unobserved utility for a given fuel choice, which is ethanol in the specification above. P (Ethanol) is the probability of choosing an ethanol stove. The model is evaluated using data gathered from the Stated Preference surveys, in which households indicate their willingness to pay and their willingness to accept the different levels of the other characteristics (smoke, safety).

References

Bailis R et al (2007) Performance testing for monitoring improved biomass stove interventions: experiences of the household energy and health project. Energy Sustain Dev 11(2):57–70

Chiwele PL, Muchelemba, Siamwiza RJ, Ellegård A (1994) Household energy and nutrition: a study of low-income households in Lusaka, Energy, environment and development series no. 37. Stockholm Environment Institute, Stockholm

Ellegård A (1997) Household energy, air pollution and health in Maputo, Energy, environment and development series no. 42. Stockholm Environment Institute, Stockholm

Heltberg R (2004) Fuel switching evidence from eight developing countries. Energy Econ 26(5):869–887

IEA (2010) World energy outlook 2010. International Energy Agency: Paris

Kaoma J, Kasali GB, Ellegård A (1994) Efficiency and emissions of coal combustion in two unvented cookstoves, Energy, environment and development series no. 40. Stockholm Environment Institute, Stockholm

Kassa M (2009) Ethanol Cooking Fuel and Stove Market Development Project Highlights, Summary Document prepared for Federal Environmental Protection Authority, Addis Ababa
Kohlin G, Gupta G (2006) Preferences for domestic fuel: analysis with socio-economic factors and rankings in Kolkata, India. Ecol Econ 57(1):107–121
Leach G (1992) The energy transition. Energy Policy 20(2):116–123
Lopes J, Ellegård A (1990) Quick and dirty: the Maputo stove project 1985–89, Energy, environment and development series no. 1. Stockholm Environment Institute, Stockholm
Malimbwi, Zahabu E (2007) Situation analysis of charcoal sector in Dar es Sa-laam; charcoal supply and consumption. Department of Forest Mensuration and Management, Faculty of Forestry and Nature Conservation, Sokoine University of Agriculture, Morogoro
McFadden D (1974) Conditional logit analysis of qualitative choice behaviour. In: Zarembka P (ed) Frontiers in econometrics. New York: Academic Press
Ramanathan V, Carmichael G (2008) Global and regional climate changes due to black carbon. Nat Geosci 1(221). doi:10.1038/ngeo156
Schlag N, Zuzarte F (2008) Market barriers to clean cooking fuels in sub-Saharan Africa: a review of literature, SEI working paper 8. Stockholm Environment Institute, Stockholm
Smith KR, Mehta S, Feuz M (2004) Indoor air pollution from household use of solid fuels. In: Ezzati M, Lopez AD, Rodgers A, Murray CJL (eds.) Comparative quantification of health risks: global and regional burden of disease attributable to selected major risk factors. WHO, Geneva
Takama T, Tsephel S, Zuzarte F, Debebe M, Lambe F, Johnson FX (2008) An approach to distinguishing socio-economic product-specific factors to understand determinants of sustainable clean bioenergy stove. Eighth international conference on ecobalance, Tokyo, 10–12 Dec 2008
Takama T, Lambe F, Johnson FX, Arvidson A, Atanassov B, Debebe M, Nilsson L, Tella P, Tsephel S (2011) Will African consumers buy cleaner fuels and stoves?: a household energy economic analysis model for the market introduction of bio-ethanol cooking stoves in Ethiopia, Tanzania, and Mozambique. SEI research report, Stockholm Environment Institute, Stockholm. ISBN 978-91-86125-25-7
UNDP (2009) The energy access situation in developing countries: a review focusing on the least developed countries and sub-Saharan Africa. United Nations Development Programme, New York
van der Horst H, Hovorka G (2008) Reassessing the energy ladder: household energy use in Maun, Botswana. Energy Policy. doi:10.1016/j.enpol.2008.05.006
WHO (2008) Indoor smoke from solid fuel use: assessing the environmental burden of disease. Update to environmental burden of disease series, no. 4, World Health Organization, Geneva

Part VI
Summary and Conclusion

Chapter 31
Opportunities and Risks of Bioenergy in Africa

Dominik Rutz and Rainer Janssen

Abstract The use of biomass for energy provision in Africa has a long history, as wood-fuels (charcoal, fuel-wood) are traditionally used since long time. Current developments foresee also the production and use of modern bioenergy for electricity, heat and transport applications. Both, traditional and modern bioenergy pose opportunities and risks for sustainable development in Africa. The present chapter describes these opportunities and risks by summarising the contributions of the book "Bioenergy for Sustainable Development in Africa" which was elaborated as a result of the COMPETE network.

Keywords Opportunities and risks • Bioenergy potential • Bioenergy options • Bioenergy policies • Sustainable development • Sustainability • Financing of bioenergy

31.1 Introduction

The use of biomass for energy provision in Africa has a long history, as wood-fuels (charcoal, fuel-wood) are traditionally used since long time. Current developments foresee also the production and use of modern bioenergy for electricity, heat and transport applications. Both, traditional and modern bioenergy pose opportunities and risks for sustainable development in Africa. The present chapter describes these opportunities and risks by summarising the contributions of the book "Bioenergy for Sustainable Development in Africa" which was elaborated as a result of the COMPETE network.

D. Rutz (✉) • R. Janssen
WIP Renewable Energies, Sylvensteinstr. 2, Munich 81369, Germany
e-mail: dominik.rutz@wip-munich.de; rainer.janssen@wip-munich.de

R. Janssen and D. Rutz (eds.), *Bioenergy for Sustainable Development in Africa*,
DOI 10.1007/978-94-007-2181-4_31,

31.2 Bioenergy Potential in Africa

Africa has a large potential for biomass production and use (Sekhwela and Kghati 2011) with large discrepancies between African countries. Today, more than one third of Sub-Sahara Africa's population rely on traditional bioenergy (wood-fuels). The use of traditional bioenergy is prominent in Africa due to the low cost and lack of available alternatives in rural areas. Projections indicate that the (relative) contribution of traditional bioenergy will decrease, but that the total use of traditional biomass energy systems will increase during the coming decades (Smeets et al. 2011).

However, still a large fraction of the bioenergy potential remains unused. At the same time many African countries import expensive fossil energy carriers. The challenge for the coming decades is to use this potential in African countries and to export bioenergy surpluses. Thereby, bioenergy needs to be used in a sustainable way without competing with food production. Simultaneously with the increasing use of biomass, also efficiency measures have to be applied for the use and conversion of biomass. This concerns simple conversion technologies such as cooking stoves (Smeets et al. 2011), as well as sophisticated conversion technologies, e.g. for the production of electricity, heat, and high-quality transport fuels. Furthermore, the efficiency of agricultural feedstock production in Africa has to be increased considerably.

The overall biomass potential in Africa includes feedstock from the following three main categories:

- Uncultivated biomass from marginal land
- Biomass waste and residues
- Dedicated energy crops

Uncultivated biomass from marginal land includes wood-fuels which are traditionally used for cooking such as wood from shrubs and trees on land that is not purposely cultivated or managed. Although this land is often not considered "productive", marginal land has high value for many rural people who collect firewood and feed their animals or grow food in small-scale subsistence systems. If shrubs and trees are not cut completely and if they have enough recreation time, the use of this feedstock can be considered renewable. However, in many regions firewood is becoming a scarce source since too much wood is used. Although the use of wood from marginal land is often neither efficient nor sustainable, it will remain an important energy source for the rural poor in the short to long term future since it presents a low cost option for cooking energy. Thus, there is need to further develop and promote energy efficient cooking stoves in order to reduce the amount of required wood-fuels.

There is a large range of **biomass waste and residues** available in Africa, which is currently unused. This includes waste from agriculture, industry, and households. Large energy potentials from the use of waste biomass in Africa are estimated e.g. for the use of molasses for ethanol production (Munyinda et al. 2011; Walter and Segerstedt 2011; Janssen and Rutz 2011; Dianka 2011; Malaviya and

Ravindranath 2011) as well as residues from sawmills and forestry (Benaiah 2011) for cogeneration and second generation biofuels. Also for example sugarcane bagasse (Munyinda et al. 2011), rice husks (Benaiah 2011), palm oil residues, nutshells, sunflower husks and other agro-industrial residues constitute a high potential. Currently, in many cases biomass wastes are still seen as a problem of waste management and not as an opportunity for electricity production or cogeneration. Often, biomass waste is burned to reduce its volume and to recover nutrients (ash), but the energy content of these wastes is not used.

Since waste from agro-industrial processing is produced centrally and in predictable amounts, its use for energy is often technically viable. However a challenge may be the seasonality of some waste materials. In comparison to waste from agro-industrial processing, the use of biomass waste from households (e.g. for anaerobic digestion) is much more challenging, since sophisticated waste management systems and logistics have to be in place.

The use of **dedicated energy crops** in Africa is far behind its large potential. Promising crops for bioenergy include jatropha (van Eijck and Smeets 2011; Sinkala and Johnson 2011), sugarcane, sweet sorghum (Munyinda et al. 2011; Chagwiza and Fraser 2011; Janssen et al. 2010), palm oil, as well as short rotation woody crops and several other crops. Continuous improvements in crop breeding and agricultural practices to increase the bioenergy output are needed to tap the potential of dedicated energy crops. A main research focus for these crops is their cultivation on land that does not compete with food production or with other land uses. As the example of jatropha shows, which was characterised some years ago as "miracle" crop, crops need intensive research before they are implemented at large scale in order to avoid wrong expectations.

Finally, an important challenge for assessing the potential of bioenergy in Africa is the influence of climate change on agricultural systems. As predictions foresee, many regions in Africa will be seriously affected by climate change. The projected climate change is likely to lead to land degradation, water stress, increased pest occurrence and ultimately reduction in yields of energy crops. If energy crops are to be cultivated on a large-scale in Africa, it is very important to develop and adopt adaptation practices and strategies for sustainable crop yields (Malaviya and Ravindranath 2011). Strategies for bioenergy activities need to be defined in order to mitigate climate risk and vulnerability as well as to improve the adaptive capacity of the population in African countries. For this, systematic analyses of bioenergy activities in different geographical and social contexts are needed (Ulsrud 2011).

31.3 Bioenergy Options in Africa

The options for bioenergy in Africa are manifold since the use of biomass is applicable at different scales, under various climatic conditions as well as under different legal, regulatory and social framework conditions. Bioenergy options in Africa range from small-scale household applications (e.g. wood-fuels for cooking) to large-scale liquid biofuel production for transport. For all scales, different mechanical, thermo-chemical and biological conversion technologies are available.

Due to the potential of different bioenergy pathways to substitute other energy sources, bioenergy could be a solution to meet all types of energy needs: cooking, lighting, electricity, heat, and transport, depending on the feedstock and on the technology. The main criterion for sustainability is related to feedstock production. Thereby waste material is usually a preferred feedstock source since it avoids land use and other sustainability conflicts, and since it contributes to sustainable waste management practices. Furthermore, inefficient systems need to be improved, in order to ensure sustainability of bioenergy chains. For instance, many African households are still cooking with inefficient three-stone stoves. By improving the stove technology, considerable amounts of wood-fuel could be saved. Furthermore, a general challenge in Africa is the low efficiency of the agricultural sector affecting current food production as well as future bioenergy production.

Although bioenergy provides a potential solution for all energy needs (cooking, light, electricity, heat, transport), other options may still better meet the needs of the local population today, since they can be more affordable. Current priority in Africa is not on sustainability of energy, but on access and availability of energy. Alternative options can be based on fossil fuels (LPG, kerosene, petrol, diesel, crude oil, natural gas, coal, lignite) or on other renewable energies (wind, solar, geothermal, hydropower).

31.4 Bioenergy Policies in Africa

Primary policy drivers for bioenergy promotion in Africa include security of energy supply, a reduction of the foreign exchange burden of oil importing countries, as well as environmental benefits such as the restoration of degraded land, reduced land abandonment, and the mitigation of greenhouse gas (GHG) emissions (Janssen and Rutz 2011). Furthermore, the development of modern bioenergy systems offers opportunities for investment and infrastructure improvements in agriculture with the promise to diversify agricultural production and to create additional employment and thus to stimulate socio-economic development (Janssen et al. 2009).

However, concerns exist that bioenergy expansion in African countries may have severe negative impacts on biodiversity and the use of natural resources through increasing competition over land and water resources. Rising prices of agricultural commodities may negatively affect food security of the poor in developing countries and the implementation of large-scale bioenergy projects may cause negative social impacts such as conflicts over land ownership and displacement of rural communities.

Several African countries have launched initiatives to establish sound policy frameworks for bioenergy in order to ensure environmentally, economically and socially sustainable production, promotion and use of bioenergy. Nevertheless, fully functional legal and regulatory frameworks have not yet been established. The most advanced regulatory frameworks for bioenergy exist in South Africa and Mozambique with the Biofuels Industrial Strategy of the Republic of South Africa (enacted in 2007) and the National Biofuels Policy and Strategy (NBPS) published in May 2009 by the Government of Mozambique. Specific activities in the field of bioenergy sustainability certification as an essential component of the regulation of the bioenergy sector are performed in Mali, Mozambique, Tanzania, and on regional level by the Southern Africa Development Community (SADC) (Janssen and Rutz 2011).

It is recommended that before African countries develop new legislation related to bioenergy well-considered and clear policies on the subject should be implemented. The existing legal framework should be analysed, the gaps and weaknesses identified and the challenges, threats and opportunities examined. Thereby, it is of crucial importance to carefully integrate policies for land use, agriculture and energy and align them with policies for rural development, transport and finance. Furthermore, bioenergy development in African countries will only find its proper environmental context and agricultural scale if convergence with biodiversity, GHG emissions, and water use policies is achieved (Janssen and Rutz 2011).

Finally, a political framework is needed that can accommodate guidelines and regulations specific to different policy levels. Policies are needed at (1) national SSA level, (2) regional SSA level, (3) Pan-African level, and (4) SSA – European level (Jumbe and Madjera 2011).

31.5 Sustainability of Bioenergy in Africa

In 1987, the United Nations released the Brundtland Report (WCED 1987), which defines sustainable development as "development which meets the needs of the present without compromising the ability of future generations to meet their own needs". Sine the use of fossil energy is based on depleting sources and since it has considerable impact on the global climate, the current global energy system is not sustainable. However, not only the need of future generations is affected by the current energy system, but also the increasing discrepancy of energy availability and access between developed and developing countries.

Although bioenergy is not per se sustainable, it has the large potential to improve the global energy system and to contribute to sustainable development. Currently more than two thirds of Africa's energy supply consists of wood-fuels which are in many cases non renewable. This means that trees and shrubs are often cut as cooking fuel without providing the opportunity for re-growth. This practice meets the basic energy needs of the poorest, but it leads to environmental degradation and even to desertification.

This example shows that all three dimensions have to be included in sustainability assessments: environment, socio-economic issues, and economy. Some authors also ask for the inclusion of cultural and policy dimensions.

Another crucial factor to be considered in sustainability evaluations is the complexity of bioenergy value chains. Thereby, the size of the systems is an important determinant (Rutz and Janssen 2011). At the community level farmers who produce dedicated energy crops can increase their incomes and grow their own supply of affordable and reliable energy for their internal needs. At the national level, cultivating biofuel crops may generate new industries, technologies, jobs and markets. At the same time, producing more biofuels will reduce energy expenditures and allow developing countries to spend more resources on health, education and other services for their neediest citizens. On the other hand, the cultivation of large energy crop monocultures for industrial use will most likely be dominated by international large-scale companies and investors. This may cause negative socio-economic impacts, especially on land tenure issues. Many large-scale economic models discourage pro-development practices. In order to prevent negative social, economic, and environmental impacts, suitable policies, enforcement of legislation and sustainability schemes are needed.

To counterbalance the possible negative effects of bioenergy, some measures (e.g. certification, accreditation, and traceability) are currently being put in place to ensure sustainability. This has an impact, either positive or negative, on the development of the bioenergy industry (Diaz-Chavez and Woods 2011). For instance, environmental and social criteria have been developed by the Roundtable on Sustainable Biofuels (RSB), the Roundtable on Sustainable Palm Oil (RSPO), the Round Table on Responsible Soy (RTRS), the Global Bioenergy Partnership (GBEP) and some country initiatives. Verification systems offer a possibility to assure sustainable production, but have not fully addressed the perception of communities and the effects at different scales.

A challenge in current sustainability certification schemes and legislation (e.g. the European Renewable Energy Directive) is that they mainly focus on environmental impacts, whereas more research is needed on socio-economic impacts of biofuels (Rutz et al. 2010). For instance an important socio-economic issue in African countries is related to gender aspects as described by Farioli and Dafrallah (2011), since the energy-poverty nexus in Africa has distinct gender characteristics. Of the approximately 1–3 billion people living in poverty, it is estimated that 70% are women, who are mainly living in female–headed households in rural areas. In Sub-Saharan Africa, women have challenging roles on the energy scene as they are in charge of supplying their households with energy amongst other subsistence activities.

31.6 Financing of Bioenergy in Africa

The implementation of bioenergy projects in Africa needs suitable framework conditions (legislation, policies, will of involved stakeholders), know-how on production systems and technologies, and access to financing or direct financial support. The type of financing depends on the size of the bioenergy system, the framework conditions, and on the organisation structure.

At the household level money for improved systems is often not available for the poor population. Poor individuals, small enterprises and small non profit organisations often lack collateral, steady employment or a verifiable credit history. Minimal qualifications to gain access to traditional loans are often not met.

For households and small enterprises microfinancing can be a good financing option. Microfinancing is the provision of financial services to low-income clients (consumers, enterprises, organisations) in developing countries, who lack access to banking and related services. This access to financing should facilitate the poor to reduce their poverty.

The most often applied tools in microfinancing are microcredits. Microcredits could be a suitable option for financing small bioenergy projects, such as the purchase of improved stoves or the electrification of villages by small biodiesel or plant oil generators.

Although microfinance seems to be a good option, long-term dependencies and exploitation by unjustified increasing interest rates which cannot be paid back by the credit holder, need to be avoided. Negative examples for exploitation in other sectors were recently reported in some countries in Asia and Africa.

At small to large company level, challenges include access to affordable loans and sound business models. At this level traditional loan financing, project financing, and financing by certificates or carbon credits is generally possible.

The nature or choice of business model for a biofuel enterprise, small or large, influences the funding requirements and the financing implications for the venture. The key differentiating characteristic of the business models is whether the venture is offering energy to customers in the form of an *energy carrier* such as biodiesel, gelfuel or bioethanol or an *energy service* such as cooked meals, heated water or lighting (Hofmann et al. 2011).

The first model focuses on the supply chain up to the sale of the energy carrier, whereas in the second model an energy service company (ESCO) encapsulates the whole supply chain for the energy service including distribution and energy conversion. The key difference in the financing requirements of these two broad categories is that the former addresses individual energy production facilities on the basis of financing specific production equipment and systems, whereas the latter is focused on an overall service business which is not dependent or based on any particular plant or equipment but rather on the overall operations required to provide the energy service (Hofmann et al. 2011).

31.7 Conclusion

Increasing fossil fuel prices and depleting fossil energy resources are currently stimulating the global debate on renewable energies. As one of the renewable energy options, energy from biomass (bioenergy) will become increasingly important in the future energy mix. In Africa, bioenergy is already a main energy source, mostly used as wood-fuels for household applications.

The production and use of bioenergy involves many different sectors, such as agriculture, energy, environment, development and poverty alleviation, trade, industry, and much more. Considering the complex nature of bioenergy, it is challenging to evaluate and judge bioenergy systems with respect to their potential to contribute to sustainable development. Social, economic and environmental aspects are site-specific and crucially depend on the specific framework conditions, and thus need to be evaluated for individual systems and applications. Generalisations are often inappropriate and may lead to wrong results and expectations.

However, it can be concluded that many bioenergy systems in Africa have the potential to contribute to sustainable development. If social, economic and environmental aspects are properly considered, bioenergy in Africa presents a huge opportunity for development and poverty alleviation. Suitable bioenergy applications exist for small to large-scale, local use or export, centralised and decentralised applications, as well as for liquid, solid and gaseous energy carriers. However, sound and supportive legal and regulatory framework conditions for sustainable bioenergy systems need to be developed and implemented in close cooperation between developing and developed countries in order to exploit the full potential of bioenergy for sustainable development in Africa.

References

Benaiah AK (2011) Modern use of solid biomass in Africa: prospects for utilization of agro-waste resources in Nigeria. In: Janssen R, Rutz D (eds) Bioenergy for sustainable development. Springer Dordrecht Heidelberg London New York

Chagwiza C, Fraser G (2011) Economic evaluation of sweet sorghum in biofuel production as a multi-purpose crop: the case of Zambia. In: Janssen R, Rutz D (eds) Bioenergy for sustainable development. Springer Dordrecht Heidelberg London New York

Dianka M (2011) Promoting biofuels in West Africa: an engine for development. In: Janssen R, Rutz D (eds) Bioenergy for sustainable development. Springer Dordrecht Heidelberg London New York

Diaz-Chavez R, Woods J (2011) Keynote introduction: sustainability considerations for biofuels production in Africa. In: Janssen R, Rutz D (eds) Bioenergy for sustainable development. Springer Dordrecht Heidelberg London New York

Farioli F, Dafrallah T (2011) Gender issues of biomass production and use in Africa. In: Janssen R, Rutz D (eds) Bioenergy for sustainable development. Springer Dordrecht Heidelberg London New York

Hofmann M, Morris G, Ballard-Treemer G, Kathun K (2011) Financing of biofuel projects in Africa. In: Janssen R, Rutz D (eds) Bioenergy for sustainable development. Springer Dordrecht Heidelberg London New York

Janssen R, Rutz D (2011) Keynote introduction: overview on bioenergy policies in Africa. In: Janssen R, Rutz D (eds) Bioenergy for sustainable development. Springer Dordrecht Heidelberg London New York

Janssen R, Rutz D, Helm P, Woods J, Diaz-Chavez R (2009) Bioenergy for sustainable development in Africa – Environmental and social aspects. In: Proceedings of the 17th European biomass conference and exhibition, Hamburg, 29 June – 3 July 2009, pp 2422–2430

Janssen R, Rutz D, Braconnier S, Reddy B, Rao S, Schaffert R, Parella R, Zaccharias A, Rettenmaier N, Reinhardt G, Monti A, Amaducci S, Marocco A, Snijman W, Terblanche H, Zavala-Garcia F (2010) Sweet sorghum – an alternative energy crop. In: Proceedings of the 18th European biomass conference and exhibition, Lyon, pp 200–206, ISBN 978-88-89407-56-5

Jumbe C, Madjera M (2011) Strategies for a sustainable Pan-African biofuels policy. In: Janssen R, Rutz D (eds) Bioenergy for sustainable development. Springer Dordrecht Heidelberg London New York

Malaviya S, Ravindranath NH (2011) Implications of climate change on sustainable biofuel production in Africa. In: Janssen R, Rutz D (eds) Bioenergy for sustainable development. Springer Dordrecht Heidelberg London New York

Munyinda K, Yamba FD, Walimwipi H (2011) Bioethanol potential and production in Africa: sweet sorghum as a complementary feedstock. In: Janssen R, Rutz D (eds) Bioenergy for sustainable development. Springer Dordrecht Heidelberg London New York

Rutz D, Janssen R (2011) Keynote introduction: socio-economic impacts of different scales of biofuel production in Africa. In: Janssen R, Rutz D (eds) Bioenergy for sustainable development. Springer Dordrecht Heidelberg London New York

Rutz D, Van Dam J, Janssen R, Hiegl W, Diaz Chavez RA, Woods J, Faaij APC, Smeets EMW, Vos J, Vis M, Bottriell K, Rettenmaier N, Reinhardt G, Nussbaum R, Gozzi A, Haye A, da Silva Walter AC, Hilbert JA, Coto C, Fallot A, Sawe EN, Togola I, Ouattare O, Jaax R (2010) Global socio-economic impact assessment of biofuel and bioproduct chains. In: Proceedings of the 18th European biomass conference and exhibition, Lyon, ISBN 978-88-89407-56-5

Sekhwela MBM, Kghati DL (2011) Biomass potential in arid and semi-arid regions in Botswana. In: Janssen R, Rutz D (eds) Bioenergy for sustainable development. Springer Dordrecht Heidelberg London New York

Sinkala T, Johnson FX (2011) Small-scale production of jatropha in Zambia and its implications for rural development and national biofuel policies. In: Janssen R, Rutz D (eds) Bioenergy for sustainable development. Springer Dordrecht Heidelberg London New York

Smeets E, Johnson FX, Ballard-Treemer G (2011) Keynote introduction: traditional and improved use of biomass for energy in Africa. In: Janssen R, Rutz D (eds) Bioenergy for sustainable development. Springer Dordrecht Heidelberg London New York

Ulsrud K (2011) Bioenergy and sustainable adaptation to climate change in Africa. In: Janssen R, Rutz D (eds) Bioenergy for sustainable development. Springer Dordrecht Heidelberg London New York

Van Eijck J, Smeets E (2011) Jatropha – a promising crop for Africa's biofuel production? In: Janssen R, Rutz D (eds) Bioenergy for sustainable development. Springer Dordrecht Heidelberg London New York

Walter A, Segerstedt A (2011) International trade of biofuels: current trends and the potential role of Africa. In: Janssen R, Rutz D (eds) Bioenergy for sustainable development. Springer Dordrecht Heidelberg London New York

WCED (1987) Our common future. World Commission on Environment and Development; Published as Annex to General Assembly document A/42/427, Development and International Co-operation: Environment 2 Aug 1987. Retrieved 14 Nov 2007

Index

R. Janssen and D. Rutz (eds.), *Bioenergy for Sustainable Development in Africa*,
DOI 10.1007/978-94-007-2181-4, © Springer Science+Business Media B.V. 2012

B

C

D

E

T

U